A Guide to Plant Poisoning

OF ANIMALS IN NORTH AMERICA

T0191478

A Guide to Plant Poisoning

OF ANIMALS IN NORTH AMERICA

Anthony P. Knight

BVSc, MS, MRCVS, DACVIM

Professor and Chair
Departmental of Clinical Sciences
College of Veterinary Medicine
Veterinary Teaching Hospital
Colorado State University
Fort Collins, Colorado

Richard G. Walter

BSAH, MABtny

Assistant Professor
Department of Biology
Colorado State University
Fort Collins, Colorado

Teton NewMedia
The Innovative Health Science Publisher
Jackson, Wyoming 83001

Executive Editor: Carroll C. Cann

Development Editor: Susan L. Hunsberger

Editor: Cynthia J. Roantree

Design and Layout: Anita B. Sykes

Printer: Colour Box, Burbank, CA

Teton NewMedia

P.O. Box 4833

125 South King Street

Jackson, WY 83001

1-888-770-3165

www.tetonnm.com

Interior photographs and illustrations by Anthony P. Knight (unless otherwise noted)

ISBN # 1-893441-11-3

Print number 5 4 3 2 1

Library of Congress Cataloging-in-Publication Data

Knight, Anthony P.
    Guide to plant poisoning of animals in North America / Anthony P. Knight, Richard Walter.
        p. cm.
    Includes bibliographical references (p.).
    ISBN 1-893441-11-3 (alk.paper)
    1. Livestock poisoning plants--North America. 2. Livestock poisoning plants--Toxicology--North America. I. Walter, Richard, 1925- II. Title.
SB617.5.N7 K57 2000
636.089'5952'097--dc21

                                                                    00-039260

DEDICATION

To the many students and animal owners who have encouraged us over the years, we dedicate this book. Our thanks and admiration to our wives Cassandra and Verda for their encouragement and patience as they tolerated endless hours of waiting while we collected and photographed plants, many of which they preceived as "just weeds." A special thanks also to Jenger Smith and Charlie Kerlee for their photographic and computer expertise and advice in helping compile the images in this book.

"What greater or better gift can we offer the republic than to teach and instruct our youth?"

– Cicero

"Where flowers degenerate man cannot live."

– Napoleon

CONTENTS

Chapter 4 Plants Affecting the Skin and Liver

Chapter 5 Plants Affecting the Blood

Chapter 6 Plants Affecting the Nervous System

Chapter 10   Plants Affecting the Mammary Gland

INTRODUCTION

Plant poisoning has plagued humans and animals throughout history, especially in North America where many immigrants and their animals were poisoned by unfamiliar plants. In the eastern states, milk sickness was a problem among those who drank the milk of cows that had been grazing white snakeroot (*Eupatorium rugosum*). Numerous deaths of people and animals due to "trembles" or "milk sickness" were reported in Indiana in the early 1800s. Once the association between snakeroot and the disease was determined, destroying the plants and preventing animal access to the plants largely controlled the disease.[1] Similarly, as settlers moved westward, they, and often their livestock, suffered fatalities from eating the bulb of death camas (*Zigadenus* spp.) mistakenly for that of the edible camas lily (*Camassia* spp.) or wild onion (*Allium* spp.).[2] A group of plants that were particularly troublesome to the early ranchers and farmers in the western states were the locoweeds (*Astragalus and Oxytropis* spp.). These widely distributed plants caused severe disease in all animals that ate them. Locoweeds continue to be the most economically important poisonous plants in North America considering their wide geographic distribution and diverse effects on livestock health.

Not only did early immigrants encounter unfamiliar toxic plants, they also inadvertently introduced weed seeds that contaminated the seed grains they brought with them from Europe and Asia. These weeds soon became established and spread rapidly because of the lack of competition in their new environment. Some of these introduced weeds such as leafy spurge (*Euphorbia esula*) have become noxious weeds in that they displaced indigenous plants and crowded out valuable forages. Other introduced weeds such as yellow star thistle (*Centaurea solstitialis*), Russian knapweed (*Centaurea repens*), and hound's tongue (*Cynoglossum officinale*) are not only noxious but also cause severe poisoning in animals that eat them when other forages are scarce.

Despite our ever-increasing knowledge about plants and their toxins the prevalence of plant poisoning is likely to increase with the influx of small acreage farmers into native rangelands. Livestock on small acreages frequently increase grazing pressure on the land, and, as a consequence, toxic plants may be consumed. Furthermore, overgrazing disturbs the normal balance of plant species, often allowing an aggressive plant or weed to invade the area. Plant poisoning is also being encountered in wildlife that are forced to concentrate their grazing in ever decreasing areas because of the encroachment of human populations on their natural range. A good example of this is the frequency with which elk develop locoweed poisoning as their population density increases in areas where human activity has curtailed the elk's normal migratory patterns.

Plant poisoning can be a significant impediment to profitable livestock management and production. In 1978, a study on the economic impact of poisonous plants on the range livestock industry in 17 western states estimated that the problem cost the industry $107 million annually.[3] In some years and in localized areas the economic losses may be proportionally much higher. Although the most obvious economic losses are those attributable to actual deaths from poisonous plants, many other aspects of plant poisoning contribute a great deal to the overall economic losses. Locoweeds, for example, exert their major effect through decreased reproductive performance including abortions, fetal deformity, and decreased fertility in male and female animals. Decreased weight gains are also common in animals consuming less than fatal doses of locoweed. Additionally, the costs for fencing, herbicides to control the plants, and decreased carrying capacity of livestock on the land invaded by these plants further adds to the economic impact of poisonous and noxious weeds.

Veterinarians and animal owners are frequently confronted with the task of determining whether or not a plant is responsible for poisoning of animals. Similarly the presence of plants other than grasses in a pasture are always a concern to livestock owners. A reference source that can help them identify toxic plants and the effects they have on animals is a necessity. Although there are many books dealing with wild flowers, weeds, and toxic plants, none adequately combine definitive color photographs of toxic plants for easy recognition with information on the effects of the plant toxins on animals. To those who are not botanists, many of the available books on toxic plants are frustrating because the illustrations are either line drawings or black and white photographs that make plant identification difficult. At present, the most comprehensive coverage of poisonous plants in North America is John M. Kingsbury's book *Poisonous Plants of the United States and Canada*. It is, however, sparsely illustrated, and since its publication in 1964, there have been numerous new documented plant poisonings that are not included in the book.

With over 50 years combined field and teaching experience, we have compiled a book that will be useful to students, livestock owners, veterinarians, and anyone who is interested in the fascinating effects of toxic plants on animals. Supplementing the book is a CD that contains numerous additional illustrations of the plants discussed in the text. Most books on plant poisoning are categorized according to the toxin in the plant. In reality however, animals with plant poisoning are encountered with one or more clinical signs that the animal owner or veterinarian must try to relate to a particular plant. With this in mind, the toxic plants covered in this book have been grouped into 10 chapters based on the most common presenting clinical signs seen in the animal. This arrangement allows the veterinarian or livestock producer who is confronted with an animal exhibiting specific clinical signs to easily review the plants that would most likely cause the problem. Those who want to know if a particular plant from their pasture, range, or hay is poisonous can match the plant to the color photographs and descriptions of the plants. Once the plant is identified, further information is provided on the toxic components of the plant, the clinical signs it produces, treatment, and management of the plant to prevent poisoning.

At the beginning of each chapter is a general description of the toxicology of the plant toxin and its effects on animals. This is followed by a description and illustration of each plant. Each chapter has a reference list for those who wish to pursue the topic further. In Chapter 1, for instance, there is a general discussion on the plants capable of causing sudden death, the mechanism of action of the various toxins, and relevant treatments, followed by a description of the individual plants. Subsequent chapters cover the most common presenting signs of plant poisoning. Some plant toxins that affect multiple organ systems, such as the pyrrolizidine alkaloids found in various plant species, are cross-referenced for the reader's benefit. A lengthy discussion of herbicide applications is intentionally omitted because of the continual availability of products, differences in the regional use of the chemicals, and the changing recommendations of the Environmental Protection Agency.

This book emphasizes the toxic plants most commonly associated with animal poisoning in North America and that are well documented in the literature. Where documentation is poor, the plant may only be mentioned briefly or listed as being a suspect or potentially poisonous plant. It is beyond the scope of this book to include the numerous poisonous house and garden plants that affect humans and pet animals. Some ornamental plants, however, are included when they have become a problem to livestock. Minimal reference is given to plant poisoning in dogs and cats and, when included, is for comparative purposes. Some plants such as *Abrus precatorius* are included because of their deadly toxins and not because they frequently cause animal poisoning. We intend this book be a useful compilation of the current information on poisonous plants that will be of benefit to those who are involved with the welfare of animals.

Special recognition and appreciation is extended to the following individuals who contributed advice and photographs or illustrations to the book: Drs John and Emily Smith, Baldwin, Georgia; Dr. Julia F. Morton, Miami, Florida; Mark Dimmit, Tucson, Arizona; Dr. Darrel N. Ueckert, San Angelo, Texas; Dr. Mike H. Ralphs, and Dr. Jim Pfister, Logan, Utah; Dr. Bob Glock, Tucson, Arizona; Dr. Mike Murphy, St Paul, Minnesota; and Dr. Gerald D. Carr, Hawaii.

A special thanks goes to our families who have encouraged us to persevere in publishing this book, and without whose support it would not have been possible.

Anthony P. Knight
Richard G. Walter

1. Wolf FA, Curtis RS, Kaup BF: A monograph of trembles or milk sickness and white snakeroot. *North Carolina Agricultural Experiment Station Bulletin* 1918;15:1-74.

2. Beath OA, Eppson HF, Draize JH, Justice RH: Three species of Zygadenus (death camas). *Wyoming Agricultural Experiment Station Bulletin* 1933;194:1-38.

3. Nielsen DB: The economic impact of poisonous plants on the range livestock industry in the 17 western states. *J Range Manage* 1978;31:325-328.

CHAPTER 1

Plants Causing
Sudden Death

Plants Containing Cyanogenic Glycosides

Cyanogenic glycosides are substances present in many plants that can produce highly toxic hydrogen cyanide (HCN) or prussic acid. Specific plant enzymes released when plant cells are damaged when chewed, crushed, wilted, or frozen, hydrolyze the glycosides to cyanide. At least 2000 plant species are known to contain cyanogenic glycosides with the potential to produce HCN poisoning.[1-6] However, relatively few of these plants are frequent causes of cyanide poisoning in humans or animals because they are infrequent food sources for humans or animals.[1,2] Plants that have been most frequently associated with cyanide poisoning in animals are listed in Table 1-1. Some of these plants are grown as food sources for humans and animals, for example, sorghum (*Sorghum* spp.), corn (*Zea mays*), clovers (*Trifolium* spp.), and manihot or cassava (*Manihot esculenta*), and can be used safely provided attention is paid to the circumstances under which these plants accumulate cyanogenic glycosides.

Most plant-induced cyanide poisoning in humans occurs in tropical countries where cassava is commonly used as food. The chronic consumption of poorly prepared cassava diets produces a disease syndrome in humans known as tropical ataxic neuropathy.[7] Pigs and goats have been similarly poisoned when fed cassava tubers and leaves.[8] Plants such as blue flax (*Linum* spp.), grown for fiber (linen) and linseed oil, will also accumulate toxic levels of HCN under the right growing conditions.[9]

Cyanide poisoning of livestock is most commonly associated with Johnson grass (*Sorghum halapense*), Sudan grass (*Sorghum vulgare*), and other forage sorghums.[10,11] Choke-cherries (*Prunus* spp.),[12] service berry (*Amelanchier alnifolia*),[13-15] and arrow grass (*Triglochin* spp.)[16-19] are less frequent but long recognized sources of cyanide poisoning. Crab apple leaves (*Malus* spp.), and sugar gums (*Eucalyptus cladocalyx*) have caused cyanide intoxication in goats,[20,21] and even wild deer have become victims of cyanide poisoning after eating service berry.[22] Occasionally cattle have consumed lethal doses of cyanogenic glycosides from eating acacia tree leaves and poison suckleya (*Suckleya suckleyana*).[23,24] Under some undetermined growing conditions, certain grasses such as tall manna grass (*Glyceria grandis*) and Indian grass (*Sorgastrum nutans*) accumulate toxic levels of cyanogenic glycoside.[25,26]

Cyanogenic glycosides in the leaves and stems of plants are not toxic unless acted on by the plant or rumen microorganism enzymes, b-glucosidase and hydroxynitrile lyase, to form HCN.[1,27] Enzymatic conversion of the glycosides is enhanced when plant cells are damaged or stressed as occurs when the plant is chewed, crushed, droughted, wilted, or frozen.[28] In the process, the glycosides, which are normally isolated in cell vacuoles, come into contact with the cell enzymes and HCN is formed. Generally most parts of the plant contain cyanogenic glycosides; the young rapidly growing portion of the plant and the seeds contain the highest concentrations. The flesh of the ripe fruits is edible. Drying the plants decreases their cyanogenic potential especially over time. Ensiling plants will reduce cyanogenic glycoside content by as much as 50 percent, with free cyanide being liberated from silage pits or silos in the curing process.[28,29]

The concentration of cyanogenic glycosides in plants varies with the stage of

Table 1-1
Plants Associated with Cyanide Poisoning

BOTANICAL NAME	COMMON NAME
Acacia spp.	Catclaw, acacia
Amelanchier alnifolia	Service, June, or Saskatoon berry
Bahia oppositifolia	Bahia
Mannihot esculentum	Cassava, manihot, tapioca
Cercocarpus montanum	Mountain mahogany
Chaenomales spp.	Flowering quince
Cynodon spp.	Star grass
Eucalyptus spp.	Eucalyptus, gum tree
Glyceria grandis	Tall manna grass
Hydrangea spp.	Hydrangea
Linum spp.	Flax
Lotus spp.	Bird's foot trefoil
Malus spp.	Crab apple
Nandina domestica	Heavenly or sacred bamboo
Phaseolus lunatus	Lima bean
Photinia spp.	Christmas berry
Prunus spp.	Choke-cherry, pin cherry
Pteridium aquilinum	Bracken fern
Sambuccus spp.	Elderberry
Sorghum spp.	Johnson, Sudan grass
Sorghastrum nutans	Indian grass
Stillingia texana	Texas queen's delight
Suckleya suckleyana	Poison suckleya
Trifolium repens	White clover
Triglochin maritima	Arrow grass
Vicia sativa	Common vetch
Zea mays	Corn, maize

growth, time of year, soil mineral and moisture content, and time of day. Cool moist growing conditions enhance the conversion of nitrate to amino acids and cyanogenic glycosides instead of plant protein. As the glycosides accumulate they further inhibit nitrite reductase in the plant, favoring the conversion of nitrate to cyanogenic glycoside rather than to amino acids.[28] Nitrate fertilization of cyanogenic plants therefore has the potential to increase the cyanogenic glycoside content of plants. Frost and drought conditions may also increase cyanogenesis in some plant species. Young plants, new shoots, and regrowth of plants after cutting often contain the highest levels of cyanogenic glycosides. Application of herbicides (2,4-dichlorophenoxyacetic acid [2,4-D]) can also increase the cyanogenetic glycoside content of plants.

At least 55 cyanogenic glycosides are known to occur in plants, many being synthesized from amino acids as part of normal plant metabolism.[1,8,30] Some of the better known glycosides include amygdalin (laetrile) from bitter almonds, peach, apricot, cherry and apple seeds;[31] prunasin from choke-cherries and service berry leaves; linamarin and lotaustralin from flax and white clover[9]; dhurrin from sorghum; and triglochinin from arrow grass. In the 1970s, laetrile received considerable attention as a potential cure for cancer, but its efficacy has never been proven, and, in fact, it was shown to have the potential to be highly toxic to humans and animals.[32,33]

Selective breeding of certain varieties of plant species with naturally low glycoside content has resulted in varieties that are low cyanogenic glycoside; and has increased their food value for humans and animals. The development of sweet almond varieties with low cyanogenic glycoside content has facilitated the human consumption of almond seeds. Similarly sorghum varieties low in cyanide have been developed that have greatly increased the safety of feeding sorghums, such as Sudan grass, to livestock.

Ruminants are more susceptible to cyanide poisoning than other animals because the normally mildly acidic to alkaline rumen contents (pH 6.5-7), high water content, and microfloral enzymes in the rumen hydrolyze the cyanogenic glycosides to HCN.[34] Water drunk after animals have eaten cyanogenic plants enhances the hydrolysis of the glycosides. Conversely, ruminants that are on high energy grain rations where the rumen is more acidic (pH 4-6) have a slower release of HCN than if they were fed a grass, hay, or alfalfa diet.[34] Humans, pigs, dogs, and horses that have a highly acidic stomach (pH 2-4) tend to have a reduced rate of glycoside hydrolysis and cyanide production in their digestive systems and therefore rarely suffer from cyanide poisoning of plant origin. Atypically, donkeys have been reported to develop acute cyanide poisoning from eating the new shoots from wild choke-cherries.[35]

Mechanism of Acute Cyanide Poisoning

Hydrogen cyanide (HCN) is highly poisonous to all animals because it rapidly inactivates cellular respiration thereby causing death.[36-38] The cyanide ion is readily absorbed from the intestinal and respiratory tracts and has a strong affinity for binding with trivalent iron of the cytochrome oxidase molecule, inhibiting its enzymatic action and preventing cellular respiration.[37,39-41] The characteristic cherry red venous blood seen in acute cyanide poisoning results from the failure of the oxygen-saturated hemoglobin to release its oxygen at the tissues because the enzyme cytochrome oxidase is inhibited by the cyanide. Normally, small quantities of cyanide are detoxified by cellular enzymes and thiosulfates in many tissues to form relatively harmless thiocyanate, which is excreted in the urine. When large quantities of cyanide are rapidly absorbed and the body's detoxification mechanisms are overwhelmed, cyanide poisoning occurs. In most species, the lethal dose of HCN is in the range of 2 to 2.5 mg/kg body weight.[36-38] However, if plenty of other plant material and carbohydrates are present in the stomach, formation and absorption of cyanide may be slowed, allowing animals to tolerate higher doses.

Chronic Cyanide Poisoning

In addition to the acute toxic effects of cyanide poisoning, low levels of cyanide will over time cause a variety of chronic effects in humans and animals. Chronic cyanide poisoning is thought to be a form of lathyrism, a neurotoxicity recognized in people in some eastern Asian countries where the seeds of certain peas (*Lathyrus* spp.) are eaten when other foods are scarce. The neurotoxin in the uncooked peas results in damage to the spinal cord leading to paralysis. Similarly, a chronic neuropathy occurs in people who consume poorly cooked cassava (*Manihot esculenta*).

The perennial sweet pea (*Lathyrus latifolius*) and the annual sweet pea (*L. odoratus*) seeds contain neurotoxins (lathyrogens) capable of producing osteolathyrism in animals, especially horses.[42] The primary lathyrogen in the annual sweet pea is b-amino proprionitrile, which causes defective cross-linking of collagen and elastin molecules.[42] This disease in horses is characterized by skeletal deformities and aortic rupture caused by defective synthesis of cartilage and connective tissue.[42] A similar syndrome of musculoskeletal deformities in foals and calves has been associated with pregnant mares and cows chronically eating Sudan grass (*Sorghum sudanense*) or sorghum hybrids containing low levels of cyanogenic glycosides.[43-48] In addition to limb deformities (arthrogryposis), calves also develop severe degeneration of the spinal cord and brain.[44] Affected animals develop posterior ataxia, urinary incontinence, and cystitis resulting from lower spinal cord degeneration.[48] The cystitis may become complicated by an ascending infection of the kidneys.

The underlying problem is the loss of the myelin sheath surrounding peripheral nerves with resulting loss of nerve function. The demyelinization of the nerves is thought to result from the conversion of the cyanide glycoside to T-glutamyl β-cyanoalanine, a known lathyrogen that interferes with neurotransmitter activity.[46] Neuronal degeneration in the brain associated with chronic cyanide poisoning may be associated with the depletion of hydroxycobalamin.[45] Animals may slowly recover if the source of the toxic aminonitrile is removed before neuronal degeneration becomes severe.

Low doses of cyanide are also goitrogenic in humans and animals. Enlargement of the thyroid gland is caused by the formation of thiocyanate, which inhibits the intrathyroidal transfer of iodine and elevates the concentration of thyroid-stimulating hormone.[8]

Pregnant ewes grazing star grass (*Cynodon* spp.) have developed goiter because of the presence of cyanide in the grass.[29]

Clinical Signs of Acute Cyanide Poisoning

Sudden death is often the presenting sign of acute cyanide poisoning. Affected animals rarely survive more than 1 to 2 hours after consuming lethal quantities of cyanogenic plants and usually die within a few minutes of developing clinical signs of poisoning.[29] If observed early, poisoned animals show rapid labored breathing, frothing at the mouth, dilated pupils, ataxia, muscle tremors, and convulsions. The heart rate is usually increased and cardiac arrhythmias may be present. Regurgitation of rumen contents occurs when ruminants become recumbent and bloating occurs. High blood ammonia levels coupled with increases in neutral and aromatic amino acids may be a significant cause for the loss of consciousness associated with terminal cyanide poisoning.[49] The mucous membranes are bright red in color because oxygen saturates the hemoglobin. Cyanosis of the mucous membranes occurs terminally when the animal's tissues become depleted of oxygen.

Postmortem Findings

Animals poisoned by cyanide may have characteristic cherry red venous blood if examined immediately after death. Generalized congestion and cyanosis of the internal organs is often seen at necropsy. The blood clots slowly and the musculature is dark and congested. Hemorrhages occur commonly in the heart, lungs, and various other organs. The smell of bitter almonds reputedly characteristic of cyanide may occasionally be detected in the rumen gas when performing a fresh postmortem examination.

Diagnosis

Cyanide is rapidly lost from animal tissues unless specimens are collected within a few hours of death and frozen for chemical analysis. Liver, muscle, and rumen contents should be collected and frozen in air-tight containers before shipment to a laboratory capable of doing cyanide analysis. Levels of cyanide in liver or blood exceeding 1 ppm, or in muscle exceeding 0.63 g/mL are considered diagnostic for cyanide poisoning.[29,50] Cyanide poisoning can be confirmed by demonstrating toxic levels of HCN in the rumen contents or the suspect plants (or both) using the sodium picrate paper test.[39] Filter paper strips soaked in yellow sodium picrate and suspended over the suspect plant material in an air-tight jar turn brick red in a few minutes if significant cyanide is present. Commercial test kits are available for testing plants and rumen contents for cyanide. More precise determination of cyanogenic glycoside and cyanide levels is possible through the use of liquid chromatography and colorimetry.[41] Plant material containing more than 20 mg HCN/100 g (200 ppm = 200 mg/kg) is potentially toxic to all animals.[36,50] Liver or blood levels greater than 1 ppm are highly suggestive of cyanide poisoning.[50] Because many factors can alter cyanide determinations in tissues, the most consistent and diagnostically valuable HCN levels can be obtained from the brain and ventricular myocardium. Levels of HCN in excess of 100 g/100 g wet tissue are diagnostic of cyanide poisoning.[51]

Treatment

Many remedies for cyanide poisoning in man and animals have been evaluated over time. In all cases the successful treatment of acute cyanide poisoning depends on the rapid inactivation and removal of cyanide by metabolizing or complexing it with other compounds to allow its excretion via the kidneys.[41,52] In animals this has been traditionally accomplished by injecting sodium nitrite and sodium thiosulfate intravenously.[53] Sodium nitrite converts some hemoglobin to methemoglobin for which cyanide has a strong affinity, producing cyanmethemaglobin.[41] This complexing of cyanide reactivates the cytochrome oxidase system essential for cellular respiration. Sodium thiosulfate in the presence of the tissue enzyme rhodanese combines rapidly with the cyanide molecule cleaved from cyanmethemaglobin to form relatively nontoxic and excretable sodium thiocyanate.[37,54]

A recommended treatment for cyanide poisoning is the intravenous administration of a mixture of 1 mL of 20 percent sodium nitrite and 3 mL of 20 percent sodium thiosulfate. This dose is given intravenously per 100 lb of body weight.[53] The dose may be repeated in a few minutes if the animal does not respond. Additional sodium nitrite should be given cautiously because excess will compound the toxicity by producing, in effect, nitrite poisoning. Better results have been obtained in sheep experimentally poisoned with cyanide by administering 660 mg/kg sodium thiosulfate and 22 mg/kg sodium nitrite intravenously.[54] Thiosulfate is the safest and most effective agent tested for treating cyanide poisoning in dogs. The effectiveness of these compounds given in combination is enhanced if oxygen is given simultaneously.[55] The oral administration of a 5 percent solution of cobaltous chloride at a dose of 10.6 mg/kg body weight further improved the effectiveness of the combination.[54] Alpha-ketoglutaric acid is an effective treatment if administered in conjunction with thiosulfate.[56,57] It is also beneficial to administer orally via stomach tube a solution of sodium thiosulfate (30 g to an adult cow) to detoxify free cyanide still present in the rumen. Animals suspected of consuming cyanogenic plants but that show no clinical signs should also receive oral sodium thiosulfate prophylactically. Administered via stomach tube, 1 gallon of vinegar diluted in 3 to 5 gallons of water will help acidify the rumen and reduce the production of hydrogen cyanide.

Prevention of Cyanide Poisoning

Appropriate pasture management to avoid exposing livestock to potentially toxic plants will help prevent animal losses. Animals should be prevented from grazing sorghums during early regrowth after the plants have been cut, droughted, or frosted, when they are likely to be most toxic.[58] Allowing sorghum forages to grow at least 2 feet high before allowing animals to graze them significantly reduces the potential for poisoning. Where uncertainty exists about the cyanogenic content of plant crops for animal forage, a sample of the plants should be tested for cyanide content. Properly curing hay and silage destroys the cyanogenic glycosides. Selecting forage sorghums and white clover varieties that have been specifically developed for low cyanogenic glycoside content further reduces the chances for poisoning and allows their safe use as forage crops.[59-61]

The selective use of herbicides in localized areas can be used to control dense stands of cyanide-containing plants such as choke-cherry. Whenever herbicides are used as a control measure, it is important to follow manufacturer recommendations for the herbicide and observe local herbicide application ordinances.

Acacia, Gregg's Catclaw Acacia, Guajillo
Acacia greggii
Fabaceae (Legume family)

Habitat

Acacia is found in the limestone cliffs and calcareous rocks of flats and valleys of the southwestern desert areas of North America.

Description

This shrub or small tree grows up to 20 feet (6 meters) in height and forms thickets (Figure 1-1A). Branches are armed with stout curved spines. Leaves are bipinnately compound with three pairs of branchlets, each divided into two to seven pairs of leaflets. Flowers are pale yellow and clustered in dense round spikes 2 to 3 inches (5 to 8 cm) long and 2 cm in diameter (Figure 1-1B). Fruits are curled, contorted leguminous pods containing numerous seeds.

Figure 1-1A Catclaw acacia (*Acacia greggii*)

Figure 1-1B Catclaw acacia flowers (*Acacia greggii*).

Principal Toxin

Cyanogenic glycoside, prunasin.[1,6]

Note: There are many species of acacia, some of which are commonly browsed by livestock especially when other forages are scarce. Being high in protein, acacias often form an important part of the diet of range animals. Some species of acacia (*A. berlanderieri*), when eaten for prolonged periods in times of drought, can also cause neuromuscular weakness and ataxia, and can adversely affect fertility in animals. These effects are possibly induced by one or more of the various amines and alkaloids including nicotine, nornicotine, mescaline, mimosine, and amphetamines that have been identified in acacias.[62]

Western Service Berry, Saskatoonberry, June Berry
Amelanchier alnifolia
Rosaceae (Rose family)

Habitat

Service berry is common on rocky slopes, canyons, and alongside streams at altitudes from 5000 to 9500 feet throughout North America.

Description

Service berry shrubs or small trees grow up to 13 feet (4 meters) tall with simple, alternate, petioled leaves, that are oval to suborbicular in shape (Figure 1-2A) The margins are coarsely serrate or dentate to the middle or below, rarely entire. The flowers are perfect and regular. The calyx has five lobes, with five white petals each up to 0.5 inches (6 to 10 mm) long (Figure 1-2B). The ovary has five styles and develops into a purple pome when ripe.

Figure 1-2A Service berry shrub in bloom (*Amelanchier alnifolia*).

Figure 1-2B Service berry flowers (*Amelanchier alnifolia*).

Principal Toxin

Cyanogenic glycoside, prunasin.[13]

Note: Livestock will browse on the plants and usually do not have problems unless they consume excessive quantities of new growth or wilted leaves in times of drought. It is not advisable to plant service berry shrubs in livestock enclosures. The ripe dark red to black fruits are edible.

Mountain Mahogany
Cercocarpus montanus
Rosaceae (Rose family)

Habitat

This woody shrub is common in the drier foothills, chaparrals, and Ponderosa pine terrain of the Rocky Mountains.

Description

The shrub or small tree is up to 6 feet (2 meters) tall (Figure 1-3A). Leaves are simple, oval to linear, entire or toothed, often prominently veined beneath. The flowers are small and inconspicuous with yellowish sepals and no petals; they are either single or in clusters. The fruit consists of a small seed inside a hairy achene with a long, twisted style (Figure 1-3B). The sharp pointed basal end of the fruit and the corkscrew style enable the seed to bore into the soil as the style coils and uncoils in response to changes in humidity.

Figure 1-3A Mountain mahogany (*Cercocarpus montanus*).

Figure 1-3B Mountain mahogany showing pointed fruits with twisted, hairy style (*Cercocarpus montanus*).

Principal Toxin

Cyanogenic glycosides.[1,2]

Note: Livestock poisoning is seldom a problem with mountain mahogany because animals rarely browse on it. Wild deer commonly use it as a food.

Wild Blue Flax
Linum lewisii
Linaceae (Flax family)

Habitat

Blue flax is widely distributed from the plains to higher mountain elevations, growing in meadows, open forest, and along roadsides.

Description

This native herbaceous perennial has slender stems and alternate very slender leaves. The light powdery-blue flowers have 5 sepals and 5 petals that drop off by midday (Figure 1-4). New flowers open each day. Fruits are capsules that have 5 to 10 compartments, each containing a flattened, shiny, black seed.

L. perenne is a perennial species introduced from Europe and tends to be a more robust plant with deep blue petals, and sepals with entire margins. Flax that is used for linen and linseed oil production comes from the annual *L. usitatissimum*. It has deep blue flowers, and inner sepals with a fringed margin. Of the approximately 200 species of flax that occur globally, some including *L. flavum* (yellow flax), *L. grandiflorum* (red flax) are popular ornamental garden plants.

Figure 1-4 Blue flax (*Linum* spp.).

Principal Toxin

Cyanogenic glycoside, linamarin (lotaustralin).[3,9] Highest concentrations of the glycosides are found in the seedling tops and seeds (910 mg/100 g).[6] There however, appears to considerable variation in the quantity of cyanogenic glycoside present depending upon the species and growing conditions of flax. Yellow pine flax (*L. neomexicana*) has been reported as poisonous to sheep, but no cyanogenic glycosides were detected.

Note: Flax is rarely a problem for animals unless they have little else to eat. Linen flax (*L. usitatissimum*), commonly grown as a crop plant in Europe, is the principal source of linseed meal (cake), a by-product of linseed oil production. Cyanide poisoning results when uncooked linseed cake is fed to cattle. Commercially produced linseed cake should be boiled for 10 minutes before feeding it to ruminants in order to destroy the enzyme linamarase that liberates cyanide from the glycoside linamarin.[63]

Western Choke-Cherry
Prunus virginiana var. *melanocarpa*
Rosaceae (Rose family)

Habitat

This shrub or small tree grows in thickets, along waterways, mountainsides below 8000 feet, and occasionally in the drier plains. It is common in the western states especially in the Rocky Mountain area.

Figure 1-5A Western choke-cherry showing cylindrical flower raceme (*Prunus virginiana*).

Figure 1-5B Western choke-cherry ripe berries (*Prunus virginiana*).

Description

The shrub or small tree growing up to 16 feet (5 meters) in height, has gray bark with obvious lenticels. The leaves are generally ovate to obovate, occasionally lanceolate, with serrate margins. The leaves are simple, glossy, and alternate with a few glands on the petiole or base of the blade. The inflorescence is a cylindrical raceme of showy white fragrant flowers appearing in early spring after the leaves have appeared (Figure 1-5A) The fruit is a dark purple drupe when ripe and is the only edible part of the plant (Figure 1-5B).

Other members of the genus known to be toxic include pin cherry (*P. pennsylvanica*) and wild black cherry (*P. serotina*) (Figure 1-5C). These species vary from choke-cherry (*P. virginiana*) in leaf shape or inflorescence, or both.

Principal Toxin

Choke-cherries contain two cyanogenic glycosides—amygdalin, which is commonly found in the seed, and prunasin, which is found in the leaves, bark, and shoots.[1,4,12]

Note: The larger succulent leaves are the most toxic with concentrations of cyanogenic glycosides reaching 368 mg/100 g of fresh leaves. Ruminants consuming 25 percent of their body weight in green leaves are likely to suffer fatal cyanide poisoning. Wilted leaves and branches and new growth are especially toxic. The ripe berries are edible, although the seeds are toxic. It is inadvisable to plant choke-cherry shrubs or trees in or adjacent to animal enclosures.

Figure 1-5C Black cherry (*Prunus serotina*).

Elderberry
Sambucus canadensis
Caprifoliaceae (Honeysuckle family)

Habitat

Elderberry plants generally prefer open areas in the rich moist soils surrounding ponds and along ditches and streams throughout North America.

Description

The woody shrubs grow to 10 feet (3 meters) in height, forming colonies from underground runners. The stems are thick and are filled with white pith. The leaves are opposite, pinnately compound with lanceolate serrated leaflets. Conspicuous, terminal, round, or flat topped clusters of numerous white, five-petaled flowers 0.25 inches (4 to 6 mm) in diameter (Figure 1-6A). Drooping clusters of dark purple (*S. canadensis*) (Figure 1-6B) or red (*S. racemosa*) (Figure 1-6C) berries with several seeds form July through September. Some species of elderberry have been hybridized for ornamental purposes.

Figure 1-6A (above)
Elderberry (*Sambuccus canadensis*).

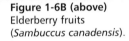

Figure 1-6B (above)
Elderberry fruits
(*Sambuccus canadensis*).

Principal Toxin

Elderberries contain a cyanogenic glycoside, sambunigrin, and an irritant alkaloid that acts as a cathartic.[2]

Note: All parts of the elderberry plant are potentially toxic to animals although animal poisoning is rarely encountered. Cultivated varieties of elderberry should be considered potentially toxic and probably should not be planted in or around animal enclosures. The purple to black ripe berries are edible if well cooked and are often used for making jams and wine.

Figure 1-6C Elderberry
(*Sambuccus racemosa*).

Johnson Grass
Sorghum halepense
Poaceae (Gramineae)(Grass family)

Habitat

Johnson grass is a common weed grass found in alluvial bottom land and along roadsides and ditches throughout most areas east and south of the Rocky Mountains. Sudan grass (broom corn, kafir corn), *Sorghum sudanense*, and its hybrids are frequently grown as a forage crops.

Description

Johnson grass is a coarse drought-resistant perennial growing 3 to 8 feet (1 to 3 meters) tall, with scaly root stalks and relatively broad leaves with a prominent mid vein (Figure 1-7A). Seeds are yellow-purple and occur in a large multi-branched panicle. Sudan grass and its hybrids are very similar in appearance but are annuals and tend to be more robust than Johnson grass, with broader leaves and thicker stems (Figure 1-7B). The seed heads are produced terminally and, in some hybrids, contain much larger seeds than the species.

Figure 1-7A Johnson grass (*Sorghum halepense*).

Principal Toxin

Cyanogenic glycoside, dhurrin.[3,4]

Note: Johnson and Sudan grasses are the most common cause of cyanide poisoning in cattle and sheep and are especially toxic when growing rapidly. Fertilization with nitrogen increases the potential for cyanide toxicity.[1,11] Regrowth of sorghums after cutting has high potential for poisoning, and there is an old saying that sorghums should not be grazed until they are above knee height. Cyanide-free hybrids of Sudan grass are available as forage crops for animal consumption. All species of sorghum may also accumulate toxic levels of nitrate, and are a common source of nitrate poisoning in cattle.

In addition to their acute cyanogenic and lathyrogenic effects, certain grain *Sorghum* species also contain tannins that can have (astringent) protein-precipitating properties. Condensed tannins impart a red color to sorghum seed that deters birds from eating the seed. Ruminants are not affected by the tannins in sorghum seed.

Figure 1-7B Sudan grass (*Sorghum sudanense*).

Poison Suckleya
Suckleya suckleyana
Chenopodiaceae (Goosefoot family)

Habitat

This small, prostrate plant grows in moist conditions preferring the edges of receding reservoirs and ponds. It also grows around the edges of irrigated fields, and where run-off water collects, it becomes a robust plant. Suckleya is found in localized areas from Montana to New Mexico.

Figure 1-8A Poison suckleya (*Suckleya suckleyana*).

Figure 1-8B Poison suckleya showing the green flowers at the leaf axils (*Suckleya suckleyana*).

Description

Suckleya is an annual, succulent, prostrate herbaceous plant with reddish fleshy stems 1 to 2 feet (30 to 60 cm) long (Figure 1-8A). The leaves are alternate, triangular, or spade shaped with dentate margins and long petioles. The inconspicuous small flowers are formed in the leaf axils (Figure 1-8B). The plants are monoecious with the male flowers at the tips of the branches and the female flowers along the remainder of the branching stems. The fruits are reddish brown and are enclosed by two papery, dark-colored scales joined at the tip.

Principal Toxin

Cyanogenic glycosides.[23,24,64]

Note: Poisoning from suckleya varies, being troublesome in some years when cattle and sheep find it quite palatable.[1,24] This may be due to the fact that it often grows around ponds, and in years of drought, livestock eat it as something green when they come to drink and congregate in the area. In some years, large populations of suckleya may appear in flood plains. The glycoside content of the plants is quite variable depending upon the growing conditions of the plants.

Arrow Grass, Pod Grass, Goose Grass
Triglochin maritima
T. palustris (Marsh arrow grass)
Juncaginaceae (Arrow grass family)

Figure 1-9A Arrow grass showing grass-like leaves and seed heads (*Triglochin maritima*).

Habitat

Arrow grass species grow throughout North America preferring alkaline soils at most elevations and flourishing in marshy ground and irrigated pastures. *T. maritima* is found mostly in the midwestern states and to the west and south; *T. striata* occurs in the southeastern states into Florida and southern California into Mexico. *T. palustris* is found in the northern states from Washington to New York and the Rocky Mountain area.

Description

These are perennial "grasslike" plants with fleshy, half rounded, dark green leaves clumped at the base of the plant (Figure 1-9A). Leaves are 6 to 12 inches (15 to 30 cm) long, linear, unjointed, and sheathed at the base. The flower is a pediceled raceme up to 4 feet (1.5 meters) in length that appears as an unbranched, unjoined flower spike. The individual flowers are inconspicuous and numerous, with a greenish, six-part perianth. The fruits are made up of capsules or pods of three to six cells that turn golden brown before splitting open. (Figure 1-9B).

Principal Toxin

Cyanogenic glycoside, triglochinin.[18]

Note: Arrow grass is not a true grass. Its long slender grass-like leaves make it difficult to recognize in a grass meadow unless the distinctive flower and seed pods are present. The plant is most toxic when green, losing its cyanogenic potential when dried. Arrow grass has long been recognized as a poison for cattle and sheep when green, but it is generally not a problem when incorporated in hay.[16]

Figure 1-9B Arrow grass seed pods (*Triglochin maritima*).

Plant-Associated Nitrate Poisoning

Nitrate poisoning is a universal and economically important problem for ruminants that is caused by the ingestion of plants that have accumulated toxic levels of nitrate.[65-74] Normally plants absorb nitrates from the soil converting them into plant proteins. Application of organic or inorganic nitrogenous fertilizers can result in excessive accumulation of nitrates in crop plants and common weeds. Livestock consuming these plants in quantity can develop nitrate poisoning. Plant nitrates, however, are rarely a problem for horses because their digestive system does not readily convert nitrate (NO_3) to toxic nitrite (NO_2). The effects on animal health of nitrate and its metabolites, nitrite and N-nitroso compounds, have been extensively reviewed.[75] The potential for nitrate poisoning is increased when water sources for livestock also contain high levels of nitrates.[76,77] Nitrate fertilizers are highly toxic chemicals capable of causing fatal poisoning in ruminants and horses that gain access to them accidentally. When animals consume nitrate fertilizers, the chemicals themselves cause severe gastrointestinal irritation, colic, and diarrhea.

Many common weeds, forage crops, and cereal grain plants have the potential for accumulating nitrate under specific growing conditions (Table 1-2). Nitrate poisoning has been reported most often in ruminants that have consumed sorghums.[74,78-80] However, sugar beet tops, turnips, kale, oats, silage, Italian rye grass, white clover, red root pigweed, and variegated thistle have also been incriminated in nitrate poisoning.[78,81-86] Nitrate levels in plants vary considerably depending on the plant species, stage of growth, water and organic content of the soil, and application of nitrogen fertilizers. Drought conditions, acidic soils, and soils deficient in sulfur, phosphorous, and molybdenum result in nitrate accumulation in plants.[73,87-89] Cool, cloudy days enhance nitrate formation in plants because the light-and warmth-dependent enzyme, nitrate reductase, is inhibited, thus allowing nitrate to accumulate in the plant.[90] Nitrate levels are therefore highest in plants at night and early morning when the nitrate-reducing enzymes are least active. Highest levels of nitrate tend to be found in the stems where nitrate reduction normally occurs, and not in the leaves.[90,91] Nitrate does not accumulate in the flowers or fruits of plants and

Table 1-2 Plants Known to Accumulate Nitrates [92,93]	
Botanical Name	**Common Name**
Ambrosia spp.	Ragweeds
Amaranthus spp	Pigweed
Avena fatua	Wild oat grass
Chenopodium spp.	Lamb's-quarter
Cirsium arvense	Canada thistle
Convolvulus arvense	Field bindweed
Datura stramonium	Jimsonweed
Echinochloa spp.	Barnyard grass
Helianthus annuus	Sunflower
Kochia scoparia	Kochia weed
Malva spp.	Cheese weed
Melilotus spp.	Sweet clover
Polygonum spp.	Smart weed
Rumex spp.	Sorrel, curly leafed dock
Salsola kali	Russian thistle
Solanum spp.	Nightshades
Solidago spp.	Goldenrods
Sorghum halapense	Johnson grass
Crop Plants	
Avena sativa	Oats
Beta vulgaris	Sugar beets
Brassica napus	Rape
Glycine max	Soybean
Linum spp.	Flax
Medicago sativa	Alfalfa
Pennisetum glauca	Pearl millet
Secale cereale	Rye
Sorghum vulgare	Sudan grass
Triticum aestivum	Wheat
Zea mays	Corn

therefore nitrate poisoning is unlikely when seeds (corn, oats, barley) are fed to live-stock. Properly prepared silage from forage crops high in nitrates reduces the nitrate content by 60 percent, while there is little reduction of nitrate in dried hay.[77] The application of herbicides such as 2,4-dichlorophenoxyacetic acid (2,4-D), not only increases the nitrate content of plants, but also the palatability of the plants thereby increasing the potential for poisoning.[87,90,92]

Nitrate Toxicology

There is considerable variation as to what constitutes a safe level of nitrate in animal feeds because of different factors that influence nitrate metabolism. Under normal circumstances, nitrate is reduced in the rumen in a series of steps from nitrate to nitrite, to ammonia, and eventually to microbial proteins. It is the rapid formation and absorption of large quantities of nitrite (NO_2) and not nitrate (NO_3) that causes poisoning. The rate at which nitrate is converted to highly toxic nitrite depends on the rate of adaptation of rumen microorganisms to nitrate, the rate and amount of nitrate ingested, and the amount of carbohydrate available in the rumen. Experimental data suggest that nitrate poisoning is more likely to occur in ruminants after several days of feeding forages high in nitrate.[94,95] Other investigators have demonstrated that nitrate-adapted rumen microflora more completely reduce nitrate beyond nitrite to ammonia thereby reducing the potential for poisoning.[96,97] Similarly, when carbohydrates such as corn and molasses are present in the rumen, nitrates are more rapidly converted to ammonia and microbial proteins without the accumulation of nitrite.[87,96,98,99] On the other hand, low-energy diets increase an animal's susceptibility to nitrite poisoning.

The amount of nitrate that can be safely consumed in forages (45 g nitrate/100 lb body weight) is three times greater than the amount of potassium nitrate (KNO_3) that can be given orally as a drench.[87] Similarly sheep can be fatally poisoned by a single oral dose of KNO_3, although the same dose has no ill effects when incorporated in the feed.[96] The lethal dose of nitrate given as a drench is 0.5 g KNO_3 kg body weight.[83] From this information it is apparent that nitrate produced in plants is far less toxic than the pure chemical present in fertilizer. Plants or hay containing more than 1 percent nitrate (10,000 ppm) dry matter are potentially toxic and should be fed with caution. Forages containing more than 1 percent nitrate should only be fed if the total nitrate intake can be reduced to less than 1 percent by diluting the nitrate forage with nitrate-free forages. Because nitrate is often reported in different units, care must be exercised in interpreting nitrate values. Conversion factors for nitrate and nitrite compounds are given in Table 1-3.

The addition of monensin to rations high in nitrate may precipitate poisoning. This has been reported in cattle fed turnips and forage high in nitrate that produced no clinical signs until monensin was given as a feed additive.[100,101]

Accumulation of nitrates in water sources frequently poisons livestock.[76,98,102] Surface water and water from shallow wells is most likely to contain nitrates, especially if there is the potential for run-off water from fertilized arable land contaminating the water source. Acute nitrate poisoning resulted when cattle were given well water containing 2790 ppm of nitrate.[102] An unusual case of nitrate poisoning occurred in cattle drinking water from shallow ponds created by blasting using the explosive mixture of ammonium nitrate fertilizer and dynamite.[103] Acceptable levels of nitrate nitrogen in water according to United States Health Service standards should not exceed 10 ppm of nitrate nitrogen (45 ppm nitrate).[104] Water nitrate levels less than 45 ppm are also desirable for livestock, but nitrate levels up to 445 ppm (100 ppm nitrate nitrogen) can be tolerated.[105,106] Levels above 200 ppm of nitrate should be considered potentially toxic to pregnant animals.[107,108] However, it has been shown experimentally that sheep can consume water with up to 667 ppm of nitrate-nitrogen without measurable adverse effects presumably because they were on a high plane of nutrition.[109] Water containing up to 100 ppm of nitrate can be considered safe for all classes of livestock assuming that the animals are on a normal diet that does not have high levels of nitrate.[98] To be safe, both the water and the forage should be analyzed to ensure that total nitrate does not exceed potentially toxic levels.

Table 1-3 Conversion Factors for Nitrate and Nitrite Compounds		
Nitrate ppm = Nitrate mg/kg = Nitrate mg/L		
% Nitrate X 10,000 = ppm		
Ppm X 0.0001 = % Nitrate		
Ppm = mg/mL		
Potassium/Sodium nitrate X 0.61 = Nitrate		
Potassium/Sodium nitrate X 0.14 = Nitrate-nitrogen		
Nitrate-nitrogen (NO-N) X 4.45 = Nitrate		
Nitrate-nitrogen (NO-N) X 3.29 = Nitrite		
Nitrate-nitrogen (NO-N) X 6.1 = Potassium or Sodium nitrate		

Monogastric animals such as horses, pigs, and dogs are unlikely to develop nitrate poisoning from plants because they cannot readily convert nitrate to nitrite in their digestive systems. However, monogastric animals that gain access to nitrate fertilizers will eat them and develop severe gastrointestinal irritation resulting in colic and diarrhea. Horses, pigs, and ruminants are equally susceptible to nitrite poisoning should they ingest nitrite salts such as sodium or potassium nitrite.[70,71,96,110] Nitrites are also potent vasodilators and cause a rapid drop in blood pressure.[111] In general all animals are susceptible to nitrate poisoning if they consume enough of the chemical.

In all animals, the nitrite ion readily reacts with hemoglobin in red blood cells, oxidizing it to form methemoglobin, which cannot transport oxygen. When over 30 to 40 percent of hemoglobin is converted to methemoglobin, clinical signs of poisoning become apparent.[95,107,111,112] Death occurs as methemoglobin levels approach 80 percent.

Clinical Signs

The first sign of nitrate poisoning is usually the sudden death of one or more animals. If observed before death, ruminants with nitrate poisoning may exhibit drowsiness and weakness, followed by muscular tremors, increased heart and respiratory rates, staggering gait, and recumbency.[65,67,78-80,86,113-116] Signs of poisoning develop within 6 to 8 hours of the consumption of a toxic dose of nitrate.[68] Stress or forced exercise will increase the severity of clinical signs and hasten death. Examination of the mucous membranes, especially the vaginal mucous membranes, may reveal a brownish discoloration depending on the quantity of methemoglobin present. This color change can be detected in the vaginal mucous membranes when 20 percent or more methemoglobin has formed.[94] This brownish discoloration occurs well before other clinical signs become evident, suggesting vaginal color changes are a good means of detecting nitrate poisoning before severe toxicity develops.[94,117] Venous blood also has a chocolate brown discoloration. Depending on the quantity and rate of absorption of nitrite from the digestive tract, and the amount of stress to which the animal is subjected, death may occur within 2 to 10 hours.[118]

Chronic Nitrate Poisoning

Sublethal doses of nitrate may induce abortion because nitrate readily crosses the placenta and causes fetal methemoglobinemia and death.[119] Severe methemoglobinemia also impairs oxygen transportation across the placenta, thus contributing to fetal hypoxia and death.[94] Fetal death and abortion may occur at any stage of gestation as a result of the combined effects of decreased placental oxygen transport and the limited ability of the fetus to metabolize nitrite.[74,77,78,80,120,121] Abortions may also result from decreased progesterone production induced by chronic nitrate poisoning interfering with luteal production of progesterone.[127]

Low levels of nitrate in the diet of cattle have also been suspected of affecting vitamin A metabolism, thyroid function, reproduction, and milk production. There are however many conflicting reports on the effects of low level nitrate poisoning. Nitrate appears to affect metabolism of carotene and vitamin A, reducing liver levels but having little effect on plasma levels.[123,124] Experimentally however, there appears to be little destruction of vitamin A in the rumen in the presence of high levels of nitrate, and nitrates do not appear to have any goitrogenic effect in cattle.[125-127] Nitrate levels in the ration exceeding 1 percent can cause a reduction in feed intake in cattle,[128] but if a ration with high levels of digestible nutrients is fed concurrently, increased weight gains occur.[128,129] These findings indicate that high-carbohydrate rations increase the utilization of nitrate in the formation of protein. Numerous studies have also shown that nitrates do not affect milk production unless the levels of nitrate induce significant methemoglobinemia.[128,130-133] Low levels of nitrate accumulate in milk of animals consuming forages high in nitrate, with maximum levels attained 2 hours after ingestion of the nitrate.[122,134,135] Reproductive efficiency, gestation length, and birth weights appear to be unaffected by low nitrate consumption.[114,119]

Treatment

Animals showing signs of nitrate poisoning should be handled carefully to avoid stress or excitement that will worsen the animal's respiratory distress. The suspected nitrate food source should be removed. The preferred treatment for nitrate poisoning is methylene blue solution administered intravenously. As a reducing agent it converts methemoglobin to hemoglobin thereby restoring normal oxygen transport by the red blood cells. The recommended dose range for methylene blue is 4 to 15 mg/kg body weight administered as a 2 to 4 percent solution.[73,136] A dose of 8 mg/kg body weight intravenously has been effective in cattle.[82,118,136] In sheep, the half-life of methylene blue is about 2 hours, indicating that small doses of the drug can be repeated as needed every few minutes to reduce methemoglobinemia to the point that the animal is not in severe respiratory distress.[108] Excessive administration of methylene blue to animals other than ruminants will result in hemolytic anemia due to formation of Heinz bodies. Horses, and especially dogs and cats, are particularly susceptible to methylene blue toxicity. Animals with severe respiratory distress can be given oxygen where possible to optimize oxygen saturation of remaining hemoglobin. In severe cases, epinephrine can be administered intravenously to counter the acute hypotensive effects of the nitrite.

The administration of mineral oil (1 gallon for a 500-kg cow) orally via stomach tube will counteract the caustic effect of the nitrates on the gastrointestinal system and will speed up the passage of the nitrates. Several gallons of cold water with added oral broad-spectrum antibiotics will further decrease nitrate reduction to nitrite by rumen microorganisms. Similarly vinegar given orally via stomach tube will help prevent nitrate reduction in the rumen.

Diagnosis

Sudden deaths in ruminants grazing Sudan grass (*Sorghum* spp.), weeds or crop stubble post harvest should raise suspicion of nitrate poisoning. Confirmation should be based on the demonstration of toxic levels of nitrate in the forage, water source, rumen contents and tissues of the animal. At necropsy, the highest concentrations of nitrite are found in the heart, lungs, and kidneys. Detection of high levels of methemoglobin (greater than 40 percent of total hemoglobin) in the animal is diagnostic of nitrate poisoning. Normal levels of methemoglobin in cattle blood are in the range of 0.1 to 0.2 g/dL.[138] Because reduction of methemoglobin occurs rapidly after death, samples should be taken and submitted for analysis as soon after death as possible. Tissue and plant samples should be frozen if they cannot be analyzed immediately. If the blood sample cannot be tested immediately, it should be diluted with 1 part blood to 20 parts phosphate buffer (pH 6.6) and frozen.[93]

If the animal has been dead for several hours or more, the best sample to submit for nitrate analysis is the aqueous humor from the eyes. Nitrate in the aqueous fluid is protected from autolytic changes that occur rapidly in other parts of the body after death. Good correlation with serum levels can be obtained for about 24 hours postmortem and the nitrate levels remain diagnostically significant for as long as 60 hours.[136-142] The normal level of nitrate in the ocular fluid of healthy cattle is 4 to 5 mg/L.[136] Nitrate levels in aqueous humor of 20 to 40 ppm should be considered suspect, and over 40 mg/L (40 ppm) could be considered diagnostic of nitrate poisoning if corroborating clinical signs are seen and evidence of high nitrate levels is found in the forage and/or water. Ocular fluid from an aborted fetus is useful for determining if nitrate is the cause of abortion provided the levels detected are interpreted in light of forage and water nitrate levels to which the dam would have had access.

As a general rule, levels of nitrate over 0.5 percent in forages and water levels exceeding 200 ppm are potentially hazardous to pregnant animals especially if fed continuously.[107] Forages containing in excess of 1 percent nitrate dry matter should be considered toxic.[108] Water levels of 1500 ppm or greater are potentially toxic to ruminants especially if consumed with forages high in nitrate.[107,108] Nitrate and nitrite can be assayed in forage, rumen contents, and water using the diphenylamine test,[136] ion-specific electrodes,[141] and high-performance liquid chromatography.[141] Presumptive diagnosis of nitrite poisoning can be made in the field using diazotization urine test strips in the aqueous humor and peritoneal and pericardial fluids of animals suspected of nitrate poisoning.[143]

Depending on the laboratory performing the test, nitrate may be reported in a variety of units that can be confusing eg: nitrate nitrogen (NO-N), ppm, NO_3 mg/kg, etc. The conversion factors are given in Table 1-3. It is important to convert the reported units to a standard nitrate unit that can be related to established normal values.

Prevention of Nitrate Poisoning

Nitrate poisoning can be prevented if the nitrate levels in forages are predetermined and managed accordingly. Forages such as sudan grass and sorghum hybrids, oat hay, and corn stalks should be tested especially if heavy nitrogen fertilization has been used or drought has affected the plants.[88,144] Hay containing high nitrate levels that is exposed to rain can have the nitrate leached out into the lower bales making them especially high in nitrates. It is also prudent to check the water of the animals to ensure it is not a source of nitrates that would be additive to any nitrate in the food. Contrary to popular belief, boiling water for prolonged periods does not decrease the level of nitrate.[145]

Forages containing 1 percent nitrate or more should be fed cautiously to ruminants. There are several strategies that can be implemented if hay and other forages are found to contain high levels of nitrates. Ideally, hay that has more than 1 percent nitrate should be diluted with hay containing no nitrates so that the total nitrate level in the ration is below 1 percent. Feeding low-nitrate forage or hay before turning cattle onto forages containing higher levels of nitrate reduces the amount of nitrate consumed. Feeding high-nitrate forages to nonpregnant cattle eliminates the risk of abortion. Products containing nitrate-reducing bacteria (*Propionibacteria* spp.) are available commercially for feeding to ruminants before exposing them to high-nitrate forages. This enables cattle to tolerate higher nitrate consumption.[146] Increasing the total energy content of the ration also enhances the metabolism of nitrate in the rumen thereby helping ruminants tolerate higher nitrate levels in their diet. It is for this reason that cattle on a good plain of nutrition are able to consume forages that have 2 percent or more of nitrates. However, sudden changes of feed from a low energy ration to one that is high in energy and nitrates may result in high mortality because the rumen microorganisms will not have had time to adapt to the high nitrate ration. Ensuring animals are on a good plain of nutrition before introducing a forage high in nitrate reduces losses.

Common Weeds that Accumulate Nitrate

Ragweed
Ambrosia **spp.**
Asteraceae (Sunflower family)

Habitat

The ragweeds are native annual or perennial weeds of most states west of the Missouri River. They grow in most soils in cultivated fields, roadsides, pastures, and rangeland.

Description

Giant ragweed is a rapidly growing annual that may attain heights of 8 to 10 feet (2 to 3 meters) in moist fertile soils. The stems and leaves are rough, the leaves usually being trilobed, but they also may have either no lobes or as many as five lobes (Figure 1-10). The leaves are alternate and large and attached with a long petiole. The male flowers (uppermost) and the female flowers (lowermost) are small and produced in terminal clusters. The seeds are about 1 cm in length, each with four to five terminal spikes.

Figure 1-10 Giant ragweed (*Ambrosia trifida*).

Principal Toxin

Nitrates are readily accumulated in the young plants especially in rich fertile soils. The pollen produced from ragweed can be a serious cause of hay fever and allergies in people and dogs.

Note: Various other members of the ragweed family also have the potential for nitrate accumulation. These include common ragweed (*A. artemisifolia*), woolly leaf bursage (*A. grayi*), and skeleton leaf bursage (*A. tomentosa*). Giant ragweed (*A. trifida*) is notorious for causing allergies in people.

Wild Oat Grass
Avena fatua
Poaceae (Grass family)

Habitat

Wild oats is a widely distributed grass, introduced from Europe, that becomes established in grain fields, pastures, and disturbed soils. It seeds are viable in the soil for at least 10 years, making it difficult to eradicate. It is considered a noxious weed in many areas.

Description

Wild oats is an annual, erect, hollow-stemmed grass that grows to 4 feet in height. It has leaf blades up to a half inch (1.5 cm) wide, with open sheaths and membranous ligules. Seedlings have leaves that twist counterclockwise. The flower is an open panicle that droops. The awn has a distinctive long bristle with a characteristic right-angle bend (Figure 1-11). The seeds are yellow to black and about 0.5 inches (1 cm) in length.

Figure 1-11 Wild oats showing distinctive bent awn (*Avena sativa*) .

Principal Toxin

Wild oats, like domestic oats, has the potential to accumulate significant levels of nitrate.

Note: Cultivated oats (*A. sativa*) can be differentiated from wild oats in that it has a straight bristle. Cultivated oats may accumulate toxic levels of nitrates especially if fertilized.

Lamb's-Quarter
Chenopodium spp.
Chenopodiaceae (Goosefoot family)

Habitat

Lamb's-quarter (*C. album*) is a common annual weed, introduced from Europe, and found growing throughout North America in cultivated fields, gardens, roadsides, and pastures. Nettleleaf goosefoot (*C. murale*) and netseed lamb's-quarter (*C. berlandieri*) are also common rapidly growing weeds with similar distribution and habitat.

Description

Lamb's-quarter is a variable annual, with erect branched stems, alternate leaves that often have a grayish, powdery undersurface. The basal leaves have a more coarsely serrated margin than the smaller upper leaves. The stems often have distinct red or purple stripes. Flowers are inconspicuous, gray green, and crowded at the axils and branch tips (Figure 1-12). Large quantities of dark seeds with a "netted" surface are produced.

Figure 1-12 Lamb's-quarter (*Chenopodiun album*).

Principal Toxins

Lamb's-quarter is capable of accumulating significant quantities of nitrate, sulfate, and oxalates.

Canada Thistle
Cirsium arvense
Asteraceae (Sunflower family)

Habitat

Introduced from Eurasia, Canada thistle has become a widely established noxious weed. It grows readily in many soil types forming dense stands in cultivated as well as waste places. It is a prolific producer of wind-borne seeds and also reproduces from an extensive root system.

Description

Canada thistle is an erect perennial colony-forming weed with an extensive horizontal root system. Stems reach 3 to 4 feet (1 meter) in height, branching above with terminal purple thistle flowers (Figure 1-13). Leaves are alternate, without petioles, and lance-shaped with spiny-tipped irregular lobes. Flowers are unisexual on separate plants, with purple and occasionally white, 0.5 inch (1 to 2 cm) diameter heads. The brownish seeds have a tuft of white hairs to aid in distribution by the wind.

Figure 1-13 Canada thistle (*Cirsium arvense*).

Principal Toxin

Canada thistle is listed as a noxious weed in many states because of its invasive nature. It will accumulate nitrates and may contain high levels of sulfates when growing in sulfate-rich soils.

Field Bindweed
Convolvulus arvensis
Convolvulaceae (Morning glory family)

Habitat
Field bindweed was introduced from Europe and has become a noxious weed throughout North America. It adapts to almost any soil type and is very invasive, drought tolerant, and difficult to eradicate.

Description
Bindweed is a perennial, prostrate vine 2 to 4 feet (1 meter) long. It forms dense mats covered with white to pink morning glory-like flowers (Figure 1-14). It readily twines around and up other plants or objects. Leaves are alternate, arrow shaped with pointed or rounded lobes at the base. The root system is extensive, penetrating to depths of 10 feet (3 meters) or more. The fruits are round capsules containing four brown seeds, flattened on two sides.

Figure 1-14 Field bindweed (*Convolvulus arvensis*).

Principal Toxins
Field bindweed will accumulate nitrates. It also contains various tropane alkaloids including pseudotropine, tropine, tropinone, and cuscohygrine.[147,148] Pseudotropine, the predominant alkaloid, is capable of affecting smooth muscle activity. Other nortropane alkaloids (calystegins) present in bindweed (*Calystegia* spp. some *Solanum* spp., and *Ipomoea* spp.) are potent glycosidase inhibitors.[149]

Note: Hedge bindweed (*C. sepium*) is similar to field bindweed but has two large bracts immediately below the flower. All bindweeds are similarly toxic.

Barnyard Grass
Echinochloa crus-galli
Poaceae (Grass family)

Habitat

Barnyard grass was introduced from Europe and is widespread in North America. It is commonly found in gardens and cultivated fields; it is a noxious weed.

Description

Barnyard grass is a vigorous, annual warm-season grass, reaching 1 to 5 feet (1 to 2 meters) in height. Bases of stems are often dark purple. Leaf blades are flat, smooth, up to 1 inch (2.5 cm) broad, and with no ligules at the junction of the sheath and blade. The flower is an open panicle, reddish brown in color (Figure 1-15). The seeds are crowded on spikelets, each with a short stiff awn. As many as 40,000 seeds may be produced by a single mature plant.

Figure 1-15 Barnyard grass seed heads (*Echinochloa crus-gali*).

Principal Toxin

Barnyard grass may accumulate toxic levels of nitrate if it is growing in fertile soils or is fertilized.

Sunflower
Helianthus annuus
Asteraceae (Sunflower family)

Habitat

Sunflowers are indigenous to North America and are common weeds of roadsides, cultivated fields, pastures, and disturbed soils.

Figure 1-16 Sunflower (*Helianthus annuus*).

Description

Sunflowers are annuals or perennials, growing from 1 to 10 feet (1 to 3 meters) tall, with erect single to many-branched, rough stems. Leaves are alternate, rough, hairy, and heart shaped with toothed edges. The flowers are showy, with yellow ray flowers and brown or yellow disk flowers (Figure 1-16). The seeds (achenes) vary from black to gray, and may be striped.

Principal Toxin

Rapidly growing sunflowers can accumulate toxic levels of nitrate.

Kochia, Mexican Fireweed
Kochia scoparia

Habitat

Kochia weed was introduced from Asia as an ornamental and now has become established as a weed in most of North America. It is versatile, growing in gardens, cultivated fields, pastures, disturbed soils, and along roadsides. It is readily grazed by livestock and can be a valuable forage when little else is available, especially in dry areas or in times of drought.

Description

Kochia is an annual growing up to 10 feet (3 meters) tall when established in fertile, moist soils (Figure 1-17A). Stems are much branched, round, hairy, and often red tinged especially in the fall (Figure 1-17B). Leaves are alternate, hairy especially below, lance shaped, smooth edged, and up to 2 inches (5 cm) in length. The flowers are small and are produced in dense, bracted spikes in the leaf axils (Figure 1-17C) Seeds are brown, slightly ribbed, and produced in large quantities.

Figure 1-17A Kochia weed growing in a corn field (*Kochia scoparia*).

Principal Toxins

Kochia weed is capable of accumulating significant levels of nitrate as a young, rapidly growing plant. It may also accumulate oxalates and sulfates and has been responsible for causing liver disease and photosensitization in some years due to an as yet undefined toxin (see Chapter 4). The variability in the reported toxicity of kochia weed is poorly understood and is probably related to the growing conditions of the plant.

Note: Kochia weed is grown in some areas with low rainfall as a drought-tolerant forage crop for livestock without evidence of toxicity.

Figure 1-17B Kochia weed immature plant (*Kochia scoparia*).

Figure 1-17C Kochia weed in flower (*Kochia scoparia*)

Mallow, Cheese Weed
Malva neglecta
Malvaceae (Mallow family)

Habitat

Mallow is a common weed introduced from Europe that has become widespread throughout North America. It successfully establishes itself in gardens and cultivated and waste areas.

Figure 1-18 Mallow (*Malva* spp.)

Description

Common mallow is an annual or perennial, with prostrate stems and erect branches that may reach 3 feet (1 meter) in length. Mallows have a substantial taproot. Leaves have long petioles and are rounded with prominent veins dividing the leaf into five to seven lobes. The flowers with fused petals range from white to pale lavender in color (Figure 1-18). Fruits consist of a circle of flat-sided, brown seeds arranged like a "cheese cake."

Little mallow (*M. parviflora*) is similar in appearance but is a more upright and robust plant with 2 to 5 inch (5 to 12 cm) broad leaves, with a red spot at their base.

Principal Toxin

Mallows have the potential for accumulating toxic levels of nitrate.

Note: The plant and seeds of mallows may cause a pink coloration to egg whites when eaten by laying hens.

Russian Thistle, Tumble Weed
Salsola iberica (S. kali)
Chenopodiaceae (Goosefoot family)

Habitat

Russian thistle was introduced from Russia and is now well established as a weed in much of western North America, especially in drier areas. It, however, will grow in cultivated areas, roadsides, and overgrazed rangeland. Russian thistle is the classic "tumble weed" that is encountered blowing across the rangelands of the west in the fall.

Figure 1-19A Russian thistle (*Salsola iberica*).

Figure 1-19B Russian thistle young plant (*Salsola iberica*).

Description

Russian thistle is a rounded, bushy, annual plant, up to 5 feet (1.5 meters) in height. Stems are often red or purple striped (Figure 1-19A). Immature plants have soft, thin, stringlike leaves, which later shorten and become stiff and tipped with a spine (Figure 1-19B). This makes the plant prickly to the touch. Flowers are small and green and are produced in the axils of the upper leaves, each accompanied by two spiny bracts.

Principal Toxin

In fertile moist soils, Russian thistle may accumulate toxic amounts of nitrates and oxalates.

Note: Each Russian thistle plant produces thousands of seeds, which are scattered widely when blown about in the fall. The seeds can remain dormant for years, germinating when conditions are optimal for growth. Russian thistle is also an undesirable weed in sugar beet fields because it is a host plant for the sugar beet leaf hopper that transmits the virus that causes "curly top" in beets.

Plants Containing Toxic Alkaloids

Larkspur Poisoning

Larkspurs in the genus *Delphinium* cause more fatal poisoning of cattle in the western United States than any other native plant species.[150,151] As early as 1897, livestock losses to larkspur were reported in Montana and even now larkspur poisoning remains a serious threat to cattle with access to the plants.[152-154] Livestock losses to larkspur in the United States have been estimated to exceed $234 million annually, making larkspurs second only to the locoweeds in terms of economic losses to the livestock industry.[155] In some areas of the intermountain states, cattle losses to larkspur poisoning average 2 to 5 percent per year and may be as high as 10 percent in some year.[156,157]

Larkspur, Poison Weed
Delphinium spp.
Ranunculaceae (Buttercup family)

Figure 1-20A Tall larkspur
(*Delphinium barbeyi*).

Habitat

There are at least 60 species of larkspur found throughout North America, the majority occurring in the western states. They range from Alaska and the Canadian Provinces south to Mexico.[151,157] Larkspurs grow in rich loamy soils of moist areas of the mountains and in the drier sandy soils of the plains and foothills.

Larkspur poisoning of cattle has been attributed to relatively few species, including *D. barbeyi, D. bicolor, D. geyeri, D. glaucescens, D. glaucum, D. nuttallianum, D. occidentale, D. tricorne,* and *D. virescens.*[158] It is probably wise, however, to assume that all larkspurs are potentially poisonous, including those cultivated as ornamentals. Larkspurs are often grouped for descriptive convenience into tall and low varieties according to their growth habit. Tall larkspurs (*D. barbeyi and D. occidentale, D. glaucescens*) (Figure 1-20A and B) grow in deep, moist, and highly organic soils at high altitudes and often reach 7 feet in height.[151,159] The tall larkspurs generally grow in montane forests, especially where snow drifts occur perennially. The plants emerge as the snows recede and once established form very long-lived dense stands. Low larkspurs (*D. nuttallianum* [Figure 1-20C], *D. nelsoni,* and *D. virescens* [Figure 1-20D]) grow at lower elevations in drier rangeland, seldom growing more than 2 to 3 feet (0.5 to 1 meter) tall; they die in early summer as the soil dries out.

Foothills larkspur (*D. geyeri*) is intermediate in its growth habit, attaining a height of 3 to 4 feet (1 meter) when in flower (Figure 1-20E and F).

Description

Larkspurs are erect herbs arising from a single or clustered, often woody root stock.

Indigenous species are perennials, whereas introduced species are annuals. Leaves clustered at the base of the plant are simple, alternate, petioled, palmately lobed into three to five divisions, and in some species further deeply divided. Stems are hollow. The showy flowers, generally blue to purple in color, but ranging from white to occasionally red, are produced on terminal erect racemes. Flowers have five sepals and four petals; the upper sepal and pair of petals are elongated to form a characteristic spur that protrudes backward (see Figure 1-20D). Seed pods are erect, three to five-celled, splitting down the inside ridge to release numerous dark brown to black seeds.

Toxicity

Young rapidly growing larkspur plants are most toxic, with the highest concentration of alkaloids in the leaves. Cattle, however, appear to not eat tall larkspur until the plants initiate and elongate flower stalks.[160,161] At this stage, the alkaloid content of the plants is generally declining, although the seed pods contain high levels of alkaloids.[161,162] Tall larkspur consumption in cattle may range from virtually nothing in the preflower stage to as much as 30 percent of the animal's diet when the plant is in the flowering stage. There is a "toxic window" of time when cattle find tall larkspur increasingly palatable, and the plants contain significant total toxic alkaloids.[160,161] (Graph 1-1) This window of time occurs as

Figure 1-20B Tall larkspur inflorescence showing flowers with characteristic spur (*Delphinium barbeyi*).

the tall larkspurs begin to elongate their flower stalks and until the flowers have been replaced with seed pods. Feeding studies with cattle using both fresh and dried tall larkspur showed no correlation between alkaloid concentration and palatability.[160] Sheep, however, will avoid eating the plants in the preflower stage when the alkaloid content is highest, but will readily eat the flower stalks and buds as they mature. In drought years the consumption of tall larkspur by cattle almost entirely ceases, and mortality from larkspur poisoning is lowest during droughts.[160,162] Another as yet unexplained phenomenon that contributes to tall larkspur poisoning is the glutinous consumption of the larkspur by cattle in a short period after a summer rain storm.[163]

The toxicity attributed to the tall larkspur species cannot be assumed to hold true for the low or foothills larkspur species. Unlike the tall larkspurs, the foothills larkspur (*D. geyeri*) is readily eaten by cattle in the early spring before it flowers, causing significant numbers of

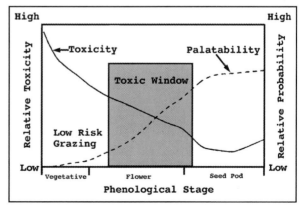

Graph 1-1

deaths in some years. By the time it flowers, the foothills larkspur is not consumed in any quantity presumably because cattle generally have plenty of other forage available by this stage.

Cattle are most susceptible to larkspur poisoning, but sheep and horses may also be poisoned if they eat sufficient quantity of the plants over a short period of time and are concurrently subjected to stress. Sheep are be able to tolerate about four times the amount of larkspur alkaloids that would be fatal to cattle.[151] Establishing the toxic dose of larkspur for cattle is difficult because the toxicity varies with the species, season, stage of growth, amount ingested, and duration over which the plant is eaten.[164] It has been estimated that cattle must eat 0.7 percent of their body weight of green tall larkspur in an hour to be fatally poisoned.[151] If one assumes a 1200-lb cow eats about 25 lb of forage a day, it would have to eat about 6 lb of green tall larkspur containing 5 mg/g of total toxic alkaloid to be fatally poisoned. This is about 25 percent of the cow's total daily feed intake.[160] The LD_{50} of total toxic alkaloid for tall larkspur has been shown experimentally to be between 25 and 40 mg/kg body weight.[164]

Figure 1-20C Low larkspur (*Delphinium nuttallianum*).

Larkspurs contain many toxic and nontoxic diterpenoid alkaloids, 40 of which have been identified in the nine larkspur species most frequently associated with poisoning in cattle.[158,165,166] The alkaloids may vary in quantity depending on the species and stage of growth.[150,167] Even within a small area, certain stands of larkspur appear to be more toxic than others. These "hot spots" are well recognized by ranchers who frequently anticipate losses when cattle are herded in these areas. The alkaloid content of tall larkspurs also appears to be consistently higher in plants growing in full sun as opposed to those growing in the shade. Studies with *D. barbeyi* have shown it to be the most toxic of the larkspurs.[151,159,162] As little as 17 g/kg body weight of the green plant of *D. barbeyi* is lethal to cattle.[162,168] A rapid electrospray mass spectrometry method has been developed that may prove useful in determining the toxic alkaloid content of larkspur species at different growth stages that will facilitate the safe management of larkspur rangeland for cattle grazing.[169]

To date the most toxic of the alkaloids isolated from tall larkspurs are 14-deactylnudicauline and methyllycaconitine (MLA).[158,170,171] Although the former alkaloid is more toxic, greater than 15 times as much MLA is produced in the plant making it the most important toxic component of the tall larkspurs.[161] The foothills larkspur (*D. geyeri*) and the low larkspur (*D. nuttallianum*) contain the highly toxic alkaloid

Figure 1-20D Prairie larkspur (*Delphinium virescens*).

Figure 1-20E Foothills larkspur (*Delphinium geyeri*) immature plant.

nudicauline.[172] The toxicity of the different larkspur species is, however, most likely due to the combined effects of the alkaloids present in the plants.

The alkaloids act principally at the neuromuscular junction (postsynaptic nicotinic and cholinergic receptors) causing a curare-like blockade with muscle weakness and paralysis.[167,173] The alkaloids reversibly bind to and block the action of nicotinic acetylcholine receptors at the neuromuscular junction thus competing with acetylcholine for the receptors. This effect is similar to that of nicotine and the snake toxin bungertoxin.[174] Binding affinity of the larkspur alkaloids to the cholinergic receptors varies with the tissue and among species.[175] Cholinergic receptors in cattle are possibly more susceptible to the alkaloids than are those of sheep, offering an explanation for the refractoriness of sheep to larkspur toxicity.

Clinical Signs

Sudden death in cattle is often the first indication of larkspur poisoning. Cattle frequently die within 3 to 4 hours of consuming a lethal dose of larkspur. Poisoned cattle initially show uneasiness, increased excitability, and muscle weakness that causes stiffness, staggering, and a base-wide stance.[176] The front legs may be affected first causing the animal to kneel before finally becoming recumbent. Muscle weakness may cause sudden collapse especially if the animal is stressed. Frequent attempts to stand are uncoordinated and result in rapid exhaustion. Muscle twitching, abdominal pain, regurgitation, and constipation are common clinical findings. Similar signs of poisoning occur in horses, sheep, and goats that have eaten larkspur except that vomiting is uncommon and fewer deaths are likely.

Bloat is common in larkspur poisoning because of the neuromuscular blocking effect of the alkaloids impairing eructation and the fact that the green larkspur is highly fermentable by rumen microorganisms, thereby increasing the rate of gas production in the rumen. The high protein content of larkspur may also facilitate the production of a stable foam in the rumen that enhances the severity

Figure 1-20F Foothills larkspur (*Delphinium geyeri*) in bloom.

of the bloat. Inhalation of regurgitated rumen contents in the recumbent animal will lead to fatal pneumonia. Cattle appear to be able to repeatedly eat a toxic dose of larkspur without marked clinical signs of poisoning provided larkspur consumption is significantly decreased for 2 to 4 day intervals.[177] This appears sufficient time for metabolism and clearance of the larkspur alkaloids to occur and reduce the cumulative effect of daily larkspur consumption.

No specific postmortem findings are seen in animals that have died of larkspur poisoning. Bloat, inhalation of rumen contents, venous congestion, and mild gastrointestinal inflammation are common secondary findings. Diagnosis of larkspur poisoning is made by searching for parts of the plants in the animal's rumen and by finding evidence of the animal having grazed larkspur. It is also possible to detect microscopically the plant cell structure of *Delphinium* spp. in the feces or rumen contents that will indicate the animal had consumed the plant in the recent past.[178]

Treatment

Ever since larkspurs were known to be poisonous to cattle, a variety of treatments have been advocated. Treatment myths abound in the early literature and have involved the use of a variety of compounds including atropine, potassium permanganate, turpentine, bacon fat, chewing tobacco, whiskey, and bleeding the animal from the tail vein.[153,179] Most of these early remedies have little scientific basis and have not been proven to be effective. The use of turpentine and bacon fat given orally may have helped reduce the severity of the bloat that develops with larkspur poisoning. Apparent successes with unique remedies have not considered the variability in toxicity of the larkspurs under different growing conditions, time of year, and quantity of total toxic alkaloid consumed. If less than a lethal dose of larkspur alkaloids is consumed, an animal will likely recover despite the treatment, unless severe bloat and regurgitation of rumen contents occurs while the animal is recumbent.

Anticholinesterase drugs that allow the accumulation of acetylcholine at the neuromuscular junction by inhibiting cholinesterase are the most appropriate for the reversal of some of the effects of larkspur alkaloids. Physostigmine (0.08 mg/kg) has been effective if given intravenously to cattle about to collapse from larkspur poisoning.[177] Cattle in sternal or lateral recumbency from tall larkspur poisoning recover after treatment with physostigmine (0.4-0.8 mg/kg body weight) given intravenously.[180] Treatment should be repeated as needed over several hours until clinical signs have abated. Neostigmine (0.04 mg/kg) appears to be as effective in reversing some of the effects of the larkspur alkaloids but is possibly not as effective as physostigmine for treating larkspur poisoning. A formulation that has been beneficial if administered early in the course of poisoning has been the injectable mixture of physostigmine salicylate, pilocarpine hydrochloride, and strychnine sulfate.[156] Some organophosphate compounds with an anticholinesterase effect may also have potential benefit in the treatment of larkspur poisoning as they will decrease the breakdown of acetylcholine.[181] Because none of these drugs are approved for use in food-producing animals they must be administered under the supervision of a veterinarian in accordance with the regulations pertaining to the use of extra-label drugs.

An early diagnosis of larkspur poisoning is essential if treatment is to be successful. Stress and excitement of the affected animal should be avoided because it will exacerbate respiratory distress and hasten death. It is often better to quietly herd affected range cattle away from the area where larkspur is being grazed and not attempt to catch and restrain an animal to treat it. Affected animals should be kept sternal if they become recumbent. Bloat should be relieved by passing a stomach tube to remove excess rumen gas and reduce respiratory difficulty when possible.

Trocarization of the rumen to relieve the bloat may be more effective than trying to pass a stomach tube because it is less stressful to the animal.

Acute larkspur poisoning in a range cow can resemble grass tetany (hypomagnesemia) and milk fever (hypocalcemia) especially when the affected animal is recumbent, and laboratory facilities are not readily available to differentiate these conditions. Magnesium solutions for the treatment of grass tetany are contraindicated in suspected larkspur poisoning as magnesium will exacerbate the effect of the alkaloids at the neuromuscular junction. Calcium gluconate in contrast will have a beneficial effect at the neuromuscular junction. However, calcium solutions should be administered very cautiously because of the effect of calcium on the heart.

Prevention of Larkspur Poisoning

Intuitively, cattle should be kept off of ranges containing large quantities of larkspur. However, this would eliminate vast areas of rangeland for livestock grazing, making such a management option uneconomical. By knowing the growth habits of tall larkspur, the times when it is most toxic, and when cattle like to eat the plant, it is possible to manage cattle so that they are kept away from the larkspur during a "toxic window" when chances for poisoning are highest (see Graph 1-1).[161] Prior to this "toxic window," cattle may be grazed on the larkspur range even though tall larkspur is most toxic at this stage, because cattle find the plant unpalatable. Similarly, after the larkspur is past flowering it is relatively less toxic and more palatable. Consequently, knowing the toxicity, palatability and growth stage of tall larkspur, it is possible to make effective use of rangeland for livestock production while minimizing the risk for tall larkspur poisoning.[182] This management strategy is only valid if there is a diversity of other forages available, and the cattle are not forced into a situation where they must eat larkspur because they are without adequate food. If early season grazing of tall larkspur is used, very close attention must be paid to the eating patterns of the cattle. As soon as the tall larkspur starts to elongate its flower stalks and cattle start to eat the flower shoots, they should be moved off of the range. Early season grazing is relevant only to tall larkspur and should not be attempted with the foothills or low larkspurs that seem to be quite palatable in the early spring when they can be highly toxic.

The fact that sheep can eat larkspur without problem makes them useful biological controls for tall larkspur. Sheep if herded into larkspur stands will eat and trample the plants thereby reducing the availability of larkspur to cattle that follow the sheep.[183,184] It is doubtful, however, if sheep effectively reduce the potential for larkspur poisoning in cattle unless large numbers of sheep are actively herded into areas where the plants are abundant.

Providing adequate calcium, phosphorus, and mineralized salts for cattle has been recommended as a preventive measure for larkspur poisoning. However, mineral supplementation of cattle grazing rangeland infested with tall larkspur had no effect in reducing the amount of larkspur consumed.[185] A balanced mineral supplement should always be provided to cattle to prevent mineral deficiencies and should not be relied on as a preventive measure for larkspur poisoning. It has been postulated that cattle may be deficient in minerals in late winter and early spring and may crave plants like foothills larkspur that are high in calcium.

Aversion to Larkspur

Cattle unfortunately do not have a natural aversion to larkspur, but they can be trained under certain management conditions to avoid eating tall larkspur. Cattle can form a strong aversion to eating larkspur if they are given intraruminal infusions of larkspur extract with lithium chloride.[186-188] Lithium is a potent irritant and emetic that induces abdominal pain that the animal associates with the last thing it was eating or was fed.

Once cattle have developed an aversion to larkspur they continue to have the aversion from year to year provided they are not exposed to cattle that are eating larkspur.[187,189,191] Socializing with cattle that are not aversed to larkspur leads to the aversed animals relearning to eat larkspur.[189] Training cattle to avoid eating larkspur can be accomplished by harvesting and feeding fresh larkspur to cattle. As soon as an animal has started eating the larkspur it is restrained and given lithium chloride (100 mg/kg body weight) via stomach tube. The subsequent abdominal discomfort is associated by the animal with the last thing it ate, namely the larkspur.

Induced aversion to eating larkspur has potential beneficial implications for some ranch enterprises. In closed herd situations where all cattle can be treated with lithium and not be exposed to non-averted cattle, larkspur aversion can be maintained in the entire herd for many years at minimal cost.[191,192]

Control of Larkspur

Although it is possible to control larkspurs with herbicides, it is economically prohibitive to do so on a wide scale.[155,193-195] However, spraying of larkspur hot-spots can be effective in reducing cattle losses. The tall larkspurs can be controlled using a variety of currently available herbicides including picloram (Tordon), metsulfuron (Escort), glyphosate (Roundup).[196,197] The most effective herbicide for all growth stages of larkspur is picloram. All are effective because they kill the root and not just the vegetative portion of the plant as is the case with 2,4-D.[196] Surfactants enhance the efficacy of herbicides because they improve the absorption of the chemicals through the waxy surface of the leaves. The most effective time to apply herbicides is in the early vegetative or leaf stage before the flower stalks begin to form.[196] A second application of the herbicide will eliminate any plants that survive the first application. Newer application methods such as vehicle mounted carpeted rollers apply the herbicides to tall larkspur without affecting lower growing useful forbs and use less chemical than conventional spraying methods. It should be noted that herbicides increase the alkaloid content of larkspur and therefore the plants should not be grazed after spraying until the plants have completely died off.[196] Herbicidal control of tall larkspur has additional benefit in that significant increases in the growth of grasses for up to 5 years after spraying increases the carrying capacity of the range.[197]

Research is currently underway to investigate the effectiveness of insect biologic controls to control tall larkspurs. The larkspur myrid (*Hopplomachus affiguratus*), which sucks on the plants reducing plant vitality, has shown potential as an insect control.[198] Plants infected by the myrids become stunted, fail to produce flowers and seeds, and appear to be unpalatable to cattle. The success of this insect as a biologic control will depend on whether or not it can sustain itself in large numbers once transplanted to new stands of larkspur.

Aconite, Monkshood
Aconitum spp.
Ranunculaceae (Buttercup family)

Habitat

Several common species of monkshood grow in North America, the most important of which include *A. columbianum* (western monkshood), *A. uncinatum* (Pennsylvania to Georgia), *A. reclinatum* (eastern states and west to Ohio), and *A. lutescens* (Idaho to New Mexico). Monkshood is generally found growing in rich, moist soils of meadows and open woods. Western monkshood (*A. columbianum*) often grows in the same areas as the tall larkspur (*Delphinium*) species found at high altitude.

Description

Monkshood plants are perennial herbaceous plants with tall leafy stems growing to 1 foot in height. The leaves are alternate, palmately lobed or parted, and similar to *Delphinium* spp. Monkshood flowers are usually deep blue-purple, but occasionally white or pale yellow, and are produced on simple racemes or panicles. The flowers are perfect, zygomorphic, with five sepals that are petal like, the upper sepal being larger and forming a characteristic helmet or hood (Figure 1-21). The two to five petals usually are concealed within the hooded sepal. Numerous stamens and two to five pistils are present in each flower. The fruit consists of three to five pods (follicles) that spread apart when mature to release the brown seeds. Monkshood can be differentiated from larkspur, if the flowers are not present, by the fact that the stems of monkshood are not hollow like those of larkspur. Wild geranium leaves can resemble those of monkshood when they first emerge, but can be differentiated by the distinctive geranium smell to the leaves when they are crushed.

Figure 1-21 Monkshood (*Aconitum columbianum*).

Principal Toxin

Highly toxic, diterpenoid alkaloids including aconitine, mesaconitine, napelline and hypaconitine form the principal toxins in monkshood. The mode and site of action of the alkaloids are similar to those found in *Delphinium* and *Garrya* species.[199] All species of monkshood including cultivated species (*A. napellus*) should be considered toxic to animals and humans.[200] All parts of the plant are toxic, with the roots, seeds, and new leaves being especially toxic. Although there is no extensive documentation of the toxic dose of monkshood, horses have been reported to be fatally poisoned after eating 0.075 percent of their body weight in green plant. In the western United States, most suspected cases of *Aconitum* poisoning are due to tall larkspur, which grows more abundantly in the same areas. Tall larkspur species are also more toxic than monkshood.[201] Human poisoning from monkshood is mostly due to the misuse of medicinal extracts of aconitine.[202] Occasionally poisoning occurs when the root of monkshood is mistakenly eaten for root of wild horseradish or other wild plants. The alkaloids can also be absorbed through the skin and can be hazardous to florists and those handling the plant.

Clinical Signs

Symptoms of monkshood poisoning resemble those of poisoning from larkspur

(*Delphinium* spp.). Symptoms begin within a few hours of ingesting the plant, and death may occur a few hours to a few days later depending on the dose of toxic alkaloids ingested. The alkaloids primarily cause cardiac conduction disturbances and arrhythmias. Affected animals initially become restless, salivate excessively, develop muscle weakness, hypotension, and have difficulty in breathing before collapsing into lateral recumbency.[2] Bloating is a common problem in ruminants once they become recumbent.

There is no proven effective treatment for monkshood poisoning. Affected animals should be stressed as little as possible, and have a better chance of survival if they are herded away from the source of the plants without stressful attempts at treatment. Symptomatic treatment with intravenous fluids and relief of rumen bloat should be administered as necessary. Activated charcoal and osmotic laxatives orally may be helpful in preventing further absorption of alkaloids from the gastrointestinal track.

Poison Hemlock, European Hemlock
Spotted Hemlock, California Fern
Conium maculatum
Apiaceae (Parsley family)

Figure 1-22A Poison hemlock (*Conium maculatum*).

Habitat

Introduced from Europe, poison hemlock is now found throughout North America, growing along roadsides, ditches, cultivated fields, and waste areas especially where the ground is moist. It is a prolific seed producer and will form dense stands if left unchecked.

Description

Poison hemlock is a coarse, erect biennial or perennial plant 4 to 6 feet (2 meters) tall (Figure 1-22A). The smooth, branching stems are hollow, with purple spots especially near the base (Figure 1-22B). The root is a simple carrot-like tap root. Leaves are alternate three to four times pinnately dissected, coarsely toothed with a fernlike appearance. The terminal inflorescence is a compound flat topped, loose umbel with multiple, small, white five-petaled flowers. Fruits are gray brown ovoid, ridged, and easily separated into two parts. The plant including the root has a strong pungent (likened to mouse urine) odor that makes it generally unpalatable. Of the four known species of *Conium*, the only one in North America is *C. maculatum*.

Principal Toxin

At least eight piperidine alkaloids have been found in various parts of the plant.[203,204] The two predominant toxic alkaloids are coniine (mature plant and seeds), and g-coniceine (young growing plant).[203,204] The mechanism of action of the conium alkaloids is complex because they have a profound effect in blocking spinal cord reflexes. After an initial stimulatory effect, the autonomic nervous system ganglia become depressed.[205] Large doses of alkaloid cause skeletal muscle stimulation followed by neuromuscular blockade and paralysis similar to that caused by nicotine on the central and peripheral nervous systems.[3,206] In small quantities, the alkaloids cause skeletal defects in the fetal calf if poison hemlock is grazed by pregnant cows.[206-208]

The leaves and stems prior to the development of seed heads are the most toxic part of the plant. The seeds themselves are highly toxic and can be a source of poisoning when they contaminate cereal grains fed to livestock. Young plants in the first year of growth are less toxic than mature plants, and those growing in the warmer southern states appear to be more toxic than those in the northern areas.[209] Poison hemlock is toxic to a wide variety of animals including man, birds, wildlife, cattle, sheep, goats, pigs, and horses.[203,206,210-219] People are usually poisoned when they mistakenly eat hemlock for plants such as yampa (*Perideridia gairdneri*), parsley

Figure 1-22B Poison hemlock showing spotted stems (*Conium maculatum*).

(*Petroselinum crispum*), wild anise (*Pimpenella anisum*), or wild carrot, Queen Anne's lace (*Daucus carota*).[215] A tea made from poison hemlock was reportedly used to kill Socrates. Livestock seldom eat hemlock because of its strong odor, but they will do so if no other forage is available or if it is incorporated in hay or silage. Cattle have been fatally poisoned by eating as little 0.5 percent of their body weight of green hemlock.[215] Experimental hemlock poisoning in cattle, pigs, and sheep has been produced by a wide range of doses suggesting that there is considerable variation in the toxic alkaloid content of the plant. Cattle dosed with1 g/kg developed clinical signs, while 5.3 g/kg of green plant was lethal.[207] In sheep, repeated *Conium* doses of 10 g/kg body weight were lethal.[18] Cattle were most susceptible to pure coniine administered by stomach tube, requiring 3.3 mg/kg body weight to induce severe poisoning. Mares required 15.5 mg/kg and sheep 44.0 mg/kg of the alkaloid to induce severe poisoning.[214]

Clinical Signs

Signs of poisoning are similar in all species and develop within an hour of hemlock consumption. If a lethal dose has been consumed, death from respiratory failure occurs in 2 to 3 hours. Salivation, abdominal pain, muscle tremors, and incoordination will occur initially, followed soon by difficulty in breathing, dilated pupils, weak pulse, and frequent urination and defecation.[215] Prolapse of the nictitating membrane across the cornea in cattle and pigs may cause temporary blindness. Cyanosis of the mucous membranes, respiratory paralysis, and coma without convulsions precede death.[215] Goats may recover from hemlock poisoning only to develop a strong craving for the plant, which ultimately proves fatal.[204] Pregnant animals that survive the acute toxicity may abort.[215]

Poison hemlock is also teratogenic causing abnormal fetal development if it is eaten by pregnant cows between the 40th and 70th days of gestation.[213] Calves and piglets may be born with crooked legs, deformed necks and spines (torticollis, scoliosis), and cleft palates that are indistinguishable from similar deformities caused by the teratogenic effects of lupines and tobacco species.[206,207,213,219,222-225] Sows consuming poison hemlock in early gestation produce litters of piglets with congenital skeletal malformations.[208,220,221] Observations in pregnant ewes using real-time ultrasound have shown that an increase in birth defects is associated with decreased fetal movements induced by ingestion of lupines, poison hemlock, and tobacco species.[226] Lambs born to ewes fed poison hemlock from 30 to 60 days of gestation showed varying degrees of excessive carpal joint flexure and lateral deviation that corrected itself by the time the lambs were 2 months old.[227] The teratogenic and toxic effects of poison hemlock appear to be most severe in cattle with sheep being most tolerant.[206] Mares are susceptible to poison hemlock poisoning, but foals born to mares fed coniine between 45 and 75 days of gestation did not develop congenital deformities.[214]

Because there is no specific treatment for hemlock poisoning, acutely poisoned animals should be given supportive treatment as necessary. Stressing the animal should be kept to a minimum. If the hemlock has been recently consumed, saline cathartics and activated charcoal are beneficial in removing the plant from the gastrointestinal tract. A dilute tannic acid solution administered via stomach tube may help to detoxify the hemlock alkaloids. All hemlock should be removed from pastures to which animals have access. Destroying the plants by mowing or with herbicides before the seed stage greatly reduces the chances of hemlock becoming an invasive weed and a problem to livestock.

Water Hemlock, Cowbane, Poison Parsnip
Cicuta douglasii
Cicuta maculata (Spotted water hemlock)
Apiaceae (Parsley family)

Habitat

Water hemlocks, as the name suggests, prefer wet meadows, riverbanks, irrigation ditches, and water edges, often growing with their roots underwater. *Cicuta maculata* is found predominantly in the eastern United States; *C. douglasii* occurs more commonly in the western states (Figure 1-23A). Water hemlocks may be found growing at altitudes as high as 8000 feet (2,438 meters) above sea level. At least eight poorly differentiated species of water hemlock occur in North America and all should be considered highly toxic.[228,229]

Synonyms for water hemlock include musquash root, muskrat weed, fever root, mock-eel root, and beaver poison.

Figure 1-23A Water hemlock (*Cicuta douglasii*).

Description

Water hemlock is a stout, erect, hairless, perennial or biennial growing to a height of 4 to 6 feet (1 to 2 meters) from a condensed bundle of two to eight thick tuberous (parsnip-like) roots (Figure 1-23B). Flowers and seeds are produced in the second year of growth. At the base of the hollow stem are a series of tightly grouped partitions that may contain an acrid, yellow fluid (Figure 1-23C). The leaves are alternate, one to three times pinnate, the leaflets are 2 to 4 inches (3 to 10 cm) long, linear-lancelate to ovate-lancelate with toothed edges (Figure 1-23D). The flowers are white in the form of a loose, compound umbel. The flower is supported by a whorl (involucre) of a few narrow bracts. There is also an involucel of several narrow bractlets. A disc-like swelling (stylopodium) is present at the base of the style. The fruits are oval, flattened laterally with prominent ribs.

Figure 1-23B Water hemlock tuberous roots showing yellow fluid on cut surfaces (*Cicuta douglasii*).

Several other members of the family *Apiaceae (Umbelliferae)* are also toxic to man and animals, including spotted hemlock (*Con-ium* spp.) and water dropwort (*Oenanthe spp*). Cowbane (*Oxypolis fendleri*), found growing along alpine streams, closely resembles water hemlock but is a much smaller, slender plant with a few, toothed leaves. (The name hemlock is also given to the evergreen trees of the genus *Tsuga*, which are not considered toxic.)

Principal Toxin

Cicutoxin ($C_{17}H_{22}O_2$), a highly unsaturated (diol), is one of the most toxic naturally occurring plant compounds known.[230,231] The alcohol derivative, cicutol ($C_{17}H_{22}O$), is relatively nontoxic.[232] The toxin is concentrated in the tuberous roots, but all parts of the plant including the fluid found in the hollow stems are toxic. The roots are highly poisonous at all times, and livestock consuming the root usually die. The roots are easily pulled up because the ground in which the plants grow is usually wet. Mature plants often have a prominent root crown that protrudes above ground making it accessible to animals even in winter. The newly emerging plant in the spring is the most toxic, whereas the mature leaves in late summer seem to have minimal toxicity to cattle that eat them. The dry stems are minimally toxic.

All animals, including humans, can be fatally poisoned by eating as little as 50 to 110 mg/kg body weight of the green water hemlock plant.[228,233-237] The lethal dose of fresh green water hemlock (*C. douglasii*) is 2 oz for adult sheep, 12 oz for adult cattle, and 8 oz for adult horses.[233] Sheep dosed with ground water hemlock tubers with 1.4, 2.8, and 6.4 g/kg of tuber developed signs ranging from mild increase in salivation, nervousness, and a few muscular tremors without seizures at the lowest dose, to seizures followed by recovery at a dose of 2.8 g/kg. Those that received 6.4 g/kg of tuber developed severe seizures and died 90 minutes after dosing.[232]

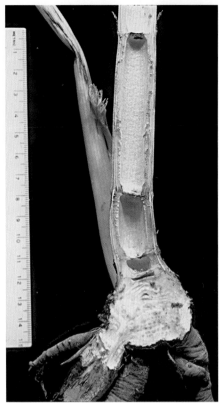

Figure 1-23C Water hemlock hollow stem with horizontal partitions (*Cicuta douglasii*).

Fatalities have occurred in children who have sucked on the hollow stems of water hemlock. In 1984, a river tour guide died after mistaking the tuberous roots of water hemlock for the edible root of common yampa (*Perideridia gairdneri*). The tuberous roots have also been mistaken for edible wild parsnip (*Pastinaca sativa*).

Clinical Signs

Cicutoxin is a potent neurotoxin capable of causing rapid onset of muscle tremors and violent convulsions.[235] Death often occurs in a matter of 2 to 3 hours after a lethal dose of water hemlock has been consumed. Excessive salivation, vigorous chewing movements, teeth grinding, frequent urination, and defecation are common. Depending on the quantity of toxin absorbed, animals become ataxic and uncoordinated and develop

Figure 1-23D Water hemlock leaf showing veins running to each notch at the leaf's margin (*Cicuta douglasii*).

grand mal seizures.[5] During the convulsions, animals may chew off their tongue. Signs may start within an hour of eating the plant or tuberous roots and progress rapidly to convulsive seizures and lateral recumbency. Poisoned animals have dilated pupils and progress to a state of coma, before dying from respiratory paralysis and asphyxia. Death may occur in about 90 minutes after ingestion of a lethal dose.[232] If animals consume a sublethal dose, they will recover if not stressed.

Sheep that die acutely from water hemlock poisoning show no gross postmortem lesions. Because of the rapid lethal effects of the cicutoxin, pieces of the water hemlock root may be found in the esophageal groove and not in the rumen at post mortem examination. Multifocal, diffuse myocardial degeneration is characteristic in acute poisoning.[232] In less acute poisoning muscle degeneration is also seen in skeletal muscle and results from the severity of the seizures. The serum enzymes lactic dehydrogenase, aspartate transaminase, and creatine kinase markedly increase in animals that have frequent seizures of long duration.[232]

Treatment

There is no specific antidote for cicutoxin. Studies in sheep have shown that the intravenous administration of sodium pentobarbital at the onset of seizures prevented lethal cardiac and skeletal muscle degeneration and resulted in complete recovery.[232] When possible, therefore, early treatment should consist of heavy sedation of poisoned animals with sodium pentobarbital to reduce the severity of the convulsions. Laxatives may be beneficial in removing the plant from the digestive system. Vomiting should be induced in dogs and people suspected of eating water hemlock. Artificial respiration should be given where possible if respiratory failure occurs. In cattle that are actually observed eating the water hemlock roots, it may be life saving to immediately perform a rumenotomy to remove the plant parts from the digestive system before the toxin is absorbed. Dilute acetic acid (vinegar) administered via stomach tube in cattle may be beneficial in neutralizing the toxin if administered very soon after the plant has been consumed.

Control of Water Hemlock

The fact that water hemlock is so poisonous makes it important to remove all water hemlock plants from livestock enclosures. The plant spreads primarily by seed which is often disseminated by water. The plant should be dug up and burned where possible. Spraying water hemlock with herbicide will kill the plant, but it should be remembered that the herbicide may make the plant more palatable before it dies off.

Copperweed
Oxytenia acerosa
Asteraceae (Composite family)

Habitat
Copperweed is an indigenous plant of the semiarid alkaline soils of southwest North America. It is often found locally in dry streambeds and canyons.

Figure 1-24 Copperweed (*Oxytenia acerosa*).

Description
Copperweed is an erect, bushy, perennial, with woody stems growing 3 to 5 feet tall (Figure 1-24). Leaves are alternate, hairy, and irregularly divided into needle-like segments. The individual flowers are small, yellow, or white and produced in dense terminal panicles that give it the appearance of a goldenrod.

Principal Toxin
The toxin has not been defined. About 0.5 percent of an animal's body weight of the green plant is a lethal dose.[238] The plant remains toxic at all times, even when dry. Cattle and sheep are susceptible to poisoning and eat the plant when other forage is unavailable.

Clinical Signs
Cattle, and less often sheep, stop eating after ingesting copperweed and become markedly depressed with some animals showing nervousness. Affected animals become comatose, and die 1 to 3 days after the onset of signs. Liver and kidney degeneration occur in fatally poisoned animals.[238]

REFERENCES

Cyanogenic glycosides

1. Conn EE: Cyanogenesis, the production of hydrogen cyanide by plants. In Effects of Poisonous Plants on Livestock. Keeler RF, Van Kampen VR, James LF, Eds. New York: Academic Press; 1978:301–310.

2. Kingsbury JM: *Poisonous Plants of the United States and Canada.* Englewood Cliffs, NJ: Prentice-Hall; 1964:23–26.

3. Siegler DS: The naturally occurring cyanogenic glycosides. *Prog Phytochem* 1977, 83-120.

4. Siegler DS: Plants of the Northeastern United States that produce cyanogenic compounds. *Economic Botany* 1976, 30:395–407.

5. Burrows GE, Edwards WC, Tyrl RJ: Toxic plants of Oklahoma cyanogenic plants. *Okla Vet Med Assoc* 1980, 35:87–90.

6. Poulton JE: Cyanogenic compounds in plants and their toxic effects. *In* Handbook of Natural Toxicants. Volume 1: Plant and Fungal Toxins. Keeler RF, Tu AT, Eds. New York: Marcel Dekker; 1983:117–157.

7. Osuntokun BO: Cassava diet, chronic cyanide intoxication and neuropathy in Nigerian Africans. *World Rev Nutr Diet* 1981, 36:141–173.

8. Tewe OO, Iyayi EA: Cyanogenic glycosides. *In* Toxicants of Plant Origin. Volume II: Glycosides. Cheeke PR, Eds. Boca Raton, FL: CRC Press;1989:44–60.

9. Cutler AJ, Conn EE: The biosynthesis of cyanogenic glycosides in *Linum usitatissimum* (linen flax) in vitro. *Arch Biochem Biophysics* 1981, 212:468–474.

10. Gibb MC, Carberry JT, Carter RG, Catalinac S: Hydrocyanic acid poisoning of cattle associated with sudan grass. *N Z Vet J* 1974, 22:127.

11. Crawford AC: The poisonous action of Johnson grass. *USDA Bulletin 90,* 1906:31–34.

12. Flemming CE, Dill R: The poisoning of sheep on mountain grazing ranges in Nevada by the western chokecherry. *University of Nevada Agricultural Experimental Station Bulletin 110,* 1928.

13. Majak W, Ubenberg T, Clark LJ, McClean A: Toxicity of Saskatoon service berry to cattle. *Can Vet J* 1980, 21:74.

14. Majak W, McDiarmid RE, Hall WJ: The cyanide potential of Saskatoon service berry (*Amelanchier alnifolia*) and choke cherry (*Prunus virginiana*). *Can J Anim Sci* 1981, 61:681.

15. Majak W, Quinton DA, Broersma K: Cyanogenic glycoside levels in Saskatoon service berry (*Amelanchier alnifolia*). *J Range Manage* 1980, 33:197–199.

16. Marsh CD, Clawson,AW: Arrow grass (*Triglochin maritima*) as a stock-poisoning plant. *USDA Bulletin 113,* 1929.

17. Beath OA, Draize JH, Eppson HF: Arrow grass-chemical and physiological considerations. *Agricultural Experimental Station University of Wyoming Bulletin 193,* 1933.

18. Majak W, McDiarmid RE, Hall JW, Van Ryswyk AL: Seasonal variation in the cyanide potential of arrow grass (*Triglochin maritima*). *Can J Plant Sci* 1980, 60:1235–1241.

19. Cutler AJ, Hosel W, Sternberg M, Conn EE: The in vitro biosynthesis of taxiphyllin and the channeling of intermediates in *Triglochin maritima*. *J Biol Chem* 1981, 256:4253–4258.

20. Shaw JM: Suspected cyanide poisoning in two goats caused by ingestion of crab apple leaves and fruits. *Vet Rec* 1986, 119:242–243.

21. Webber JJ, Roycroft CR, Callinan JD: Cyanide poisoning of goats from sugar gums (*Eucalyptus cladocalyx*). *Austr Vet J* 1985, 62:28.

22. Quinton DA: Saskatoon service berry toxic to deer. *J Wild Life Manage* 1985, 49:362–364.

23. Stout EN: *Suckleya suckleyana*. A poisonous plant. *USDA Bulletin* 359-A, 1939.

24. Thorp F, Deem AW, Harrington HD, Tobiska JW: *Suckleya suckleyana*. A poisonous plant. *Colorado Experimental Station Bulletin* 22, 1937.

25. Puls R, Newschwander FP, Greenway JA: Cyanide poisoning from *Glyceria grandis* S Wats ex Gray (tall mannagrass) in a British Columbia beef herd. *Can Vet J* 1978, 19:264–265.

26. Vogel KP, Haskins FA, Gorz HJ: Potential for hydrocyanic acid poisoning of livestock by Indian grass. *J Range Manage* 1987, 40:506–509.

27. Cheeke PR, ed: *Natural Toxicants in Feeds, Forages, and Poisonous Plants*, edn 2. Danville, IL: Interstate Publishers; 1997:150–153.

28. Pickerel JA, Oehme FW, Hichman SR: Drought increases forage nitrate and cyanide. *Vet Hum Toxicol* 1991, 33:247–251.

29. Radostits OM, Blood DC, Gay CC, eds: *Veterinary Medicine*, edn. 8. Bailliere Tindall, Philadelphia 1994:1532–1585.

30. Conn EE: Cyanogenic compounds. *Annu Rev Plant Physiol* 1980, 31:433–451.

31. Holzbecher MD, Moss MA, Ellenberger HA: The cyanide content of laetrile preparations, apricot, peach, and apple seeds. *Clin Toxicol* 1984, 22:341–347.

32. Macy DW: Amygdalin (laetrile) and veterinary medicine. *J Am Vet Med Assoc* 1977, 171:284–286.

33. Schmidt ES, Newton GW, Sanders SM, Lewis JP, Conn EE: Laetrile toxicity studies in dogs. *JAMA* 1978, 239:943–947.

34. Majak W, McDiarmid RE, Hall JW, Cheng KJ: Factors that determine rates of cyanogenesis in bovine ruminal fluid in vitro. *J Anim Sci* 1990, 68:1648–1655.

35. Jackson T: Cyanide poisoning in two donkeys. *Vet Human Toxicol* 1995, 37:567–568.

36. Moran EA: Cyanogenic compounds in plants and their significance in animal industry. *Am J Vet Res* 1954, 15:171–175.

37. Hatch RC: Poisons causing respiratory insufficiency. *In* Veterinary Pharmacology and Therapeutics, edn 4. Jones ML, Booth NH, McDonald LE, Eds. Ames IA: Iowa State University Press; 1977:1163–1166.

38. Salkowski AA, Penney DG: Cyanide poisoning in animals and humans: a review. *Vet Human Toxicol* 1994, 36:455–466.

39. Humphries DJ: *Veterinary Toxicology*, edn 3. Bailliere Tindall, Philadelphia 1988, 188-191.

40. Radeleff RD, ed: *Veterinary Toxicology*, edn 2. Philadelphia: Lea & Febiger, 1977:50–58.

41. Way JL: Cyanide intoxication and its mechanism of antagonism. *Ann Rev Pharmacol Toxicol* 1984, 24:451–481.

42. Cheeke PR, ed: *Natural Toxicants in Feeds, Forages, and Poisonous Plants*, edn 2. Danville, IL: Interstate Publishers; 1997:193–197.

43. Pritchard JT, Voss JL: Fetal ankylosis in horses associated with hybrid Sudan pasture. *J Am Vet Med Assoc* 1967, 150:871–873.

44. Seaman JT, Smeal MG, Wright JC: The possible association of a sorghum (*Sorghum sudanense*) hybrid as a cause of developmental defects in calves. *Aust Vet J* 1981, 57:351–352.

45. Smith ADM, Duckett S, Waters AH: Neuropatholgical changes in chronic cyanide intoxication. *Nature* 1963, 4902:179–181.

46. Van Kampen KR: Sudan grass and sorghum poisoning of horses: a possible lathyrogenic disease. *J Am Vet Med Assoc* 1970, 156:629–630.

47. McKenzie RA, McMicking LI: Ataxia and urinary incontinence in cattle grazing sorghum. *Aust Vet J* 1977, 53:496–497.

48. Adams LG, Dollahite JW, Romane WM, *et al*.: Cystitis and ataxia associated with sorghum ingestion in horses. *J Am Vet Med Assoc* 1969, 155:518–524.

49. Yamamoto H: Hyperammonemia, increased brain neutral and aromatic amino acid levels, and encephalopathy induced by cyanide in mice. *Toxicol Appl Pharmacol* 1989, 99:415–420.

50. Osweiler GD, Carlson TL, Buck WB, Van Gelder GA: *Clinical and Diagnostic Veterinary Toxicology*. Dubuque, IA: Kendall Hunt Publishing; 1985:455–459.

51. Ballantyne B: Artifacts in the definition of toxicity by cyanides and cyanogens. *Fundam Appl Toxicol* 1983, 3:400–408.

52. Burrows GE: Cyanide intoxication in sheep: therapeutics. *Vet Human Toxicol* 1981, 23:22–28.

53. Buck WB, Osweiler GD, Van Gelder GA, eds: *Clinical and Diagnostic Veterinary Toxicology,* edn 3. Kendall-Hunt Publishing, Dubuque, IA.1988: 455-459.

54. Burrows GE, Way JL: Cyanide intoxication in sheep: enhancement of efficacy of sodium nitrite, sodium thiosulfate and cobaltous chloride. *Am J Vet Res* 1979, 40:613–617.

55. Burrows GE, Way JL: Cyanide intoxication in sheep: therapeutic value of oxygen or cobalt. *Am J Vet Res* 1977, 38:223–227.

56. Dalvi RR, Sawant SG, Terse PS: Efficacy of alpha-ketoglutaric acid as an effective antidote in cyanide poisoning in dogs. *Vet Res Commun* 1990, 14:411–414.

57. Dulaney MD, Brumley M, Willis JT, Hume AS: Protection against cyanide toxicity by oral alpha-ketoglutaric acid. *Vet Human Toxicol* 1991, 33:571–575.

58. Wattenbarger DW, Gray E, Rice JS, Reynolds JH: Effects of frost and freezing on hydrocyanic potential of sorghum plants. *Crop Sci* 1968, 8:526–528.

59. Nass HG: Cyanogenesis: its inheritance in sorghum bicolor, *Sorghum sudanense, lotus,* and *Trifolium repens*—a review. *Crop Sci* 1972, 12:503–506.

60. Gorz HJ, Haskins FA, Vogel KP: Inheritance of dhurrin content in mature sorghum leaves. *Crop Sci* 1986, 26:65–67.

61. Haskens FA, Gorz HJ, Johnson BE: Seasonal variation in leaf hydrocyanic potential of low and high-dhurrin sorghums. *Crop Sci* 1987, 27:903–906.

62. Clement BA, Goff CM, Forbes TDA: Toxic amines and alkaloids from *Texas acacias. Proceedings 5th International Symposium on Poisonous Plants*. San Angelo, TX; 1997:91.

63. Humphreys DJ: *Veterinary Toxicology*, edn 3. Bailliere Tindall, Philadelphia 1988:245.

64. Barr CG, Reuszer HW, Thorp F: The chemical composition of *Suckleya suckleyana. Science* 1939, 90:497.

Nitrate

65. O'Hara PJ, Fraser AJ: Nitrate poisoning in cattle grazing crops. *N Z Vet J* 1975, 23:45–53.

66. Knott SG: Nitrite poisoning in livestock. *Queensland Agric J* 1971, 485–489.

67. McIlwain PK, Schipper IA: Toxicity of nitrate nitrogen to cattle. *J Am Vet Med Assoc* 1963, 142:502–505.

68. Jones TO, Jones DR. Nitrate/nitrite poisoning of cattle from forage crops. *Vet Rec* 1977;10I:266-267.

69. Singer RH: The nitrate poisoning complex. *Proceedings from the 76th Annual Meeting of the U.S. Animal Health Association*. 1972:310–322.

70. Dollahite JW, Holt EC: Nitrate poisoning. *Vet Med/Small Anim Clinician* 1972, 257–260.

71. Dollahite JW, Rowe LD: Nitrate and nitrite intoxication in rabbits and cattle. *Southwestern Vet* 1974, 27:246–248.

72. Guerink JH, Malestein A, Kemp A, Klooster TV: Nitrate poisoning in cattle. The relationship between nitrate intake with hay or fresh roughage and the speed of intake on the formation of methemaglobin. *Neth J Agric Sci* 1979, 27:268–276.

73. Burrows GE: Nitrate intoxication. *J Am Vet Med Assoc* 1980, 177:82–83.

74. Carrigan MJ, Gardner IA: Nitrate poisoning in cattle fed sudax (*Sorghum* spp. hybrid) hay. *Aust Vet J* 1982, 59:155–157.

75. Brunning-Fann CS, Kaneene JB: The effects of nitrate, nitrite, and N-nitroso compounds on animal health. *Vet Hum Toxicol* 1993, 35:237–253.

76. Olson JR, Oehme FW, Carnahan DL: Relationship of nitrate levels in water and livestock feeds to herd health problems on 25 Kansas farms. *Vet Med/Small Anim Clinician* 1972, 67:257–260.

77. Hibbs CM, Stencel EL, Hill RM: Nitrate toxicosis in cattle. *Vet Hum Toxicol* 1978, 20:1–2.

78. Vermunt J, Visser R: Nitrate toxicity in cattle. *N Z Vet J* 1987, 35:136–137.

79. Brown CM, Burrows GE, Edwards WC: Nitrate intoxication. *Vet Hum Toxicol* 1990, 32:481–482.

80. Haliburton JC, Edwards WC: Nitrate poisoning in Oklahoma cattle during the winter of 1977–1978. *Vet Hum Toxicol* 1978, 20:401–403.

81. Savage A: Nitrate poisoning from sugar beet tops. *Can J Comp Med* 1949, 13:9–10.

82. Cawley CD, Collings DF, Dyson DA: Nitrate poisoning. *Vet Rec* 1977, 101:305–306.

83. Bradley WB, Eppsom HF, Beath OA: Nitrate as the cause of oat hay poisoning. *J Am Vet Med Assoc* 1939, 94:541–542.

84. Bjornsen CB, McIlwain P, Eveleth DF, Bolin FM: Sources of nitrate intoxication. *Vet Med* 1961, 56:198–200.

85. Dodd DC, Coup MR: Poisoning of cattle by certain nitrate containing plants. *N Z Vet J* 1957, 5:51–54.

86. Kendrick JW, Tucker J, Peoples SA: Nitrate poisoning in cattle due to ingestion of variegated thistle *Silybum marianum*. *J Am Vet Med Assoc* 1955, 126:53–56.

87. Crawford RF, Kennedy WK, Davison KL: Factors influencing the toxicity of forages that contain nitrate when fed to cattle. *Cornell Vet* 1966, 56:3–17.

88. Pickrell JL, Oehme FW, Hickman SR: Drought increases forage nitrate and cyanide. *Vet Hum Toxicol* 1991, 33:247–251.

89. Sokolowski JH, Carrigus US, Hatfield EE: Effects of inorganic sulfur on KNO_3 utilization in lambs. *J Anim Sci* 1961, 20:953.

90. Wright MJ, Davison KL: Nitrate accumulation in crops and nitrate poisoning in animals. *Adv Agron* 1964, 16:197–247.

91. Whitehead EL, Moxson AL: Nitrate poisoning. *South Dakota Agricultural Experiment Station Bulletin 254*:1952.

92. Kingsbury J M: *Poisonous Plants of the United States and Canada*. Englewood Cliffs, NJ: Prentice-Hall, 1964:38–43.

93. Ruhr LP, Osweiler GD: Nitrate accumulators. *In* Current Veterinary Therapy— Food Animal Practice 2. Howard JL, Ed. Philadelphia: WB Saunders; 1981:433–435.

94. Van't Klooster AT: Nitrate intoxication in cattle. *Proceedings X11th. World Congress on Diseases of Cattle*. The Netherlands; 1982:398–403.

95. Kemp A, Guerink JH, Malestein A: Nitrate poisoning in cattle. 2. Changes in nitrate in rumen fluid and methemoglobin formation in blood after high nitrate intake. *Neth J Agric Sci* 1977, 25:51–62.

96. Sinclair KB, Jones DIH: Nitrite toxicity in sheep. *Res Vet Sci* 1967, 8:65–70.

97. Farra PA, Satter LD: Manipulation of the ruminal fermentation. III. Effect of nitrate on ruminal volatile fatty acid production and milk production. *J Dairy Sci* 1971, 54:1018–1024.

98. Emerick RJ: Consequences of high nitrate levels in feed and water supplies. *Fed Proc* 1975, 33:1183–1187.

99. Holtenius P: Nitrite poisoning in sheep, with special reference to the detoxification of nitrite in the rumen: an experimental study. *Acta Agricul Scand* 1957, 22:357–372.

100. Malone P: Monensin sodium toxicity in cattle. *Vet Rec* 1978, 103:477–478.

101. Slenning BD, Galey FD, Anderson M: Forage related nitrate toxicoses possibly confounded by non-protein nitrogen and monensin in the diet used at a dairy heifer replacement operation. *J Am Vet Med Assoc* 1991, 98:867–870.

102. Campbell JR, Davis AN, Myhr PJ: Methaemoglobinaemia of livestock caused by high nitrate contents of well water. *Can J Vet Sci* 1954, 18:93–101.

103. Yong C, Brandow RA, Howlett P: An unusual case of nitrate poisoning in cattle. *Can Vet J* 1990, 31:118.

104. National Academy of Sciences—National Research Council, Assembly of Life Sciences. *The Health Effects of Nitrate, Nitrite and N-nitroso Compounds.* Washington, DC: National Academy of Sciences; 1981.

105. Carson TL: Water quality for livestock. *In* Howard JL, ed. Current Veterinary Therapy 2, Food Animal Practice. Philadelphia: WB Saunders;1986:381–383.

106. National Academy of Sciences—National Research Council, Subcommittee on Nutrient and Toxic Elements in Water: *Nutrients and Toxic Substances in Water for Livestock and Poultry.* Washington, DC: National Academy of Sciences; 1974.

107. Buck WB: Diagnoses of feed related toxicoses. *J Am Vet Med Assoc* 1970, 156:1434–1443.

108. Osweiler GD, Carson TL, Buck WB, Van Gelder GA: *Clinical and Diagnostic Veterinary Toxicology*, edn 3. Dubuque, IA: Kendall/Hunt Publishing; 1985:460–467.

109. Seerley RW, Emerick RJ, Embry LB, Olsen OE: Effect of nitrate and nitrite administered continuously in drinking water for swine and sheep. *J Anim Sci* 1965, 24:1014–1019.

110. Schneider NR, Yeary RA: Nitrite and nitrate pharmacokinetics in the dog, sheep, and pony. *Am J Vet Res* 1975, 36:941–947.

111. Ashbury AC, Rhode EA: Nitrite intoxication in cattle: the effects of lethal doses of nitrite on blood pressure. *Am J Vet Res* 1964, 25:1010–1013.

112. Winter AJ: Studies on nitrate metabolism in cattle. *Am J Vet Res* 1962, 23:500–505.

113. Newsom IE, Stout EN, Thorp F, Barber CW, Groth AH: Oat hay poisoning. *J Am Vet Med Assoc* 1937, 43:66–75.

114. Davison KL, Hansel WM, Krook L, McEntee K, Wright MJ: Nitrate toxicity in dairy heifers. 1. Effects on reproduction, growth, lactation, and vitamin A nutrition.

J Dairy Sci 1964, 47:1065–1073.

115. Thorp F: Further observations on oat hay poisoning. *J Am Vet Med Assoc* 1938, 92:159–170.

116. Nichols Ashbury AC, Rhode EA: Nitrite intoxication in cattle: the effects of lethal doses of nitrite blood pressure. *Am J Vet Res* 1964, 25:1010–1013.

117. Kemp A, Guerink JH, Haalstra RT, Malestein A: Discoloration of the vaginal mucous membrane as aid in the prevention of nitrate poisoning in cattle. *Stikstof* 1976, 19:40–48.

118. Beath OA, Gilbert CS, Eppson HS, Rosenfeld I: Oat-hay and oat-straw poisoning. *Wyoming Agricultural Experimental Station Bulletin* 1953, 324:46–48.

119. Winter AJ, Hokanson JF: Effects of long-term feeding of nitrate, nitrite, or hydroxylamine on pregnant dairy heifers. *Am J Vet Res* 1988, 125:353–361.

120. Simon J, Sund JM, Douglas FD, Wright MJ, Kowalczyk T: The effect of nitrate or nitrite when placed in the rumens of pregnant dairy cattle. *J Am Vet Med Assoc* 1959, 135:311–314.

121. Malestein A, Geurink JH, Schuyt G, *et al.*: Nitrate poisoning in cattle. 4. The effect of nitrite dosing during parturition on the oxygen capacity of maternal blood, and the oxygen supply of the unborn calf. *Vet Q* 1980, 2:149–159.

122. Duthu GS, Sertzer SG: Effect of nitrite on rat liver mixed function oxidase activity. *Drug Metab Dispos* 1979, 7:263–269.

123. Sebaugh TP, Lane AG, Campbell JR: Effects of two levels of nitrate and energy on lactating cows receiving urea. *J Anim Sci* 1944, 31:142–144.

124. Goodrich RD, Emerick RJ, Embry LB: Effect of sodium nitrate on the vitamin A nutrition of sheep. *J Anim Sci* 1964, 23:100–104.

125. Mitchell CE, Little CO, Hayes BW: Pre-intestinal destruction of vitamin A by ruminants fed nitrate. *J Anim Sci* 1967, 26:827–829.

126. Roberts WK, Sell JL: Vitamin A destruction by nitrite in vitro and in vivo. *J Anim Sci* 1963, 22:1081–1085.

127. Jainudeen MR, Hansel W, Davison K: Nitrate toxicity in dairy heifers. 3. Endocrine response to nitrate ingestion during pregnancy. *J Dairy Sci* 1965, 48:217–221.

128. Wallace JD, Raleigh RJ, Weswig PH: Performance and carotene conversion in Hereford heifers fed different levels of nitrate. *J Anim Sci* 1964, 23:1042–1045.

129. Hale WH, Hubbert F, Taylor RE: Effect of energy level and nitrate on hepatic vitamin A and performance of fattening steers. *Proc Soc Exp Biol Med* 1962, 109:289–290.

130. Stewart CA, Merilan CP: Effect of potassium nitrate intake on lactating dairy cows. *University of Missouri Agricultural Experimental Station Bulletin* 650:1958.

131. Kahler LW, Jorgensen NA, Satter LD *et al.*: Effect of nitrate in drinking water on reproductive and productive efficiency in dairy cattle [abstract]. *J Dairy Sci* 1975;58:771.

132. Morris M P, Cancel B, Gonzalez-mas A: Toxicity of nitrates and nitrites to dairy cattle. *J Dairy Sci* 1972, 41:694–696.

133. Murdock FR, Hodgson AS: Utilization of nitrates by dairy cows. *J Dairy Sci* 1972, 55:640–642.

134. Geissler C, Steinhofel O, Ulbrich M: Nitrate contents in milk. *Arch Anim Nutr* 1991, 41:649–656.

135. Samol S, Sokolowski M: Poisoning with nitrates and nitrites in cattle. *Med Vet* 1980, 36:477–479.

136. Boerhmans HJ: Diagnosis of nitrate toxicosis in cattle using biological fluids and a rapid ion chromatographic method. *Am J Vet Res* 1990, 51:491–495.

137. Radostits OM, Blood DC, Gay CC: *In* Veterinary Medicine, edn 8. Bailliere Tindall, Philadelphia 1994:1536–1539.

138. Schneider NR, Yeary RA: Measurement of nitrite and nitrate in blood. *Am J Vet Res* 1973, 34:133–135.

139. Diven RH, Pistor WJ, Reed RE, Trautman RJ, Watts RE: The determination of serum or plasma nitrate and nitrite. *Am J Vet Res* 1962, 23:497–499.

140. Carlson MP, Schneider NR: Determination of nitrates in forages by using selective ion electrode. Collaborative study. *J Assoc Off Anal Chem* 1986, 69:196–198.

141. Smith GS: Diagnosis and causes of nitrate poisoning. *J Am Vet Med Assoc* 1965, 147:365–366.

142. Lincoln SD, Lane VM: Post mortem chemical analysis of vitreous humor as a diagnostic aid in cattle. *Mod Vet Pract* 1985, 66:883–886.

143. Montgomery JF, Hum S: Field diagnosis of nitrite poisoning in cattle by testing aqueous humor samples with urine test strips. *Vet Rec* 1995, 137:593–594.

144. Clay BR, Edwards WC, Peterson DR: Toxic nitrate accumulation in the sorghums. *Bovine Practitioner* 1976, 11:30–32.

145. Dalefield RR, Oehme FW: Stability of water nitrate levels during prolonged boiling. *Vet Hum Toxicol* 1997, 39:313.

146. Cheeke PR: *In* Natural Toxicants in Feeds, Forages and Poisonous Plants, edn 2. Danville, IL: Interstate Publishers; 1998:231.

147. Schultheiss PC, Knight AP, Traub-Dargatz JL, *et al.:* Toxicity of field bindweed (*Convolvulus arvensis*) to mice. *Vet Hum Toxicol* 1995, 37:452–454.

148. Todd FG, Stermitz FR, Schultheiss PC, *et al.:* Tropane alkaloids and toxicity of *Convolvulus arvensis*. *Phytochem* 1995, 39:301-303.

149. Molyneux RJ, Pan YT, Goldmann A, *et al.:* Calystegins, a novel class of alkaloid glycosidase inhibitors. *Arch Biochem Biophys* 1993, 304:81-88.

Larkspur (*Delphinium* spp.)

150. Aiyar VN, Benn MH, Hanna T, *et al.:* The principle toxin of *Delphinium brownii Rydb*, and its mode of action. *Experientia* 1979, 35:1367–1368.

151. Cronin EH, Nielsen DB. Tall larkspur and cattle on high mountain ranges. *In* Effects of Poisonous Plants on Livestock. Keeler RF, Van Kampen KR, James LF, Eds. New York: Academic Press; 1978:521–534.

152. Wilcox EV: Larkspur poisoning of sheep. *Montana Experimental Station Bulletin* 1897, 15:37–51.

153. Chestnut VK, Wilcox EV: The stock-poisoning plants of Montana. *USDA Bulletin* 1901, 26:65–80.

154. Glover GH: Larkspur and other poisonous plants. *Agricultural Experimental Station Fort Collins, Colorado Bulletin* 1906,113:1–24.

155. Nielsen DB, Rimbey NR, James LF: Economic considerations of poisonous plants on livestock. *In* The Ecology and Economic Impact of Poisonous Plants on Livestock Production. James LF, Ralphs MH, Nielsen DB, Eds. Boulder, CO: Westview Press; 1988:5–16.

156. Marsh CD, Clawson AB, Marsh H: Larkspur or poison weed. *USDA Farmer's Bulletin* 1923, 988:1–15.

157. Kingsbury JM: *Poisonous Plants of the United States and Canada*. Englewood Cliffs, NJ: Prentice-Hall; 1964:131–140.

158. Olsen JD, Manners GD: Toxicology of diterpenoid alkaloids in rangeland larkspur (*Delphinium* spp.). *In* Toxicants of Plant Origin. Volume 1. Alkaloids. Cheeke PR, Ed. Boca Raton, FL: CRC Press; 1989:291–326.

159. Olsen JD: Tall larkspur poisoning in cattle and sheep. *J Am Vet Med Assoc* 1978,173:762–765.

160. Pfister JA, Ralphs MH, Manners GD, *et al.*: Relationships between tall larkspur toxicity and consumption by cattle. Larkspur Symposium Proceedings. Colorado State University Cooperative Extension, 1996:13–16.

161. Pfister JA, Manners GD, Ralphs MH, *et al.*: Effects of phenology, site, and rumen fill on tall larkspur consumption by cattle. *J Range Manage* 1988, 41:509–514.

162. Pfister JA, Manners GD, Gardner GD, Ralphs MH: Toxic alkaloid levels in tall larkspur (*Delphinium barbeyi*) in western Colorado. *J Range Manage* 1994, 47:355–358.

163. Ralphs MH, Jensen DT, Pfister JA, *et al.*: Storms influence cattle to graze larkspur: an observation. *J Range Manage* 1994, 47:275–278.

164. Pfister JA, Panter KE, Manners GD: Effective dose in cattle of toxic alkaloids from tall larkspur (*Delphinium barbeyi*). *Vet Hum Toxicol* 1994, 36:10–11.

165. Olsen JD, Manners GD, Pelletier SW: Poisonous properties of larkspur (*Delphinium* spp.) *Collectanea Botanica (Barcelona)* 1990, 19:141–151.

166. Grina JA, Schroeder DR, Wydallis ET, Stermitz FR: Alkaloids from *Delphinium geyeri*. Three new C20 Ä diterpenoid alkaloids. *J Organic Chem* 1986, 51:390–396.

167. Pfister JA, Ralphs MA, Manners GD: Cattle grazing tall larkspur on Utah mountain rangeland. *J Range Manage* 1988, 41:118–122.

168. Olsen JD: Larkspur toxicosis: a review of current research. *In* Effects of Poisonous Plants on Livestock. Keeler RF, Van Kampen KR, James LF, Eds. New York: Academic Press; 1978:535–543.

169. Gardner DR, Panter KE, Pfister JA, Knight AP: Analysis of toxic norditerpenoid alkaloids in Delphinium species by electrospray atmospheric pressure chemical ionization, and sequential tandem mass spectrometry. *J Agric Food Chem* 1999, 47:5049–5058.

170. Manners GD, Panter KE, Ralphs MH, *et al.*: The occurrence and toxic evaluation of norditerpenoid alkaloids in tall larkspur (*Delphinium* spp.) *J Food Agric Chem* 1993, 41:96–100.

171. Manners GD, Panter KE, Pelletier SW: Structure-activity relationships of norditerpenoid alkaloids occurring in toxic larkspur (Delphinium) species. *J Nat Prod* 1997, 58:863–869.

172. Manners GD, Panter KE, Ralphs MH, *et al.*: The toxic evaluation of norditerpenoid alkaloids in three tall larkspur (Delphinium) species. *In* Plant Associated Toxins: Agricultural, Phytochemical and Ecological Aspects. Colegate SM, Dorling PR, Eds. Wallingford, Oxford, UK: CAB International; 1994:178–183.

173. Nation PN, Benn MH, Roth SH, Wilkens JL: Clinical signs and studies of the site of action of purified larkspur alkaloid, methyllycaconitine, administered parenterally to calves. *Can Vet J* 1982, 23:264–266.

174. Kukel CF, Jennings KR: *Delphinium* alkaloids as inhibitors of α-bungeratoxin

binding to rat and insect neural membranes. *Can J Physiol Pharmacol* 1994, 72:104–107.

175. Ward JM, Cockcroft VB, Lunt GG, *et al.:* Methyllycaconitine: a selective probe for neuronal a-bungeratoxin binding sites. *FEBS Lett* 1990, 270:45–48.

176. Olsen JD: Larkspur poisoning: as we now know it and a glance at the future. *Bovine Practitioner* 1994, 28:157–163.

177. Olsen JD, Sisson DV: Description of a scale for rating the clinical response of cattle poisoned by larkspur. *Am J Vet Res* 1991, 52:488–493.

178. Potter RL, Ueckert DN: Epidermal cellular characteristics of selected livestock-poisoning plants in North America. *Texas Agric Exp Station* 1997. pp 20.

179. Knight AP, Pfister JA: Larkspur poisoning in livestock: myths and misconceptions. *Rangelands* 1997, 19:10–13.

180. Pfister JA, Panter KE, Manners GD, Cheney CD: Reversal of tall larkspur (*Delphinium barbeyi*) poisoning in cattle with physostigmine. *Vet Hum Toxicol* 1994, 36:511–514.

181. Stegelmeier BL, James LF, Panter KE, *et al.:* Mechanisms and treatment of larkspur poisoning. *Larkspur Symposium Proceedings*. Colorado State University Cooperative Extension. 1996:7–12.

182. Pfister JA et al: Larkspur (*Delphinium* spp.) poisoning in livestock. *J Natural Toxins* 8:81-94, 1999.

183. Ralphs MH, Browns JE. Utilization of larkspur by sheep. *J Range Manage* 1991, 44:619–622.

184. Ralphs, MH, Olsen JD. Prior grazing by sheep reduces waxy larkspur consumption by cattle: an observation. *J Range Manage* 1992, 45:136–139.

185. Pfister JA, Manners GD: Mineral salt supplementation of cattle grazing tall larkspur infested range land during drought. *J Range Manage* 1991, 44:105–111.

186. Olsen JD, Ralphs MH: Feed aversion induced by intraruminal infusion of larkspur extract in cattle. *Am J Vet Res* 1986, 47:1829–1831.

187. Ralphs MH, Olsen JD: Comparison of larkspur alkaloid extract and lithium chloride in maintaining cattle aversion to larkspur in the field. *J Anim Sci* 1992, 70:1116–1120.

188. Olsen JD, Ralphs MH, Lane MA: Aversion to eating larkspur plants induced in cattle by intraruminal infusion of lithium chloride. *J Animal Sci* (Suppl 1) 1987, 65:218–224.

189. Lane MA, Ralphs MH, Olsen JD, Provenza FD, Pfister JA: Conditioned taste aversion: potential for reducing cattle loss to larkspur. *J Range Manage* 1990, 43:127–131.

190. Ralphs MH, Olsen JD: Adverse influence of social facilitation and learning context in training cattle to avoid eating larkspur. *J Anim Sci* 1990, 68:1944–1952.

191. Ralphs MH, Cheney CD: Influence of cattle age, lithium chloride dose level, and food type in the retention of food aversions. *J Anim Sci* 1993, 71:373–379.

192. Ralphs MH: Continued food aversion: training livestock to avoid eating poisonous plants. *J Range Manage* 1992, 45:46–51.

193. Nielsen DB, Cronin EH: Economics of tall larkspur control. *J Range Manage* 1997, 30:434–438.

194. Nielsen DB, Ralphs MH, Evans JO, Call CA: Economic feasibility of controlling tall larkspur on range lands. *J Range Manage* 1994, 47:369–372.

195.   Cronin EH, Nielsen D, Madson N: Cattle losses, tall larkspur, and their control. *J Range Manage* 1976, 29:364–367.

196.   Ralphs MH, Evans JO, Dewey SA: Timing of herbicide applications for control of larkspurs (*Delphinium* spp.). *Weed Sci* 1992, 40:264–269.

197.   Ralphs MH: Long term impact of herbicides on larkspur and associated vegetation. *J Range Manage* 1994, 48:459–464.

198.   Ralphs MH, Jones W, Mower K, Quimby C: Biological agents to control larkspur or reduce risk of poisoning. *Larkspur Symposium Proceedings*. Colorado State University Cooperative Extension. 1996:28–29.

199.   Stern ES: The diterpenoid alkaloids from *Aconitum*, *Delphinium*, and *Garrya* species. *In* The Alkaloids, Volume VII. Manske RHF, Ed. New York: Academic Press; 1960:473–501.

200.   Kingsbury JM: *Poisonous Plants of the United States and Canada*. Englewood Cliffs, NJ: Prentice Hall; 1964:125–140.

201.   Olsen JD, Manners GD, Pelletier SW: Poisonous properties of larkspur (*Delphinium* spp.) *Collectanea Botanica (Barcelona)* 1990, 19:141–151.

202.   Chang TY, Tomlinson B, Tse LK, Chan JC, Chan WW, Critchley JA: Aconitine poisoning due to Chinese herbal medicines: a review. *Vet Hum Toxicol* 1994, 36:452–455.

203.   Panter KE, Keeler RF, Baker DC: Toxicoses in livestock from the hemlocks (*Conium* and *Cicuta* spp.). *J Anim Sci* 1988, 66:2407–2413.

204.   Panter KE, Keeler RF: Piperidine alkaloids of poison hemlock (*Conium maculatum*). *In* Toxicants of Plant Origin. Volume 1. Alkaloids. Cheeke PR, Ed. CRC Press, Boca Raton, FL; 1989:109–132.

205.   Bowman WC, Snaghvi IS: Pharmacologic actions of hemlock (*Conium maculatum*) alkaloids. *J Pharm Pharmacol* 1963, 15:1–25.

206.   Lopez TA, Cid MS, Bianchini ML: Biochemistry of hemlock (*Conium maculatum L*) alkaloids and their acute and chronic toxicity in livestock. A review. *Toxicon* 1999, 37:841-865.

207.   Keeler RF: Coniine, a teratogenic principle from Conium maculatum producing congenital malformations in calves. *Clin Toxicol* 1979, 7:195–206.

208.   Keeler RF, Balls LD: Teratogenic effects in cattle of *Conium maculatum* and conium alkaloids and analogs. *Clin Toxicol* 1978, 12:49–64.

209   Panter KE, Keeler RF, Buck WB, Shupe JL: Toxicity and teratogenicity of *Conium maculatum in swine. Toxicon Suppl* 1983, 3:333–336.

210.   Dyson DA, Wrathall AE: Congenital deformities in pigs possibly associated with exposure to hemlock (*Conium maculatum*). *Vet Rec* 1977, 100:241–243.

211.   Frank AA, Reed WM: Conium maculatum (poison hemlock) toxicosis in a flock of range turkeys. *Avian Dis* 1987,31:386–388.

212.   Jessup DA, Boermans HJ, Kock ND: Toxicosis in Tule Elk caused by the ingestion of poison hemlock. *J Am Vet Med Assoc* 1986, 189:1173–1175.

213.   Keeler RF: Alkaloid teratogens from lupinus, conium, veratrum and related genera. *In* Effects of Poisonous Plants on Livestock. Keeler RF, Van Kampen KR, James LF, Eds. New York: Academic Press; 1978:397–408.

214.   Keeler RF, Balls LD, Shupe JL, Crowe MW: Teratogenicity and toxicity of coniine in cows, ewes and mares. *Cornell Vet* 1980, 70:19–26.

215. Kingsbury JM: *Poisonous Plants of the United States and Canada*. Englewood Cliffs, NJ: Prentice-Hall; 1964:379–383.

216. MacDonald H: Hemlock poisoning in horses. *Vet Rec* 1937, 49:1211–1212.

217. Penney RHC: Hemlock poisoning in cattle. *Vet Rec* 1953, 65:669.

218. Widmer WR: Poison hemlock toxicosis in swine. *Vet Med* 1984, 79:405–408.

219. Shupe JL, James LF: Teratogenic plants. *Vet Human Toxicol* 1983, 25:415–421.

220. Edmonds LD, Selby LA, Case AA: Poisoning and congenital malformations associated with consumption of poison hemlock by sows. *J Am Vet Med Assoc* 1972, 160:1319–1324.

221. Panter KE, Keeler RF, Buck WB: Induction of cleft palate in new born pigs by maternal ingestion of poison hemlock (*Conium maculatum*). *Am J Vet Res* 1985, 46:1368–1371.

222. Panter KE, Keeler RF, Buck WB: Congenital skeletal malformations induced by maternal ingestion of poison hemlock (*Conium maculatum*) in newborn pigs. *Am J Vet Res* 1985, 46:2064–2066.

223. Crowe MW, Swerczek TW: Congenital arthrogryposis in offspring of sows fed tobacco stalks (*Nicotiana tabacum*). *Am J Vet Res* 1974, 35:1071–1073.

224. Panter KE, Keeler RF, Bunch TD, Callan RJ: Congenital skeletal malformations and cleft palate induced in goats by ingestion of *Lupinus, Conium*, and *Nicotiana* species. *Toxicon* 1990, 28:1377–1385.

225. Panter KE, Keeler RF: Induction of cleft palate in goats by *Nicotiana glauca* during a narrow gestational period and the relation to the reduction in fetal movements. *J Natural Toxins* 1992, 1:25–32.

226. Panter KE, James LF, Keeler RF, Bunch TD: Radio-ultrasound observations of poisonous plant-induced fetotoxicity in livestock. *In* Poisonous Plants. *Proceedings 3rd International Symposium*. James LF, Keeler RF, Bailey EM, *et al.*: Eds. Ames, IA: Iowa State University Press; 1992:481–487.

227. Panter KE, Bunch TD, Keeler RF: Maternal and fetal toxicity of poison hemlock (*Conium maculatum*) in sheep. *Am J Vet Res* 1988, 49:281–283.

228. Kingsbury JM: *Poisonous Plants of the United States and Canada*. Englewood Cliffs, NJ: Prentice-Hall; 1964:373–379.

229. Marsh DC, Clawson AB: Cicuta, or water hemlock. *USDA Bulletin* 1914, 69:1.

230. Anet EFLJ, Lythgoe B, Silk MH, Trippett S: Oenanthotoxin and cicutoxin. Isolation and structures. *J Chem Soc* 1953, 66:309–322.

231. Payonk GS, Segelman AB: Analytical and phytochemistry of the American water hemlock, *Cicuta maculata* (*Umbelliferae*). *Vet Hum Toxicol* 1980, 22:367.

232. Panter KE, Baker DC, Kechele PO: Water hemlock (*Cicuta douglasii*) toxicosis in sheep: pathologic description and prevention of lesions and death. *J Vet Diagn Invest* 1996, 8:474–480.

233. James LF, Ralphs MH: Water hemlock. *Utah Sci* 1986, 47:67–69.

234. Warwick BL, Runnels HA: Water hemlock poisoning of livestock. *Ohio Agricultural Expimental Station* 1929, 14:35.

235. Panter KE, Keeler RF, Baker DC: Toxicosis in livestock from the hemlocks (*Conium* and *Cicuta* spp.) *J Anim Sci* 1988, 66:2407–2413.

236. Fleming CE, Peterson NF: The poison parsnip or water hemlock (*Cicuta occidentalis*). *University of Nevada Agricultural Experimental Station Bulletin 100*. 1920:1.

237. Smith RA, Lewis D: Cicuta toxicosis in cattle: case history and simplified analytical method. *Vet Hum Toxicol* 1987, 29:240–241.

238. Kingsbury JM. *Poisonous Plants of the United States and Canada*. Englewood Cliffs, NJ: Prentice Hall; 1964:419–420.

CHAPTER 2

Plants Affecting the Cardiovascular System

A large group of unrelated plants have toxic compounds that have a direct effect on the heart and blood vessels and can cause the death of animals that consume them. The most recognized of these compounds are the cardiac glycosides, of which digoxin, found in foxglove (*Digitalis* spp.), is best known. The toxic and pharmacologic properties of digoxin have been known for a long time. Because of its effects on the heart at therapeutic levels, it is routinely used to treat congestive heart failure in humans and animals.[1] Other cardiotoxic compounds found in plants such as yew (*Taxus* spp.), rhododendrons (*Rhododendron* spp.), laurels (*Kalmia* spp.), and the avocado (*Persea* spp.) are responsible for poisoning in animals and humans.

Cardiac Glycosides

Cardiac glycosides are found in 11 plant families including the Apocynaceae, Asclepiadaceae, Celastraceae, Brassicaceae, Lilaceae, Moraceae, Fabaceae, Ranunculaceae, Scrophulariaceae, Sterculiaceae, and Tiliaceae.[2] The glycosides that have specific effects on the heart are present in at least 34 genera of these plant families.[2,3] There are 2 basic groups of glycosides in plants, cardenolides and bufadienolides, both of which have direct effects upon cardiac function. The best known cardiac glycosides (cardenolides) are digoxin and digitoxin, derived from the parent glycoside digitalis. The therapeutic value of digitalis in treating heart failure was first recognized in 1775[4] and it is now produced commercially for therapeutic purposes from foxglove (*Digitalis lannata*).[5] Immunologic methods have enabled detection of glycoside concentrations in a variety of other plant genera.[6] Other cultivated plants that contain cardenolides include:, *Euonymus* spp., Star of Bethlehem (*Ornithogalum* spp.), lily of the valley (*Convallaria majalis*), and oleander (*Nerium* spp.). Bufadienolides being found in hellebores (*Helleborus* spp.), lily of the valley (*Convallaria majalis*), and squill (*Drimia* or *Urginea* spp.).

Relatively few of the plants containing cardiac glycosides are a significant cause of animal poisoning in North America.[7] Milkweeds (*Asclepias* spp.) are the most frequent cause of cardiac glycoside poisoning in livestock in North America.[8-10] Oleander is becoming of increasing significance as a source of poisoning because of the frequency with which it is planted as an attractive, drought-resistant, ornamental flowering shrub. Foxglove (*D. purpurea*), oleander (*Nerium oleander*), and lily of the valley (*Convallaria majalis*) are widely grown as ornamental plants in North America and have escaped to become established in the wild. Dogbane or Indian hemp (*Apocynum cannabinum*) is an indigenous plant containing cardiac glycosides but is rarely a problem to livestock. In Australia and southern Africa various Kalanchoe (*Bryophyllum* spp.) have killed cattle as a result of their cardiac glycoside content.[11-15] These succulent perennial plants are now sold in the United States as house plants or garden ornamentals and can be grown in the warmer, mild climates of the southern States where they have the potential of becoming a threat to animals that eat them.[12,13] Another exotic plant containing cardiac glycosides with the potential of being introduced to North America is pheasant's eye (*Adonis microcarpa*).[16] Only plants that pose a risk to livestock in North America will be discussed.

Other common cultivated plants that contain cardiac glycosides include: Hellebores (*Helleborus* spp.), squill (*Urginea* spp.), *Euonymus* spp., Star of Bethlehem (*Ornithogalum* spp.), lily of the valley (*Convallaria majalis*), oleander (*Nerium* spp.), blue eyed grass (*Sisyrinchium* spp.), hyacinth (*Hyacinthium* spp.) and periwinkle (*Vinca* major).

Toxicity of Cardiac Glycosides

Cardiac glycosides are found in all plant parts, especially the leaves. Generally, only very small quantities of the plants must be ingested to produce poisoning. Drought and freezing temperatures may cause livestock to consume more of the toxic plants.[17] In cattle and horses, as little as 0.005 percent body weight of green oleander leaves is reportedly lethal.[7] Oleander leaves administered experimentally via nasogastric tube at 40 to 80 mg/kg body weight consistently caused gastrointestinal and cardiac toxicosis.[18] Although reduced, toxicity is retained in the dried plants.

Animals consuming plants containing cardiac glycosides develop primarily heart and digestive disturbances before death.[19] The glycosides act directly on the gastrointestinal tract causing hemorrhagic enteritis, abdominal pain, and diarrhea.[7,20-22] Cardiac glycosides are cardenolides (a steroid nucleus with an attached lactone group) that inhibit the cellular membrane sodium-potassium pump (Na^+K^+ adenosine triphosphatase [ATPase] enzyme system) with resulting depletion of intracellular potassium and an increase in serum potassium.[3,23] This results in a progressive decrease in electrical conductivity through the heart causing irregular heart activity and eventually completely blocking cardiac activity. In low doses, the glycosides have a beneficial therapeutic effect on the heart by increasing the force of contraction, slowing the heart rate, and increasing cardiac output. Toxic doses of the glycosides cause a variety of severe dysrhythmias and conduction disturbances through the myocardium that results in decreased cardiac output and death. Cattle on rations containing ionophore feed additives such as monensin have increased susceptibility to the cardiac glycosides.

Clinical Signs of Poisoning

Cattle especially, and less often horses consuming cardiac glycoside-containing plants are often found dead because of the profound cardiac effects of the toxins. A variety of cardiac arrhythmias and heart block, including ventricular tachycardia and first- and second-degree heart block, may be encountered with cardiac glycoside poisoning.[18] Abdominal pain (colic) and diarrhea are also signs commonly seen in animals poisoned with cardiac glycosides. If observed early in the course of poisoning, animals will exhibit rapid breathing, cold extremities, and a rapid, weak, and irregular pulse. The duration of symptoms rarely exceeds 24 hours before death occurs. Convulsions before death are not common.

In acute poisoning from cardiac glycosides as characterized by oleander poisoning, the postmortem findings include hemorrhages, congestion, edema, and cell degeneration of the organs of the thoracic and abdominal cavities. In less acute but fatal poisoning, multifocal myocardial degeneration and necrosis is often present.[24]

Treatment

No specific treatment is available for counteracting the effects of the cardiac glycosides. Gastric lavage or vomiting should be induced in dogs and cats as soon as possible. Cattle and horses should be given adsorbents such as activated charcoal (2-5 g/kg body weight) orally to prevent further toxin absorption.[13] In ruminants known to have eaten oleander, a rumenotomy to remove all traces of the plant from the rumen may be lifesaving. The cardiac irregularities may be treated using antiarrhythmic drugs such as potassium chloride, procainamide, lidocaine, dipotassium EDTA, or atropine sulfate.[1,3,13] The use of fructose-1,6-diphosphate (FDP) has effectively reduced serum potassium levels and irregularities of the heart and will improve cardiac function in dogs experimentally poisoned with oleander.[25] The mechanism of action of FDP is not known, but it apparently restores cell membrane Na$^+$ and K$^+$ ATPase function.[25] Because hyperkalemia is a common feature of oleander poisoning, the use of potassium in intravenous fluids should be avoided and serum potassium levels should be monitored closely.

Intravenous fluids containing calcium should not be given because calcium augments the effects of the cardiac glycosides. Poisoned animals should be kept as quiet as possible to avoid further stress on the heart. The use of digoxin-specific antibodies to treat digoxin toxicity, although possible in humans, has not found application in animal poisoning as yet.[26]

Diagnosis

A diagnosis of cardiac glycoside poisoning may be made if an animal is found dead and evidence indicates that the animal had access to plants known to contain cardiac glycosides. Similarly, detection of cardiac dysrhythmias, heart block, and ventricular escape rhythms is suggestive of cardiac glycoside or grayanotoxin poisoning. Detection of cardiac glycosides in the serum, urine, tissues, and stomach contents is possible using high-performance liquid chromatography.[27,28] Oleander glycosides will cross-react with digoxin radioimmunoassays.[29]

Postmortem findings in oleander poisoning may include focal pale areas and hemorrhages in the myocardium. Multifocal areas of necrosis and hemorrhage may be seen microscopically.[18] The clinical cardiac abnormalities, sudden deaths and the lesions present in the heart typical of cardiac glycoside toxicity also closely mimic poisoning due to monensin, a feed additive in cattle rations.

Dogbane, Indian Hemp
Apocynum cannabinum
Apocynaceae (Dogbane family)
Spreading Dogbane
A. androsaemifolium

Habitat

Dogbane is a common plant of open spaces, especially along streams, irrigation ditches, and roadsides throughout North America. Spreading dogbane is more common at higher altitudes.

Figure 2-1B (above) Spreading dogbane (*A. androsaemifolium*).

Figure 2-1A Dogbane showing distinctive long double seed pods (*Apocynum cannabinum*).

Description

Perennial, erect herbs grow from spreading root stalks. The stems are red, smooth, tough, branched, and 4 to 6 feet (2 to 3 meters) in height. The plant contains a white milky sap. The leaves are opposite, oblong in shape, hairless, and entire margined. The small green-white flowers are produced in clusters at the ends of the branches. The fruits are characteristically 3 to 8 inches (10 to 20 cm) long, narrow pods hanging in pairs (Figure 2-1A) The seeds are long and narrow, each having a tuft of white hairs similar to milkweed seeds.

Spreading dogbane is a smaller plant up to 3 feet (1 meter) in height, with larger pink flowers (Figure 2-1B).

Principal Toxin

Dogbanes contain a resin, apocynin, and two known glycosides, cymarin, and apocynein. Livestock are rarely poisoned by dogbane, and eat these plants only when they lack other forages.

Lily of the Valley
Convallaria majalis
Liliaceae (Lily family)

Although this plant is not indigenous to North America, it is commonly planted as ground cover in shady gardens. It is a hardy plant and when abandoned can escape to establish large stands. The plant is potentially toxic to animals if they are carelessly allowed to graze it or are fed garden clippings.

Description

Lily of the valley is a perennial plant arising from a deep underground rhizome. The plant forms dense spreading colonies. The leaves are hairless, glossy, green, parallel-veined, and sheath the flower stem. The white, fragrant, drooping, bell-shaped flowers are on a raceme (Figure 2-2A). The fruits ripen into conspicuous red berries (Figure 2-2B).

Figure 2-2B (above) Fruits of Lily of the valley

Figure 2-2A Lily of the valley (*Convallaria majalis*).

Principal Toxin

The cardiac glycosides (cardenolides) convallerin and convallamarin amongst at least 15 others, are found throughout the plant and have similar cardiac effects to digitalis glycosides. The seeds have the highest concentration of cardenolides, but the flesh of the fruit is minimally toxic. The skin of the fruit and the flowers also contain saponins that cause abdominal pain and diarrhea. Poisoning has been reported in dogs.[30]

Foxglove
Digitalis purpurea
Scrophulariaceae (Figwort family)

Habitat

A native plant of Europe, the foxglove was introduced to North America where it has escaped cultivation and now is quite widespread in the northwest. It prefers rich soils growing along roadsides, fences, and disused areas.

Description

Foxglove is a perennial herb growing 3 to 6 feet (1 to 2 meters) tall with alternate toothed, hairy, basal leaves. The characteristic purple or white tubular pendant flowers with conspicuous spots on the inside bottom surface of the tube (Figure 2-3A and B).

Figure 2-3A Foxglove plant in bloom (*Digitalis purpurea*).

Principal Toxin

Several cardiac glycosides, the most important of which are digoxin, digitoxin, and digitonin, are found in all parts of the plant, and especially in the seeds. Livestock are infrequently poisoned, but will eat the plant occasionally either fresh or in hay. All spears of digitalis and their hybrids should be considered toxic until proven otherwise.

Figure 2-3B Foxglove shows the characteristic purple spots in the flowers (*D. purpurea*) .

Oleander
Nerium oleander
Apocynaceae (Dogbane family)

Habitat

Introduced from the Mediterranean area, oleander is an evergreen showy flowering shrub, growing commonly from California to Florida. It is drought tolerant and is extensively used in landscaping along highways. Oleander is also grown as a potted house plant in northern climates.

Description

Oleander is a perennial, evergreen shrub or small tree up to 25 feet (10 meters) tall with whorled, simple, narrow, sharply pointed, leathery leaves 3 to 10 inches (6 to 20 cm) long. Two primary parallel veins run perpendicular to the mid rib of the leaf. The showy white, pink, or red flowers with five or more petals are produced in the spring and summer (Figure 2-4A and B). Fruit pods contain many seeds, each with a tuft of brown hairs (Figure 2-4C).

Figure 2-4A Oleander (*Nerium oleander*).

Figure 2-4B Oleander (*N. oleander*). White variety shows the characteristic appendages at base of the petals.

Principal Toxin

Oleandrin and neriine are two potent cardiac glycosides (cardenolides) found in all parts of the plant. Red flowered varieties of oleander appear to be more toxic. Oleander remains toxic when dry and is very poisonous to humans, many animals, and birds.[7,17,18,20,22,31] A single leaf can be lethal to a child eating it, although mortality is generally low in humans.[32] The lethal dose of the green oleander leaves for cattle and horses is 0.005 percent of the animal's body weight.[7] The minimum lethal dose of oleander for cattle is 50 mg/kg body weight.[24] Horses given 40 mg/kg body weight of green oleander leaves via nasogastric tube consistently developed severe gastrointestinal and cardiac signs of poisoning.[18] Livestock are usually poisoned when they are allowed to graze in places where oleander is abundant or when prunings are carelessly thrown into animal pens.

Figure 2-4C Oleander pod and seeds.

Yellow Oleander, Be-still Tree, Tiger Apple, Lucky Nut
(*Cascabela thevetioides, Thevetia thevetioides*)
Apocynaceae (Dogbane family)

Figure 2-5A Yellow oleander (*Thevetia thevetioides*).

Habitat

Native to tropical America, this plant is widely cultivated in the southern United States and Hawaii, and most tropical areas of the world.

Description

Yellow oleander is a perennial branched shrub or tree growing to 30 feet (12 meters) tall with dark green, glossy, alternate linear leaves up to 6 inches (15 cm) long and 0.5 inch (1 to 2 cm) wide, with milky sap. The yellow showy, tubular flowers are produced in clusters at the ends of branches (Figure 2-5A). Fruits are fleshy drupes turning yellow to black when ripe.

Thevetia peruviana is a similar, but smaller species that is a large shrub or small rounded tree growing to 15 feet (7 meters) in height. The leaves are light green and glossy, often rolling under at the edges. The funnel-shaped flowers are smaller than *T. thevetioides* and orange-yellow to salmon-pink in color depending on the variety (Figure 2-5B). The spherical fruits have rounded corners and are green, turning black when ripe. Each fruit contains two black seeds.

Principal Toxin

Thevitin A and B and thevetoxin are potent cardiac glycosides found in all parts of the plant and are concentrated in the fruits.[33,34]

Figure 2-5B Yellow oleander (*T. peruviana*).

Milkweeds
Asclepias spp.
Asclepiadaceae (Milkweed family)

Habitat

Milkweeds are widely distributed throughout the world and much of North America. They cause poisoning in sheep, goats, cattle, horses, and domestic fowl.[8,35-38] Greatest losses have occurred in sheep on western rangeland, but all animals are susceptible to poisoning especially when other forages are scarce or milkweeds are incorporated in hay.[8,39] Milkweeds grow in open areas along roadsides, waterways, and disturbed areas, preferring sandy soils of the plains and foothills. Overgrazing will enhance the encroachment of milkweeds.

Description

Milkweeds are erect perennial herbs that have either 6 to 12 cm broad veined leaves or narrow linear leaves less than 0.5 inches (1 cm) wide, arranged either alternately or in whorls (Figures 2-6A and 2-6B). Most species (except *A. tuberosa*) contain a milky sap or latex (Figure 2-6C). The flowers are produced in terminal or axillary

Figure 2-6A Showy milkweed (*Asclepias speciosa*).

umbels consisting of two, five-parted whorls of petals, the inner one being modified into a characteristic hornlike projection (Figure 2-6D) The color of the flowers varies among species from greenish white to red. The characteristic follicle or pod contains many seeds each with a tuft of silky white hairs that aids in its wind-borne dispersal.

Figure 2-6B Whorled milkweed flowers and narrow leaves (*A. subverticillata*).

Although milkweeds characteristically have a milky sap that is readily seen when the stem, leaves, or pods are ruptured, they should be differentiated from other plant genera that also have milky sap. These include the spurges (*Euphorbia* spp.), dogbane (*Apocynum* spp.), and wild lettuce (*Lactuca* spp.). The presence, therefore, of a milky sap or latex is not limited to milkweeds.

Principal Toxin

Asclepias species contain various toxic cardenolides (cardiac glycosides).[40-42] Acute death from milkweed poisoning results from the cardiotoxic effects of the cardenolides that act like ouabain, a digitalis glycoside.[43] The cardenolides act by inhibiting Na^+K^+-ATPase, thereby affecting myocardial conduction and contractility.[42,43] Milkweeds are most toxic during rapid growth and retain their toxicity even when dried in hay. Toxicity varies with the species and growing conditions.[44] However, all milkweeds should be considered potentially poisonous, especially the narrow-leafed species (see Figure 2-6B and E). The highest concentration of cardenolides occurs in the latex, with the lowest concentrations in the roots.[44]

The relative toxicity of the more common milkweeds is shown in Table 2-1. Fatal poisoning of an adult horse (450 kg) may occur with the ingestion of as little as 1.0 kg of green milkweed plant material.[36] As little as 0.1 to 0.2 percent body weight of plant on a dry matter basis of *A. labriformis* and *A. subverticillata*, respectively, induced fatal poisoning in sheep.[37] In addition to the cardiotoxic effects of the cardenolides common to most milkweeds, other glycosides and resinoids identified in milkweeds have direct effects on the respiratory, digestive, and nervous systems causing dyspnea, colic and diarrhea, muscle tremors, seizures, and head pressing.[36,37]

The presence of cardenolides in milkweeds, as many as 20 in *A. eriocarpa*, is apparently a defense mechanism for the plant to discourage most animals and insects from feeding on the plant.[44,45]

Figure 2-6C Butterfly weed (*A. tuberosum*).

Some insects, however, including the larvae of the monarch butterfly (*Danaus plexippus*) have the ability to feed on milkweeds and store the cardenolides in their own tissues as a protective mechanism.[40,44,45] The adult monarch butterfly retains the cardenolides as a defense mechanism. Birds that feed on the monarch butterfly that has fed on toxic milkweeds will experience the intense emetic effects of the cardenolides, and by association, avoid eating the insect in the future.

Figure 2-6D Woolypod milkweed (*A. eriocarpa*).

Clinical Signs

Signs of poisoning usually begin within 8 to 10 hours of the milkweed plants being eaten; the severity of symptoms depends on the quantity of plant consumed. In acute milkweed poisoning the animal may be found dead without any prior symptoms.

Figure 2-6E One of the smallest toxic milkweeds (*A. pumilla*).

Poisoned sheep show a labored and slow respiratory rate, pain and inability to stand, muscular tremors, staggering gait, a weak, rapid pulse, bloating, colic and dilated pupils prior to death.[8,19,35,37] A variety of cardiac dysrhythmias may be detected using electrocardiography. Once recumbent, the poisoned animals exhibit periods of tetany and chewing movements.[35,37] Postmortem signs in animals poisoned by milkweeds consist of nonspecific congestion of the lungs, stomach, and intestines, with hemorrhages present on the surfaces of the lungs, kidneys, and heart.

Table 2-1 Common Toxic Milkweeds[36, 38]		
COMMON NAMES	SCIENTIFIC NAMES	TOXICITY*
Labriform milkweed	Asclepias labriformis	0.05
Western whorled milkweed	A. subverticillata	0.2
Eastern whorled milkweed	A. verticillata	0.2
Woolypod milkweed	A. eriocarpa	0.25
Spider antelopehorn milkweed	A. asperula	1–2
Plains or dwarf milkweed	A. pumila	1–2
Swamp milkweed	A. incarnata	1–2
Mexican whorled milkweed	A. mexicana	2.0
Showy milkweed	A. speciosa	2–5
Broad leaf milkweed	A. latifolia	1.0
Narrow-leafed milkweed	A. stenophylla	----
Butterfly Weed	A. tuberosa	1.5
Antelope horn milkweed	A. viridis	1
* Amount of green plant as a percent of the animal's body weight that is lethal.		

Treatment and Prevention

No specific treatment is available for milkweed poisoning. Animals that have not consumed a lethal dose of the plants recover over several days. Affected animals should be moved from the source of the plants and given fresh water, good quality hay, and shade. Sedatives, laxatives, and supportive intravenous fluid therapy with addition of potassium chloride, if carefully monitored, may help to control the severity of signs. Intravenous fluids containing calcium are contraindicated.

Because milkweeds generally tend to grow singly or in relatively small stands, they can be controlled by digging out individual plants or selectively spraying the plants with a herbicide such as 2,4-dichlorophenoxyacetic acid (2,4-D). Care should be taken during the hay-making process to avoid incorporating the narrow-leafed milkweeds in the hay.

Yew
Taxus species
Taxaceae (Yew family)

Habitat

Several species of yew grow naturally or as ornamentals in North America, generally preferring more humid, moist environments. Western yew (*T. brevifoliat*) and American yew (*T. canadensis*) are two indigenous species.[46] English yew (*T. baccata*) and Japanese yew (*T. cuspidata*) are commonly cultivated species in North America.

Description

Yews are evergreen shrubs or small trees with glossy, rigid, dark green, linear leaves 1.5 to 2 inches (4 to 5 cm) long with pointed ends, closely spaced on the branches. Inconspicuous axillary male and female flowers are produced on separate plants, forming showy red to yellow fruits (aril) containing a single seed (Figure 2-7A).

Podocarpus (*Podocarpus macrophyllus*), a nonpoisonous tree found commonly in milder climates of the southern portion of North America, is frequently but incorrectly called yew. It has dark green lance-shaped leaves 3 to 4 inches (7 to 10 cm) long, and the fruits are greenish purple that help differentiate it from yew (Figure 2-7B).

Figure 2-7A Japanese yew leaves and fruits (*Taxus japonicus*).

Principal Toxin

The toxicity of yews to humans and animals has been known for many years.[46-48] Yews contain a group of 10 or more toxic alkaloids, the most toxic of which are taxine A and B, collectively referred to as taxine. Taxine inhibits normal sodium and calcium exchange across the myocardial cells, depressing cardiac depolarization and causing arrhythmias.[51,52] Taxol is a different diterpenoid alkaloid found in T. brevifolia and other yews that has anticancer activity.[51,52] Yews also contain nitriles, ephedrine, and irritant oils.[46,48] All parts of the plant, green or dried, except the fleshy part of the aril surrounding the seed are toxic.[47,48] Livestock are frequently poisoned when fed prunings from cultivated yews. The highest concentration of the alkaloids is generally found in the leaves in winter time. All domestic animals including birds are susceptible to the cardiotoxic effects of the alkaloid. Adult cattle and horses have been fatally poisoned with as little as 8 to 16 oz of yew leaves or 0.1 to 0.5 percent of their body weight. Drying of the leaves does not appreciably decrease their toxicity.[47] As little as 200 g of dried leaves fed to a 550 kg steer proved fatal.[49] Animals generally will not eat yew if they are fed a balanced diet.[50] Interestingly, deer appear to be able to eat yew without problem.

Figure 2-7B Yew (left) and *Podocarpus* (right) leaves.

Although not proven, the yew alkaloids are not apparently excreted through the milk of cows fatally poisoned.[49] However, it is wise to not use the milk from lactating animals that survive yew poisoning for at least 48 hours and to withhold animals from slaughter for 35 days.[53]

Clinical Signs

Sudden onset of muscle trembling, incoordination, nervousness, difficulty in breathing, slow heart rate, vomiting, diarrhea, convulsions, and death are characteristic of yew poisoning in animals.[46,50-62] Sudden death may be the only observed sign in many cases.[56] Deaths may however occur several days after the yew was eaten.

No postmortem lesions are diagnostic of yew poisoning. Diagnosis must be based on eliminating other causes of sudden death, evidence of access to yew, and the presence of yew leaves in the rumen and stomach contents of the animal. Finely chewed leaves may have to be examined microscopically to positively identify them as yew. Identification of taxine from chewed plant material and rumen contents using mass spectrometry affords a more precise means of confirming yew poisoning.[63,64]

There is no specific treatment or antidote for acute yew poisoning. If an animal is observed eating yew, immediate veterinary attention is indicated. Activated charcoal (2 g/kg body weight) and magnesium sulfate (2 g/kg body weight) as a cathartic should be given via stomach tube to decontaminate the rumen.[53] A rumenotomy to remove the yew leaves from the rumen of cattle in early confirmed cases of yew consumption may be lifesaving. Atropine sulfate is reportedly effective in counteracting the slow heartbeat and heart failure but should be used with caution.[49] When possible, intravenous fluid therapy and other supportive measures should be instituted to support the cardiovascular system.[53]

Yews should never be planted as hedges around animal enclosures, and prunings from yews should not be given to animals.

Avocado
Persea americana
Lauraceae (Laurel family)

Habitat

Avocados are grown in southern areas of North America and throughout tropical areas of the world for their edible fruits. As early as 1942, avocado poisoning was reported in California,[65] and since then a variety of species including cattle, horses, goats, rabbits, canaries, budgerigars, cockatiels, ostriches, and fish have been poisoned by eating the leaves and fruits of the avocado tree.[66-72] The leaves, bark, seeds, and skin of the fruit are toxic. The leaves remain toxic when dried. Both the Guatemalan and the Mexican varieties of avocado are toxic to animals and birds, with most poisoning being associated with the consumption of the Guatemalan and Nabal varieties of avocado.[73-75] (Figure 2-8). There is some variability in toxicity between different varieties of avocado, the Mexican variety being considered the least toxic.[74] However, hybrid varieties of avocados are toxic to birds.[74,75]

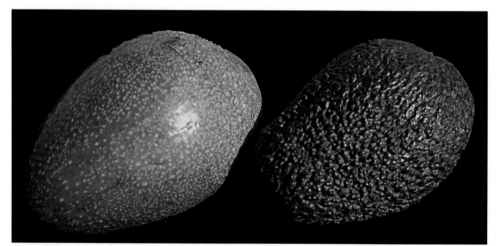

Figure 2-8 Avocado fruits (*Persea americana*). Guatemalan varieties have a dark rough skin.

Principal Toxin

The toxin present in avocados, as yet unidentified, has a direct toxic effect on the myocardium and tissues of the lactating mammary gland.[72-75]

Clinical Signs

Horses develop edematous swelling of the lips, mouth, eyelids, head, and neck, which can cause upper respiratory distress.[70] Colic is seen in some horses. The serum enzymes creatine phosphokinase (CPK) and aspartate aminotransferase (AST) are generally elevated, indicating muscle necrosis and in particular myocarditis.[70] Edematous swelling of the head and neck area is suggestive of heart failure. Brisket and neck edema, and acute pulmonary edema as a result of cardiomyopathy and heart failure, have also been reported in rabbits and goats that have died from avocado poisoning.[72] The cardiomyopathy is manifested as an acute noninfectious degeneration and necrosis of myocardial tissue, which, if severe, will result in acute death of the animal.[72-75]

Ostriches eating the leaves and immature fruits of the Guatemalan variety of avocado have developed fatal cardiomyopathy and congestive heart failure with 96 hours of eating the plants.[69] In other birds, avocado poisoning is characterized by respiratory distress and sudden death. Deaths usually occur 1 to 2 days after consumption of the avocado.

Cattle, horses, goats, and rabbits develop noninfectious mastitis after eating avocado leaves.[72,74] Goats fed as little as 31 g/kg body weight of avocado leaves showed dramatic reduction in milk production and developed hard swollen udders 24 hours after they had eaten the leaves. The milk was of a cheesy consistency and contained clots. The milk somatic cell counts also became markedly elevated. If no further avocado leaves were fed, the udder edema regressed and milk production returned, but not to the levels before feeding the avocado leaves.[72] The serum liver enzyme AST was increased in intoxicated goats.[91] Generalized necrosis of the mammary gland epithelium with no significant cellular inflammatory response and sloughing of necrotic cells are characteristic microscopic findings.[72,75]

At postmortem examination, generalized congestion of the lungs, hydropericardium, and subcutaneous edema in the pectoral area are typical.[68,69] Microscopic examination of the tissues reveals scattered nonsuppurative inflammation in the liver, heart, and kidneys and eosinophilic material in the cytoplasm of many Kupffer cells.[72-75] Similar noninflammatory necrosis of the myocardium and lactating mammary gland were observed in goats and mice fed avocado leaves.[73-75]

Horses and other livestock should not be allowed into avocado orchards and the trees should not be planted adjacent to livestock enclosures. The fruits and seeds should not be fed to pet birds.

Death Camas
Zigadenus spp.
Liliaceae (Lily family)

Habitat

Approximately 15 species of death camas occur in North America, their habitat ranging from moist mountain valleys to dry sandy hills and plains. Death camas appears in early spring, often growing among wild onion from which it can be readily differentiated on the basis of the smell.

Description

Death camas is a herbaceous, hairless perennial with grasslike, linear, V-shaped, parallel-veined leaves arising basally from an onion-like bulb 6 to 8 inches below the soil surface. The leaves are not hollow like those of onions, nor do they smell like an onion. The bulb is covered with a black membranous outer layer. The inflorescence, a terminal raceme or panicle, has small, perfect, greenish white to yellow flowers (Figure 2-9A). A showy species is mountain death camas (*Z. elegans*) (Figure 2-9B). The six perianth segments are separated, each with a gland at its base (Figure 2-9C).

Figure 2-9A Death camas in bloom (*Zigadenus venenosus*).

Figure 2-9B Showy death camas
(*Z. elegans*).

The ovary is superior or partly inferior forming a trilobed, capsuled seed pod.

Principal Toxin

Members of the genus *Zigadenus* contain several steroidal alkaloids, the best known being zygacine and zygadenine.[76] The alkaloids are similar to those found in false hellebore (*Veratrum californicum*) and decrease blood pressure by dilating arterioles, constricting veins, and slowing the heart rate. The plants vary considerably in their toxicity, with most toxic species seldom growing at elevations above 8000 feet. All parts of the plant are toxic, especially the bulbs, which have caused severe gastrointestinal disease, hypotension, and even death in people when mistakenly eaten as wild onion.[77-79] Sheep are most frequently poisoned by death camas, but cattle, horses, and pigs may be affected.[80-85] Deer and other wild ruminants may be more tolerant of the toxic effects of death camas.[86] Poisoning is most likely to occur in early spring when few other plants are available and the succulent shoots are especially enticing.

Clinical Signs

Sheep show signs of poisoning after eating as little as 0.5 lb of the green plants. Salivation, nausea, vomiting, muscular weakness, and staggering gait are typically seen. Convulsions, coma, and death soon occur if sheep eat 2 to 2.5 lb of green plant per 100 lb body weight.[82-84] No specific lesions are visible on autopsy.

Diagnosis of death camas poisoning is therefore usually made by eliminating other causes of sudden death, the presence of *Zigadenus* species in the animal's environment, and the detection of *Zigadenus* alkaloids in the rumen contents using thin-layer chromatography.[76,85]

Treatment

In most cases of death camas poisoning, little can be done to reverse the toxicity of alkaloids. The subcutaneous injection of 2 mg atropine sulfate and 8 mg picrotoxin per 100 lb body weight is reported to be effective in treating early poisoning of sheep.[86,87] Supportive therapy with intravenous fluids is helpful in combating the hypotensive effects of the death camas.

Figure 2-9C Showy death camas flowers
(*Z. elegans*).

Bloated animals should be kept in a sternal position and a stomach tube passed to relieve rumen pressure.

Keeping animals out of the areas where the plant is present, especially in the early spring when other forages are not available, can prevent poisoning from death camas. Herbicides can be selectively used to control dense stands of the death camas in problematic areas.

Plants Causing Similar Clinical Signs

Similar signs and death losses have occurred in sheep after eating fly poison or stagger grass (*Amianthium muscaetoxicum*, synonym *Chrosperma muscaetoxicum*), a plant similar in appearance to death camas, and found in the eastern United States from New York to Florida, and west to Arkansas.[87] Fly poison has basal grasslike leaves and showy white flowers borne on multiple dense racemes (Figures 2-10A and 2-10B). It prefers moist soils, emerging in the early spring when it can induce death in cattle and sheep that graze it before other forages are available. The name fly poison was given to this plant because at one time the ground up bulb was used as a fly poison.

Figure 2-10A Fly poison or stagger grass plant in bloom (*Amianthium muscaetoxicum*).

Star of Bethlehem (*Ornithogalum umbellatum*) was introduced from the Mediterranean area and has become established as a common pasture plant of eastern North America. This onion-like perennial herb, growing from a bulb, forms clumps of showy white star-shaped flowers (Figure 2-11). The bulbs appear to be the most toxic and result in rapid death of animals, particularly sheep, that consume them.[88] Members of the genus Ornithogalum are some of the most poisonous plants found in South Africa.[89]

The Atamasco or rain lily (*Zephranthes atamasco*) found in wooded areas or pastures of southeastern North America has been associated with deaths in cattle and horses.[90] The 1 inch (2.5 cm) diameter bulb is the most poisonous part of the plant. Several bluish green leaves 6 to 10 inches (15 to 25 cm) long emerge from the bulb in the spring and are followed by showy star-shaped, white flowers (Figure 2-12). The faded flowers turn pink. Within 24 hours of eating the bulbs, affected animals develop hemorrhagic diarrhea, staggering gait, and coma before death.

Figure 2-10B Fly poison or stagger grass flowers (*A. muscaetoxicum*).

Figure 2-11 Star of Bethlehem (*Ornithogalum umbellatum*).

Figure 2-12 Atamasco or rain lily (*Zephranthes atamasco*).

REFERENCES

Cardiac Glycosides

1. Adams HR: Digitalis and vasodilator drugs. *In* Veterinary Pharmacology and Therapeutics, edn 7. Adams HR, Ed. Ames, IA: Iowa State University Press; 1995:451–481.

2. Hollma A: Plants and cardiac glycosides. *Br Heart J* 1985, 54:258–261.

3. Joubert JPJ: Cardiac glycosides. *In* Toxicants of Plant Origin Volume 2. Cheeke PR, Ed. Boca Raton, FL: CRC Press; 1989:61–96.

4. Krikler DM: Withering and the foxglove: the making of a myth. *Br Heart J* 1985, 54:256–257.

5. Mastenbroek C: Cultivation and breeding of *Digitalis lanata* in the Netherlands. *Br Heart J* 1985, 54:262–268.

6. Radford RD, Cheung K, Urech R, Gollogly JR, Duffy P: Immunologic detection of cardiac glycosides in plants. *Aust Vet J* 1994; 71:236–238.

7. Kingsbury JM: *Poisonous Plants of the United States and Canada.* Englewood Cliffs, NJ: Prentice-Hall; 1964:262–267.

8. Marsh CD: Poisonous properties of whorled milkweeds *Asclepias pumila* and *A. verticillata* var. *Geyeri. USDA Bulletin* 1921, 942:1–14.

9. Glover GH, Newsom IE, Robbins WW: A new poisonous plant. *Agricultural Experimental Station Bulletin* 1918, 246:116.

10. Seiber JN, Lee SM, Benson JM: Cardiac glycosides (*cardenolides*) in species of Asclepias (*Asclepiadaceae*). *In* Handbook of Natural Toxins, Volume 1, Plant and Fungal Toxins. Keeler RF, Tu A, Eds. New York: Marcel Dekker; 1983:43–83.

11. Anderson LAP, Anitra SR, Joubert JPJ *et al.*: Krimpiekte and acute cardiac glycoside poisoning in sheep caused by bufadienolides from the plant *Kalanchoe lanceolata* Forsk. *Onderstepoort J Vet Res* 1983, 50:295–300.

12. McKenzie RA, Dunster PJ: Hearts and flowers: Bryophyllum poisoning of cattle. *Aust Vet J* 1986, 63:222–227.

13. McKenzie RA, Dunster PJ: Curing experimental *Bryophyllum tubiflorum* poisoning of cattle with activated carbon, electrolyte replacement solution and antiarrhythmic drugs. *Aust Vet J* 1987, 64:211–214.

14. Reppas GP: *Bryophyllum pinnatum* poisoning of cattle. *Aust Vet J* 1995, 72:425–427.

15. Masvingwe S, Mavenyengwa M: *Kalanchoe lanceolatum* poisoning in Brahaman cattle in Zimbabwe: the first field outbreak. *J South African Vet Assoc* 1997, 68:18–20.

16. Davies RL, Whyte PBD: *Adonis microcarpa* (pheasant's eye) toxicity in pigs fed field pea screenings. *Aust Vet J* 1989, 66:141–143.

17. Reagor JC: Increased oleander poisoning after extensive freezes in South/Southeast Texas. *Southwest Vet* 1985, 36:95.

18. Siemens LM, Galey FD, Johnson B, Thomas WP: The clinical, cardiac, and pathophysiological effects of oleander toxicity in horses. *J Vet Intern Med* 1995, 9:217.

19. Benson JM, Seiber JN, Bagley CV, *et al.*: Effects on sheep of the milkweeds *Asclepias eriocarpa* and *A. labriformis* and of cardiac glycoside-containing dirivitive material. *Toxicon* 1979, 17:155–165.

20. Szabuniewicz M, Schwartz WL, McCrady JD *et al.*: Experimental oleander poisoning in the dog, monkey (*Cebus apella*), and cat. *Proceedings 19th World Vet Congress*, Mexico City 1971, 2:729.

21. Szabuniewicz M, Schwartz WL, McCrady JD, *et al.*: Experimental oleander poisoning and treatment. *Southwestern Vet* 1972, 25:105–114.

22. Mahin L, Marzou A, Huart A: A case report of *Nerium oleander* poisoning in cattle. *Vet Hum Toxicol* 1986, 26:303–304.

23. Cheeke PR, ed: *Natural Toxicants in Feeds, Forages and Poisonous Plants*, edn 2. Danville, IL: Interstate Publishers; 1998:390–409.

24. Oryan A, Maham M, Rezakhani A, Maleki M: Morphological studies on experimental oleander poisoning in cattle. *Zentralblatt für Vet Med* 1996, 43:625–634.

25. Markov AK, Hume AS, Rao MR, *et al.*: Fructose-1,6-diphosphate in the treatment of oleander toxicity in dogs. *Vet Human Toxicol* 1999, 41:9–15.

26. Safadi R, Levy I, Caraco Y: Beneficial effect of digoxin-specific fab antibody fragments in oleander intoxication. *Arch Intern Med* 1995, 155:2121–2125.

27. Tor ER, Holstege DM, Galey FD: Determination of oleander glycosides in biological matrices by high-performance liquid chromatography. *J Agric Food Chem* 1996, 44:2716–2719.

28. Galey FD, Holstege DM,Plumlee KH *et al.*: Diagnosis of oleander poisoning in livestock. *J Vet Diagn Invest* 1996, 8:358–364.

29. Osterloh J, Harold S, Pond S: Oleander interference in the digoxin radioimmunoassay in a fatal ingestion. *JAMA* 1982, 247:596–597.

30. Moxley RA, Schneider NH, Steinegger DH, Carlson MP: Apparent toxicosis associated with lily of the valley (*Convalaria majalis*) ingestion in a dog. *J Am Vet Med Assoc* 1989, 195:485–487.

31. Alfonso HA, Sanchez LM: Intoxication due to *Nerium oleander* in geese. *Vet Hum Toxicol* 1994, 36:47.

32. Langford SD, Boor PJ: Oleander toxicity: an examination of human and animal toxic exposures. *Toxicology* 1996, 109:1–13.

33. Ansford AJ, Morris H: Oleander poisoning. *Toxicon* 1983, 3:15–16.

34. Pawha R, Chatterjee VC: The toxicity of yellow oleander (*Thevetia neriifolia Juss*) seed kernels to rats. *Vet Hum Toxicol* 1990, 32:561–564.

35. Kingsbury JM: *Poisonous Plants of the United States and Canada.* Englewood Cliffs, NJ: Prentice-Hall;1964:267–270.

36. Edwards WC, Burrows GE, Tyrl RJ: Toxic plants of Oklahoma: milkweeds. *J Okla Vet Med Assoc* 1984; 34:4–79.

37. Ogden L, Burrows GE, Tyrl RJ, Ely RW: Experimental intoxication in sheep by Asclepias. *In* Poisonous Plants. James LF, Bailey EM, Cheeke PR, Hegarty M, Eds. *Proceedings of the Third International Symposium.* Ames, IA: Iowa State University Press; 1992:495–499.

38. Clarke JG: Whorled milkweed poisoning. *Vet Human Toxicol* 1979; 21:31.

39. Glover GH, Newsom IE, Robbins WW: A new poisonous plant. *Agricultural Experimental Station Bulletin* 1918; 246:116.

40. Duffey SS, Scudder GGE: Cardiac glycosides in north American *Asclepiadaceae*, a basis for unpalatability in brightly colored *Hemiptera* and *Coleoptera*. *J Insect Physiol* 1972;18:3.

41. Seiber JN, Roeske CN, Benson JM: Three new cardenolides from the milkweeds *Asclepias eriocarpa* and *A. labriformis*. *Phytochem* 1978;17:967.

42. Benson JM, Seiber JN, Keeler RF, Johnson AE: Studies on the toxic principle of *Asclepias eriocarpa* and *A. labriformis*. *In* Effects of Poisonous Plants on Livestock.

Keeler RF, Van Kampen KR, James LF, Eds. New York: Academic Press; 1978:273–284.

43. Cheeke PR, ed: *Natural Toxicants in Feeds, Forages and Poisonous Plants*, edn 2. Danville, IL: Interstate Publishers; 1998:390–409.

44. Nelson CJ, Seiber JN, Brower LP: Seasonal and intraplant variation of cardenolide content in the California milkweed, *Asclepias eriocarpa*, and implications for plant defense. *J Chem Ecology* 1981, 7:981.

45. Brower LP, *et al.*: Plant-determined variation in the cardenolide content, thin layer chromatography profiles, and emetic potency of monarch butterflies, *Daunus plexippus* reared on the milkweed, *Asclepias eriocarpa* in California. *J Cheal Ecology* 1982, 8:579.

46. Kingsbury JM. *Poisonous Plants of the United States and Canada*. Englewood Cliffs, NJ: Prentice-Hal; 1964:121–123.

47. Bryan-Brown T: The pharmacological actions of taxine. *J Pharm Pharmacol* 1932; 5:205–219.

48. Clarke ML, Harvey DG, Humphreys DJ: *Veterinary Toxicology*, edn 2. London: Bailliere Tindall; 1981:256–257.

49. Raisbeck MF, Kendall JD: Taxus poisoning. *In* Current Veterinary Therapy 3. Food Animal Practice. Howard JL, Ed. Philadelphia: WB Saunders; 1993:371.

50. Tekol Y: Acute toxicity of taxine in mice and rats. *Vet Hum Toxicol* 1991, 33:337–338.

51. Witherup KM, Look SA, Stasko MW, *et al.*: *Taxus* spp. needles contain amounts of taxol comparable to the bark of *Taxus brevifolia*: analysis and isolation. *J Natural Products* 1990, 53:1249–1255.

52. Rowinski EK, Cazenave LA, Donehower RC: Taxol: a novel investigational antimicrotubule agent. *J Natl Cancer Inst* 1990, 82:1247–1259.

53. Hare WR: Bovine yew (*Taxus* spp.) poisoning. *Large Anim Pract* 1998, 19:24–28.

54. Alden CL, Fosnaugh JF, Smith JB, Mohan R: Japanese yew poisoning of large domestic animals in the Midwest. *J Am Vet Med Assoc* 1977, 170:314–316.

55. Lowe JE, Hintz HF, Schyrver HF, Kingsbury JM: *Taxus cuspidata* (Japanese yew) poisoning in horses. *Cornell Vet* 1970, 60:36–39.

56. Thompson GW, Barker IK: Japanese yew (*Taxus cuspidata*) poisoning in cattle. *Can Vet J* 1978, 19:320–321.

57. Kerr LA, Edwards WC: Japanese yew: a toxic ornamental shrub. *Vet Med Small Anim Clin* 1981, 76:1339–1340.

58. Rook JS: Japanese yew toxicity. *Vet Med* 1994, 89:950–951.

59. Casteel SW, Cook WO: Japanese yew poisoning in ruminants. *Mod Vet Pract* 1985, 66: 875–877.

60. Ogden L: Taxus (yews)—a highly toxic plant. *Vet Hum Toxicol* 1988, 30:563–564.

61. Veatch JK, Reid FM, Kennedy GA: Differentiating yew poisoning from other toxicosis. *Vet Med* 1988, 83:298–300.

62. Maxie G: Another case of Japanese yew poisoning. *Can Vet J* 1991, 32:370.

63. Kite GC, Lawrence JJ, Dauncey EA: Detecting taxus poisoning in horses using liquid chromatography/mast spectrometry. *Vet Hum Toxicol* 2000, 42:151-154.

64. Lang DG, Smith RA, Miler RE: Detecting taxus poisoning using GC/MS. *Vet Hum Toxicol* 1997, 39:314.

65. Appleman D: Preliminary report on toxicity of avocado leaves. *California Society Year Book* 1944, 37.

66. Kingsbury JM: *Poisonous Plants of the United States and Canada*. Englewood Cliffs,

NJ: Prentice-Hall;1964:124–125.

67. Craigmill AL, Eide RN, Schultz TA, Hedrick K: Toxicity of avocado (*Persea americana* var.) leaves: review and preliminary report. *Vet Hum Toxicol* 1984, 26:381–383.

68. Hargis AM, Stauber E, Casteel S, Eitner D: Avocado (*Persea americana*) intoxication in caged birds. *J Am Vet Med Assoc* 1989, 194:64–66.

69. Burger WP, Naude TW, Van Rensburg IB, *et al.*: Cardiomyopathy in ostriches (*Struthio camelus*) due to avocado (*Persea americana* var *guatamalensis*) intoxication. *J South African Vet Assoc* 1994, 65:113–118.

70. Mckenzie RA, Brown OP: Avocado (*Persea americana*) poisoning of horses. *Aust Vet J* 1991, 68:77–78.

71. Hurt LM: Avocado poisoning. *Los Angeles County Livestock Dept Annu Rep* 1942, 43–44.

72. Craigmill AL, Seawright AA, Matilla T, Frost AJ:. Pathological changes in the mammary gland and biochemical changes in milk of the goat following oral dosing with leaf of the avocado (*Persea americana*). *Aust Vet J* 1989, 66:206–211.

73. Sani Y, Atwell RB, Seawright AA: The cardiotoxicity of avocado leaves. *Aust Vet J* 1991, 68:150–151.

74. Sani Y, Seawright AA, NG JC, *et al.:* The toxicity of avocado leaves (*Persea americana*) for the heart and lactating mammary gland of the mouse. *In* Plant Associated Toxins. Colegate SM, Dorling PR. Exeter, UK, Eds. CAB International, Short Run Press; 1994:552–556.

75. Grant R, Basson PA, Booker HH, *et al.*: Cardiomyopthy caused by avocado (*Persea americana Mill.*) leaves. *J South African Vet Assoc* 1991, 62:21–22.

76. Majak W, McDiarmid RE, Cristofoli W, *et al.*: Content of zygacine in *Zygadenus venosus* at different stages of growth. *Phytochem* 1992, 31:3417–3418.

77. Spoerke DG, Spoerke SE: Three cases of Zigadenus poisoning. *Vet Hum Toxicol* 1979, 21:346–347.

78. Heilpern KL: Zigadenus poisoning. *Ann Emerg Med* 1995, 25:259–262.

79. Wilcox VK, Wilcox EV: The stock-poisoning plants of Montana. *USDA Bulletin* 1901, 26:51–64.

80. Marsh CD, Clawson AB: The stock poisoning death camas. *USDA Farmers Bulletin* 1922, 127:1–10.

81. Beath OA, Epson HF, Draize JH, Justice RS: Three species of *Zygadenus* (death camas). *University of Wyoming Agricultural Experimental Station* 1933, 194:1–38.

82. Kingsbury JM: *Poisonous Plants of the United States and Canada.* Englewood Cliffs, NJ: Prentice-Hall; 1964:461–466.

83. Smith RA, Lewis D: Death camas poisoning in cattle. *Vet Hum Toxicol* 1991, 33:615–616.

84. Panter KE, Ralphs MH, Smart RA, Duelke B: Death camas poisoning in sheep: a case report. *Vet Hum Toxicol* 1987, 29:45–48.

85. Collet S, Grotelueschen D, Smith R, Wilson R: Deaths of 23 adult cows attributed to intoxication by the alkaloids of *Zigadenus venosus* (meadow death camas). *Agri-Practice* 1996, 17:5–9.

86. Longland WS, Clements C: Consumption of a toxic plant (*Zigadenus paniculatus*) by mule deer. *Great Basin Naturalist* 1995, 55:188–189.

87. Marsh DC, Clawson AB, Marsh H: Stagger grass (*Chrosperma muscaetoxicum*) as a poisonous plant. *USDA Bulletin* 710, 1926:1–14.

88. Kingsbury JM. *Poisonous Plants of the United States and Canada.* Englewood Cliffs, NJ: Prentice-Hall; 1964:456–457.

89. Kellerman TS, Coetzer JAW, Naude TW: *Plant Poisonings and Mycotoxicoses of Livestock in Southern Africa*. Cape Town, South Africa: Oxford University Press; 1988:140–144.

90. Kingsbury JM. *Poisonous Plants of the United States and Canada*. Englewood Cliffs, NJ: Prentice Hall; 1964:469.

CHAPTER 3

Plants Affecting the Digestive System

Numerous toxic plants cause digestive problems in animals either through direct effects of the plant on the gastrointestinal system or indirectly by affecting other organs with a secondary effect on the digestive system. In this chapter, only plants that have a direct effect on the digestive system will be discussed. Excessive salivation, abdominal pain (colic), impactions of the stomach or intestines, bloat, constipation, and diarrhea are all signs that can be attributed to plants affecting the digestive system.

Excessive Salivation

Figure 3-1A Prickly pear (*Opuntia* spp.).

Excessive salivation, characterized by drooling or frothy saliva around the lips, may be caused by a variety of things including chemical irritants fungal toxins, virus and bacterial diseases affecting the mouth, teeth problems, choking caused by esophageal obstruction, and various toxic or injurious plants. Grazing animals that eat sharp grass awns, spiny plants such as the prickly pear cactus (*Opuntia* spp.) (Figure 3-1A), or those with burs such as burdock (*Arctium minus*) (Figure 3-1B) and cocklebur (*Xanthium* spp.) (Figure 3-1C) may injure the oral mucosa. Some common grasses such as foxtail barley (Figure 3-2A), bristle grass (Figure 3-2B) and sandbur (Figure 3-2C) have seeds with sharp awns that can become embedded in the tongue and gums of animals eating them. Initially excessive salivation may be noticed, but in time the grass awns or spines that are imbedded in the mucosae create large ulcers. The awns are not easily visible in the ulcers because they become embedded in the granulation tissue filling the ulcer. It is not uncommon for some sharp grass awns to penetrate the skin of animals, migrate through the tissues, and act as a foreign body causing abscesses and draining wounds far from the site of penetration.[1] Grasses with sharp awns that commonly become embedded in the mouth, ears, and skin of animals are included in Table 3-1.

Profuse salivation has been observed in horses and other livestock eating clover or alfalfa pasture or hay that is infected with the fungus *Rhizoctonia leguminicola*. The mycotoxin responsible for the "slobbering" that animals exhibit has been identified as slaframine, an indolizidine alkaloid produced most commonly by the fungus *R. leguminicola* growing on red clover.[2,3] Slaframine is chemically similar to the

Figure 3-1B Burdock (*Arctium minus*).

Figure 3-1C Cocklebur (*Xanthium strumarium*).

alkaloid swainsonine produced by plants of the genera *Astragalus* and *Oxytropis* that are responsible for causing locoism. Under wet or humid conditions the fungus grows on the leaves producing black or brown spotting. After they eat the fungus-infected clover for several days, horses begin to salivate excessively and lose weight; pregnant mares may abort if they continue to consume the infected clover. Recovery occurs rapidly once horses are removed from the infected hay. Problem pastures can be used for animals if they are mowed, the affected hay is removed, and the regrowth has no brown spotting on the leaves.

Miscellaneous plants cause irritation to the mouth of animals through the action of oxalate crystals present in the plants leaves and stems. The oxalate crystals become embedded in the mucous membranes of the mouth causing severe inflammation and swelling, excessive salivation, and difficulty in swallowing. In addition the soluble oxalates in the plants can induce kidney failure (see Chapter 7).

Table 3-1 Grasses That Cause Trauma	
Common Name	**Scientific Name**
Foxtail barley	*Hordeum jubatum* (Figure 3-2A)
Needle grasses	*Stipa* spp.
Squirrel tail	*Sitanion hystrix*
Bristle grass	*Setaria* spp. (Figure 3-2B)
Sandbur	*Cenchrus longispinus* (Figure 3-2C)
Medusahead rye	*Taentherum asperum*
Prairie three-awn	*Aristida oligantha*
Tanglehead	*Heteropogon contortus*

Figure 3-2A Foxtail barley (*Hordeum jubatum*).

Protoanemonins

Plants including buttercups (*Ranunculus* spp.) and related species such as marsh marigold (*Caltha palustris*) (Figure 3-3A), clematis (*Clematis* spp.) (Figure 3-3B), anemone, and baneberry (*Actaea arguta*) contain substances that cause mouth irritation, excessive salivation, and diarrhea if eaten in quantity.[4] The buttercup family (*Ranunculus* spp.) contains an oily glycoside, ranunculin, that when converted by plant enzymes to protoanemonin becomes a strong irritant.[5-7] It primarily causes irritation to the mucous membranes of the mouth and digestive system and can blister the skin and cause liver damage and secondary photosensitization.[8-9] Sheep have been fatally poisoned by bur buttercup (*Ceratocephalus testiculatus*)[10] (see Figure 3-5). Protoanemonin is rapidly polymerized to anemonin, which is nontoxic, and accounts for the fact that hay-containing plants with protoanemonins are not toxic. Feeding trials with cattle and sheep have shown that buttercups vary in their toxicity and do not seriously affect animals under most circumstances.[11,12]

Figure 3-2B Bristle grass
(*Setaria* spp.).

Figure 3-2C Sandbur
(*Cenchrus longispinus*).

Clinical Signs

Protoanemonins are toxic to all animals and cause reddening of the oral mucous membranes, salivation, gastroenteritis, colic, and diarrhea. In lactating animals, especially cows, protoanemonins produce a bitter taste to the milk and because of their irritating effects may cause blood to appear in the milk and urine.

Treatment

Because there is no specific treatment, severely affected animals should be given intestinal protectants until the toxic plant material is eliminated from the digestive tract.

Figure 3-3A Marsh
marigold (*Caltha palustris*).

Figure 3-3B Clematis (*Clematis ligusticifolia*).

Blister Buttercup
Ranunculus sceleratus
Ranunculaceae (Buttercup family)

Habitat

Blister buttercup is a plant of early spring that prefers borders of lakes, ponds, and streams and in general wet areas, from Alaska south to Iowa, New Mexico, and California.

Description

Blister buttercups are perennial herbaceous plants with fibrous roots. Stems are erect, stout, and glabrous or nearly so. The basal leaves are kidney-shaped, long petioled, and three-parted. The upper leaves are sessile or short-petioled. Flowers are few, sepals five, petals yellow, stamens number 10 to many, pistils many, and the fruit is an achene.

Principal Toxin

An oily glycoside, ranunculin, is converted to protoanemonin by the action of plant enzymes released when the plant is chewed. The protoanemonin irritates the mouth, causing excessive salivation and intestinal irritation that may result in diarrhea.

Signs of Poisoning

Signs of poisoning include excessive salivation, reddening of oral mucous membranes, and diarrhea. Cattle that consume excessive quantities of certain buttercup species may develop a severe gastroenteritis and hemorrhagic diarrhea that can be fatal. The bitter taste of the protoanemonin can also be passed into the milk of lactating animals.

Among the buttercups considered more toxic to animals than others are *R. sceleratus, R. flammula, R. parviflora, R. acris, R. abortivus, R. cymbalaria, and R. repens* (Figure 3-4).

Figure 3-4 Creeping buttercup (*Ranunculus repens*).

Bur Buttercup
Ceratocephalus testiculatus
Ranunculaceae (Buttercup family)

Habitat

Introduced from Europe, bur buttercup has become established commonly along roadsides and in disturbed areas in northwestern North America. It is a plant of spring and early summer.

Description

Bur buttercups are annual plants with stems 1 to 3 inches (3 to 10 cm) tall (Figure 3-5). The flower-bearing stems have simple, three-parted, gray-green, hairy, basal leaves with lateral segments, cleft into linear divisions and a winged petiole. The five sepals are green and hairy. The five petals are yellow. The fruits are burlike, consisting of numerous individual seed capsules, each three-chambered, but with the two lateral chambers empty.

Figure 3-5 Bur buttercup (*Ceratocephalus testiculatus*) (Courtesy of Dr. Michael H. Ralphs, USDA Poisonous Plants Research Laboratory, Logan, Utah).

Principal Toxin

The glycoside, ranunculin, which is converted to protoanemonin by the action of plant enzymes released when the plant is chewed, is responsible for oral irritation causing excessive salivation and intestinal irritation that may result in diarrhea. Sheep are particularly susceptible to bur buttercup poisoning, and large numbers of sheep have died from eating as little as little as 500 g of the plant.[10-12] Affected sheep develop watery diarrhea, labored breathing, and weakness and cannot rise when approached. Hypocalcemia is not the cause of the muscle weakness and recumbency. Postmortem findings include inflammation of the rumen; congestion of the lungs, liver, and kidneys; and excessive fluid in the thoracic and abdominal cavities.[10]

Baneberry
Actaea rubra
Ranunculaceae (Buttercup family)

Habitat

Baneberry prefers rich, moist soils of woodlands, often in deep shade of trees. It is found in most areas of North America except in the desert southwest.

Description

Baneberry is an erect perennial herbaceous plant with thick root stalks. The leaves are large with the lower ones petioled and the upper nearly sessile. They are ternate with the divisions long petioled and pinnate. The leaflets are ovate-lanceolate and serrate, with three to five lobes. The raceme elongates in the fruit, often up to 4 inches (10 cm). The petals are shorter than the stamens and are white. Stamens are numerous with the filaments flattened. The pistil has one locule and is sessile. The fruiting pedicels are elongated up to 2 cm (20 mm) long with red or white berries (Figures 3-6A and 3-6B).

Figure 3-6A Baneberry *(Actaea rubra).*

Principal Toxin

As a member of the buttercup family, baneberries contain the glycoside ranunculin and as yet other unidentified irritant compounds. Human or livestock fatalities associated with baneberry have not been recorded in the United States, but in Europe references are found concerning the death of children who have eaten the conspicuous red or white berries.

Figure 3-6B White baneberry *(Actaea rubra)*

Tung Nut
Aleurites fordii
Candle nut *(A. moluccana)*
Lumbang nut *(A. trisperma)*
Euphorbiaceae *(Spurge family)*

Habitat

The tung tree was originally imported from China and grown in Florida and along the Gulf Coast to east Texas for the purpose of producing tung oil. Although the tung oil industry is no longer of major significance, many of the trees have persisted and are grown as ornamentals.

Description

The tung tree may attain a height of 50 feet (20 meters) with stout branches, often in whorls. The leaves are simple, alternate deciduous, and palmately veined, reaching 12 inches (30 cm) in length (Figure 3-7). Ivory-colored male and female flowers appear in the spring before the leaves. The lumbang nut tree (*A. trisperma*) has white flowers tinged with red. The fruits are conical and 2 to 3 inches (5 to 7 cm) long, turning brown when ripe. The thin shell splits open to reveal three to four round dark brown seeds that are white on the inside.

Principal Toxin

Saponins and possibly a toxalbumin are found in highest concentration in the tung nuts and leaves but not in the oil.[13-15] The toxicity of the various species is variable, the tung nut being the most toxic. Experimentally 0.35 percent of an animal's body weight of fresh macerated leaves ind-uced a hemorrhagic diarrhea. Cattle do not readily eat the leaves off of the tree, but they will eat prunings. Most poisoning has been reported in people who have eaten the white-fleshed seeds.[16]

Figure 3-7 Tung nut (*Aleurities*).

Clinical Signs

Cattle become depressed and anorexic and develop an atonic rumen. A hemorrhagic diarrhea often develops after several days of consuming either the tung nuts or the leaves.[17] Chronic watery diarrhea and emaciation often develop in animals that do not die and can mimic chronic bovine virus diarrhea.

On postmortem examination the most common finding is hemorrhagic gastroenteritis, the abomasum and proximal small intestine being most severely affected.

Treatment

There is no specific treatment for tung oil poisoning. Affected animals should be removed from the source of the plant and given supportive oral and intravascular fluid therapy.

Chinaberry, Paraiso
White cedar, syringa berry, Persian lilac
Melia azadarach
Meliaceae (Mahogany family)

Texas umbrella tree - *M. azadarach* var. *umbraculiformis*

Habitat
The chinaberry tree is a fast growing tree that was introduced from China and has become established in the Southern United Sates from Florida to Southern California, Hawaii, and Mexico. The tree has escaped from cultivation and is commonly found growing in hedgerows and waste areas where birds have dropped the seeds.

Description
The chinaberry tree is a fast growing tree up to 40 feet (20 meters) in height, with compound, opposite leaves, 1 to 3 feet (0.5 to 1 cm) in length, each leaflet being serrated and 1 to 3 (2 to 6 cm) inches long. The tree is deciduous, with the leaves turning yellow before falling. The white to lavender inch-sized flowers with 5 to 6 petals are produced in clusters and are heavily and pleasantly fragrant. The smooth green fruits (drupe) hang in clusters, turn yellow when ripe, and have a ridged pit surrounded by a little pithy flesh. Fruit clusters often remain hanging on the tree even in winter (Figure 3-8). Fruit production is especially heavy in high rainfall years.

Figure 3-8 Chinaberry (*Melia azedarach*).

Principal Toxin
Uncertainty exists as to the exact nature of the toxic compounds found in chinaberry, but saponins, alkaloids(azaradine, margosine, mangrovin), and tetranortriterpenes (meliatoxins) are thought to be responsible for the gastroenteric and neurologic signs.[18-21] All parts of the tree are poisonous, but most poisoning is associated with the consumption of the fruits. Humans as well as pigs, cattle, sheep, goats, rabbits, guinea pigs, poultry and dogs are susceptible to poisoning from chinaberries.[18-20] Pigs are most frequently fatally poisoned after consuming as little as 100 g of the fruits.

Clinical Signs

The usual signs associated with chinaberry poisoning are associated with either the gastrointestinal system (vomiting , constipation, hemorrhagic diarrhea, colic) or the nervous system (weakness, muscle trembling, ataxia, and generalized paresis leading to recumbency, coma, respiratory failure).[18-21] Rumen impaction may develop in cattle eating large quantities of the ripe fruits that drop to the ground. Signs develop within a few hours to a day or two later depending on the quantity of fruits consumed. Death may occur 1 to 2 days after the onset of signs if a lethal dose of chinaberries was consumed.

Chinaberry poisoning has no known specific treatment and affected animals must be treated symptomatically. This may include inducing vomiting or gastric lavage in people and dogs to remove the berries from the stomach, oral administration of activated charcoal, and intravenous fluid therapy to counteract shock and maintain renal function.

At postmortem examination, the fruits and seeds may be evident in the stomach, which is often congested and hemorrhagic. The intestinal tract is usually congested and hemorrhagic. The meninges of the brain are often congested. Histologically the liver, kidneys, and heart show varying degrees of degeneration and coagulative necrosis.

Chinaberry trees should not be planted in or around livestock enclosures because of the high risk of poisoning.

Pokeweed, Pokeberry
Phytolacca americana
Phytolaccaceae (Pokeweed family)

Habitat and Description

Pokeweed grows mostly in the eastern and southern United States. It is a perennial branching herb from 3 to 10 feet (1 to 3 meters) tall, with a large tap root, green or purple stems, and large alternate, petioled, and ovate leaves. The flowers are small, white in color and without petals. The distinctive fruits are shiny purple-black berries carried on red stems (Figure 3-9).

Figure 3-9 Pokeweed (*Phytolacca americana*).

Principal Toxin

All parts of the plant contain saponins, oxalates, and the alkaloid phytolacine with greatest concentrations in the roots and seeds. Pokeberry also contains a protein lectin, (a mitogen) that can have wide effects on the immune system. Pokeweed mitogen affects cell division and stimulates B- and T-cell lymphocyte proliferation. The plant should be handled with gloves because the mitogen can be absorbed through cuts and abrasions on the skin.

Clinical Signs

Depending on the amount of the plant consumed, animals may show mild to severe colic and diarrhea. Fatalities are rare unless large quantities of the plant are consumed. Other species of *Phytolacca* found in South America and Africa have been associated with higher mortality rates.[22] Humans appear to be more severely poisoned by pokeweed and develop mouth irritation, stomach cramps, vomiting, and diarrhea. Death may occur in children eating large amounts of the plant or berries.

Sheep, cattle, goats, horses, pigs, and poultry are susceptible to the toxic effects of pokeweed.[23-26] The signs of poisoning are varied and include oral irritation, excessive salivation, vomiting, colic, bloody diarrhea, depression, prostration, and death. Mild to severe gastroenteritis with ulceration of the gastric mucosa are common nonspecific findings on postmortem examination.

Treatment

Intestinal protectants and other supportive treatments should be given as appropriate.

Corn Cockle
Agrostemma githago
Caryophyllaceae (Pink family)

Habitat

Corn cockle was introduced from Europe and has become a common weed of wheat fields throughout North America. Its seeds become a problem to livestock when they contaminate grains fed to them. Wheat screenings are particularly hazardous to livestock because the corn cockle seeds tend to be concentrated in the screenings.

Description

The corn cockle is an erect annual, with hairy stems, branching above and growing to 3 feet (1 meter) in height. The leaves are alternate, lanceolate, and grayish green due to the heavy covering of hairs. Flowers are showy, terminal, and purplish red in color with green sepals longer than the colored petals (Figure 3-10) The urn-like seed capsules contain relatively large 0.2 to 0.3 cm (2 to 3 mm) black seeds, rounded with a flat side and covered by rows of sharp tubercles.

Figure 3-10 Corn cockle (*Agrostemma githago*).

Principal Toxin

A saponin, githagenin, which can comprise 5 to 7 percent by weight of the seeds, appears to be the toxin that causes gastrointestinal irritation and diarrhea.[27] Poultry, cattle, sheep, pigs, and horses are susceptible to corn cockle poisoning. Similar saponins are found in members of the genus *Drymaria (Alfombrilla)* common in Mexico.

Habitat and Description

Widely distributed throughout the United States, *Saponaria* spp. are perennial erect herbs with jointed stems, opposite, lanceolate, simple leaves, which are sessile. The leaves are prominently veined. The white or pink flowers are produced in terminal clusters, with a calyx of five sepals that is green and tubular (Figure 3-11) The flower has five petals, 10 stamens, and two styles.

Figure 3-11 Bouncing bet (*Saponaria officinalis*).

Principal Toxin

Saponins are the primary toxins present in *Saponaria* spp. If eaten in sufficient quantity, the saponins may cause acute liver degeneration and death.[28] The seeds, which are especially toxic, may be a contaminant of cereal crops. Cow cockle (*Vacaria pyramidata*, or *Saponaria vacaria*) is an annual weed introduced from Europe that is common in grain fields and has similar toxic properties.

Sesbania, Coffee Weed, Bladderpod
Sesbania spp. *(Daubentonia and Glottidium)*
Fabaceae (Legume family)

Habitat

Various species of *Sesbania* are found in North America, most of which grow in the warmer climates from Florida to California and Hawaii. Some are vigorous annuals that form dense stands in the damp soils along streams. *Sesbania punicea*, red or purple sesbania (Figure 3-12A), *S. drumondii* (coffee bean, rattlebox) (Figure 3-12B), and *S. vesicaria* (annual bladderpod) are some of the more prevalent species.[29] Synonyms for the genus *Sesbania* include *Daubentonia* and *Glottidium*.

Figure 3-12A Sesbania (*Sesbania punicea*).

Description

Sesbania plants are perennial shrubs or small trees growing to 12 feet (3.5 meters) tall. Leaves are alternate, pinnately compound with 12 to 40 leaflets. Red, orange, and yellow flowers are produced in showy pendant racemes. Pods are hairless, four-winged, and contain glossy, kidney-shaped seeds.

Principal Toxin

Saponins are believed to be the principal toxins present in all parts of the plants and especially the seeds. The green seeds are most poisonous, and the seeds remain toxic for years. The leaves appear to be least toxic. Birds, with the exception of ducks, are especially susceptible to poisoning. Sheep, cattle, and goats are also susceptible to poisoning, sheep being fatally poisoned after eating less than 2 oz of seeds per hundred weight.[29] As few as 10 seeds of *S. punicea* are lethal to poultry. The primary effects of the toxin are severe gastrointestinal irritation and liver degeneration.[30] Hemolysis may also occur when large quantities of seed are consumed.

Clinical Signs

Severe hemorrhagic diarrhea is often the presenting sign of *Sesbania* poisoning in cattle and sheep. Anorexia, abdominal pain, dehydration, and prostration may also be observed before death. Poultry develop similar signs of poisoning.

Postmortem findings include a severe hemorrhagic abomasitis, enteritis, and liver and kidney degeneration with necrosis. Diagnosis of poisoning is based on the signs and evidence that the plant or the seeds have been consumed. The finding of Sesbania seeds in the rumen, coupled with the clinical and post mortem signs is strongly supportive of a diagnosis *Sesbania* poisoning.

Figure 3-12B Rattlepod sesbania (*Sesbania drummondii*).

Treatment

Laxatives to evacuate the intestinal tract, activated charcoal via stomach tube, and intravenous fluids are helpful in treating the effects of the hemorrhagic diarrhea. The prognosis is poor if extensive liver and kidney degeneration occurs.

Prevention

Mowing of the plants before the formation of seed pods is the best way to contain the plants. Herbicides (2,4-dichlorophenoxyacetic acid) are effective in controlling young plants.

Gastrointestinal Impaction and Obstruction Caused by Plants

Adult horses and cattle may swallow poorly chewed or entire fruits and seeds of some plants that can become lodged in the pharynx or esophagus. The resulting esophageal obstruction is referred to as "choke." Fruits that may cause esophageal obstruction include apples, sugar beets, turnips, onions, and persimmons. Animals that are choked generally make frequent attempts at swallowing and drool saliva profusely because they are unable to swallow it. In cattle that are choked, severe bloat usually develops because they cannot belch to allow rumen gases to escape.

If large fruits or masses of seeds are swallowed successfully, they may cause an impaction of the stomach or obstruct some portion of the intestinal tract. Plants that can cause rumen impaction and intestinal obstruction include chinaberry (*M. azedarach*), mesquite pods and beans (*Prosopis* spp.), mescal bean pods and beans (*Sophora secundiflora*) (see Chapter 6), and persimmon fruits (*Diospyros virginiana*). Occasionally sheep may eat large quantities of pasque flowers (*Anemone patens*) that are covered with poorly digestible hairs. The fibrous hairs can form a large mass in the rumen that can cause rumen impaction.

Mesquite
Prosopis glandulosa (Honey mesquite)
Mimosaceae (Mimosa family)

Habitat

Various species of mesquite are found in the drier areas of the southwestern states from California and Mexico to Texas and Oklahoma.

Description

Mesquite plants are shrubs or small trees to 15 to 20 feet (5 to 6 meters) in height. Leaves are alternate, bipinnate, with 6 to 15 leaflet pairs. Mature branches often have stout spines at the nodes. Flower spikes, 3 to 4 inches (7 to 10 cm) long, are produced in the axils. The fruits are leathery pods 6 to 10 inches (15 to 25 cm) in length, constricted between the seeds (Figure 3-13). Mesquite beans are reddish brown.

Principal Toxin

A specific toxin has not been identified in mesquite. In times of drought cattle may eat excessive quantities of mesquite pods and beans that result in rumen impaction. The seeds and pods form a sticky mass when in contact with water which contributes to their poor digestion. The high carbohydrate content of the beans may also result in rumen acidosis. At other times livestock eat the beans in the process of foraging and do well on them.

In some areas of the world where sugar cane and mesquite are fed together to cattle, an enzyme in the mesquite releases cyanide from glycosides in the sugar cane that causes cyanide poisoning.[31]

Clinical Signs

After consuming large quantities of mesquite over a period of time, cattle may show weight loss and poor appetite. Affected animals have frothy salivation, continuous chewing movements, and loss of rumen activity; the tongue may protrude between the lips. Ketosis may be present in the chronically affected animal. Horses eating mesquite pods may develop intestinal obstruction due to the formation of an indigestible mass (phytobezoar) consisting of the pods and seeds.

Valuable cattle that are not in the advanced stages of emaciation may benefit from a rumenotomy to rid the rumen of the impacted mesquite beans. Intestinal obstructions in horses require surgical intervention to remove the phytobezoar.

Prolonged consumption of mesquite pods (*P. juliflora*) by goats and cattle will cause a neurologic syndrome characterized by difficulty in chewing and eating, salivation, and tongue protrusion due to masseter muscle atrophy resulting from fine vacuolar degeneration of the trigeminal nuclei.[32]

Figure 3-13 Mesquite leaves and pods (*Prosopsis juliflora*).

American Persimmon
Diospyros virginiana
Ebenaceae (Ebony family)

Description

The American persimmon is a large deciduous tree of warm temperate regions, with distinctive "alligator-hide" bark. The leaves are glossy, alternate, and elliptical to oblong, turning red in the fall. Flowers are greenish yellow, both male and female flowers on the same plant or on separate plants. The 1.5 to 2 inches (4 to 5 cm) fruits are conical shaped, yellow when ripe, fleshy, and sweet. The seeds are oblong and flat, with one straight edge, the other rounded; they have a pale brown, hard, wrinkled coat.

Principal Toxin

Persimmon fruits contain water soluble tannins, which precipitate in the acidity of the stomach to form a sticky coagulum of fruit skin, pulp, seeds and gastric protein that becomes a solid mass or phytobezoar. Once formed, the phytobezoar is abrasive and can lead to ulcers and even rupture of the stomach of horses that have eaten large quantities of ripe persimmon fruits.[33-35] Severe colic results when impaction of the stomach occurs, or when the phytobezoar causes an intestinal obstruction.

Clinical Signs

Intermittent colic and weight loss are often the non specific presenting signs associated with persimmon ingestion in horses. The severity of the colic depends on the degree of obstruction or impaction. Persimmon phytobezoars can be difficult to diagnose, but can be suspected in the fall when the fruits are ripe, the horse has had access to the fruits, and the persimmon seeds can be visualized in the stomach using an endoscope.

Treatment for phytobezoars is aimed at softening the mass with mineral oil and dioctyl sodium sulfosuccinate to allow its passage through the gastrointestinal system. Surgical intervention becomes necessary when medical treatment fails or if colic is severe and unrelenting.

Persimmon trees should not be planted in animal enclosures where animals could have access to the fruits.

Vomiting

A group of about 40 related plants commonly referred to as bitterweeds or sneezeweeds cause a syndrome of sneezing and vomiting in livestock that consume them. Sheep and goats are most frequently affected, but the sneezeweeds are also poisonous to cattle and horses. Sneezeweeds and bitterweeds that have caused the greatest sheep losses include *Hymenoxys odorata*, *Hymenoxys richardsonii*, *Helenium autumnale*, *Helenium amarum*, and *Dugaldia hoopsei*.[36-41] Other species of bitterweed and sneezeweed can be assumed to be toxic if eaten in sufficient quantity. Desert baileya (*Baileya multiradiata*) also contains hymenoxon, the principal toxin in the bitterweeds and sneezeweeds.[42]

Principal Toxin

The sneezeweeds and bitterweeds contain sesquiterpene lactones, which are highly irritating to the nose, eyes, and gastrointestinal tract. The primary toxins isolated from members of the *Helenium* and *Hymenoxys genera* are hymenovin (dugaldin), helenalin, helenanolide, tenulin, and hymenoxon, all of which have similar effects.[38] The amount of dried bitterweed to induce severe poisoning in sheep is 2.9 to 8.5 g/kg body weight.[43] Other than the direct irritant effects on the digestive system, the lactones (hymenoxon) have a profound effect on metabolism through their ability to bind with sulfhydryl groups, resulting in metabolic acidosis and hypoglycemia.[44,45] In light of this, the toxicity of the lactones can be reduced if sulfur-containing amino acids (cysteine, methionine) or antioxidants such as butylated hydroxyanisole are fed before the bitterweed is consumed.[46,47]

Clinical Signs

Orange sneezeweed (*Dugaldia hoopesii*), bitter sneezeweed (*Helenium amarum*), common sneezeweed (*Helenium autumnale*), and bitterweed, or pingue (*Hymenoxys richardsonii*), induce sneezing, vomiting, and diarrhea. "Spewing sickness" is a name given to the syndrome of projectile vomiting associated with sheep eating orange sneezeweed over a period of weeks. If vomiting is not observed, affected sheep often have green rumen contents around the mouth and nostrils indicative of vomiting.[40] Anorexia, bloating, and teeth grinding indicative of abdominal pain may be observed. Muscle weakness, tremors, and severe weight loss often accompany the gastrointestinal signs. Liver degeneration as indicated by elevations in serum enzymes may be seen in severe cases.[48] Coughing may be an indication that inhalation of rumen contents and pneumonia have occurred. In such cases the prognosis is poor and the mortality rate is high.

The lactones in bitterweeds and sneezeweeds may also impart a bitter taste to milk, rendering it unpalatable.[41]

Treatment

Affected animals should be removed from the pasture containing the sneezeweed and given good quality hay or pasture. Antibiotics are indicated if pneumonia is present. If a large amount of the sneezeweed or bitterweed has been consumed in the previous few hours, activated charcoal and osmotic laxatives may be helpful in eliminating the plant from the digestive system. The use of L-cysteine, a sulfur containing amino acid, is only of benefit counteracting the effect of the toxic lactones if it is given before clinical signs appear.[46,47]

Postmortem findings in sneezeweed/bitterweed poisoning include severe glomerulonephritis and degeneration of the kidney cortex and medulla. Congestion of the abdominal organs and gaseous and fluid distention of the forestomachs, abomasum, and cecum are common. The liver often shows vacuolar degeneration around the central vein.[48]

Prevention

The best way to reduce sheep losses to sneezeweed/ bitterweeds is to avoid herding sheep in areas heavily infested with the plant so that the sheep do not overgraze an area to the extent that they are forced to eat the weed. Moving the herd frequently prevents heavy consumption of the plant and reduces overgrazing an area that facilitates the proliferation of the sneezeweed or bitterweeds.[49]

Orange Sneezeweed
Dugaldia hoopesii (Helenium hoopesii)
Asteraceae (Sunflower family)

Habitat

Orange sneezeweed is a perennial plant of high mountain rangelands from Montana and eastern Oregon south to New Mexico. It prefers moist areas at altitudes between 5000 and 10,000 feet and will become invasive in overgrazed mountain meadows.

Description

The broad, basal clasping leaves are up to 12 inches (30 cm) in length. Flowers are large and showy with three to five toothed yellow-orange ray flowers. The disc is rounded and yellow in color (Figure 3-14).

Figure 3-14 Orange sneezeweed (*Dugaldia hoopesii*).

Principal Toxin

Orange sneezeweed contains sesquiterpene lactones, which are irritating to the nose, eyes, and gastrointestinal tract. The primary toxin is hymenovin (dugaldin), but other lactones including helenalin, helenanolide, tenulin, and hymenoxon, probably play a role in the toxicity.[38] Other than the direct irritant effects on the digestive system, the lactones have a profound effect on metabolism through their ability to bind with sulfhydryl groups.[44,45] In light of this, the toxicity of the lactones can be reduced if sulfur-containing amino acids (cysteine, methionine) or antioxidants such as butylated hydroxyanisole are fed before the bitterweed is consumed.[46,47]

Lambs with sneezeweed poisoning develop a syndrome of weakness that results in them "falling behind" the flock specially when they are being herded.

Bitter Sneezeweed
Helenium amarum (H. tenuifolium)
Asteraceae (Sunflower family)

Habitat and Description

A common weed preferring sandy soils in fields, woods, and waste areas of the south-eastern States to Texas, *H. amarum* is an annual growing to 3 feet (1 meter) in height. The stems are not winged and branch terminally. The leaves are numerous, linear, and 1 to 3 inches (2.5 to 7 cm) in length. The flowers are produced terminally, the ray flowers being yellow and the disc flowers either yellow or purple (Figure 3-15A).

Figure 3-15A Sneezeweed (*Helenium amarum*).

Principal Toxin

Bitter sneezeweed contains sesquiterpene lactones, which are irritating to the nose, eyes, and gastrointestinal tract. Tenulin is the most poisonous of the lactones found in bitter sneezeweed, but other lactones including, hymenovin, helenalin, helenanolide, and hymenoxon appear to contribute to the toxicity of the plant. Tenulin appears to be the principle compound that imparts a bitter taste to milk. Other than the direct irritant effects on the digestive system, the lactones (hymenoxon) have a profound effect on metabolism through their ability to bind with sulfhydryl groups causing a metabolic acidosis and hypoglycemia.[44,45] Consequently, the toxicity of the lactones can be reduced if sulfur-containing amino acids (cysteine, methionine) or antioxidants such as butylated hydroxyanisole are fed before the bitterweed is consumed.[46,47]

Sneezeweed, Bitterweed
Helenium autumnale
Asteraceae (Sunflower family)

Habitat

Various forms of *H. autumnale* are found throughout North America except in the very north and southwestern states. Its preferred habitat is moist areas in meadows, waste areas, and roadsides.

Figure 3-15B Bitterweed
(*Helenium autumnale*).

Description

Sneezeweed is a perennial growing to 4 feet in height. Stems arise from a fibrous root system, are winged, and branch terminally. The leaves are lanceolate, sparsely toothed, and 2 to 6 inches (5 to 15 cm) in length. The flowers are produced singly at the ends of branches. The ray flowers are yellow with three lobes and reflexed at maturity; the disc flowers are yellow (Figure 3-15B).

Principal Toxin

Sneezeweed contains sesquiterpene lactones, which are irritating to the nose, eyes, and gastrointestinal tract. The primary toxin is helenalin, but other lactones including helenanolide, tenulin, hymenovin, and hymenoxon, all of which have similar effects on the digestive system, contribute to its toxicity. Weight loss, vomiting, and secondary inhalation pneumonia are common clinical signs seen in animals with bitterweed poisoning.

Sneezeweed
Helenium microcephalum
Asteraceae (Sunflower family)

Figure 3-15C Small head bitterweed
(*Helenium microcephalum*).

Habitat and Description

Primarily found in Texas and Mexico, it prefers moist soils of pastures and open woodlands. It is an annual herb, growing to 3 feet (1 meter) in height. The stems are winged and leaves are lanceolate, sparsely toothed below, entire above. The flowers are produced terminally and are smaller than those of *H. amarum* and *H. autumnale*. The yellow ray flowers are equal or shorter than the diameter of the reddish brown disc (Figure 3-15C).

Principal Toxin

Sneezeweed contains sesquiterpene lactones (dugaldin, helenalin, helenanolide, tenulin, and hymenoxon), which are irritating to the nose, eyes, and gastrointestinal tract. Weight loss, vomiting, and secondary inhalation pneumonia are common clinical signs seen in animals with bitterweed and sneezeweed poisoning.

Colorado Rubberweed, Pingue
Hymenoxys richardsonii
Asteraceae (Sunflower family)

Habitat
Preferring drier sandy soils, Colorado rubberweed is found from Saskatchewan to Texas and westward to California.

Description
Rubberweed is a shrub-like perennial growing to 2 feet (0.5 meters) in height. Stems branch upward from a woody base. The stems have a conspicuous mass of white or brown hairs among the lower leaves. The leaves are linear, one to three times alternately divided. The basal leaves are densely grouped around the stem base. The yellow flowers, one to five per stem, are about 0.5 inch (1 cm) in diameter with 6 to 10 three-lobed ray florets, and 60 or more disc florets (Figure 3-16A).

Principal Toxin
Sesquiterpene lactones, which are irritating to the nose, eyes, and gastrointestinal tract are similar to those found in bitter sneezeweeds. The principle lactone involved with poisoning is hymenovin, with other compounds playing a role.

Figure 3-16A Rubberweed, pingue (*Hymenoxys richardsonii*).

Bitterweed
Hymenoxys odorata
Asteraceae (Sunflower family)

Habitat and Description

A common weed of semiarid rangeland from Texas to Southern California, *H. odorata* is an erect, branching, annual weed growing up to 2 feet (0.75 meters) in height, and forming large stands. (Figure 3-16B) Leaves are alternate, once to three times divided into very narrow, hair-covered divisions. Showy flower heads are formed at tips of branches, bright yellow, up to 0.5 inch (1 cm) in diameter, with 6 or more ray florets, each trilobed.

Figure 3-16B Bitterweed (*Hymenoxys odorata*) (Courtesy Dr. Darrell N. Ueckert, Texas Agricultural Research and Extension Station, San Angelo, Texas.)

Principal Toxin

Bitterweed contains sesquiterpene lactones, which are irritating to the nose, eyes, and gastrointestinal tract. The primary toxins are dugaldin, helenalin, helenanolide, tenulin, and hymenoxon, all of which have similar effects. Other than the direct irritant effects on the digestive system, the lactones (hymenoxon) have a profound effect on metabolism through their ability to bind with sulfhydryl groups. Consequently, metabolic acidosis and hypoglycemia develop rapidly.[44,45] In light of this, the toxicity of the lactones can be reduced if sulfur-containing amino acids (cysteine, methionine) or antioxidants such as butylated hydroxyanisole are fed before the bitterweed is consumed.[46,47]

Desert Baileya, Desert Marigold
Baileya multiradiata
Asteraceae (Sunflower family)

Habitat and Description

Common to the semiarid region of the southwestern states and Mexico, desert baileya is an annual or in some areas a perennial. The plant is wooly, extensively branched from its base, growing to 18 inches (46 cm) in height. The leaves are numerous, alternate, with those at the base having long petioles and those higher up being smaller without a petiole. Showy yellow flowers are produced on long stems, each with one flower (Figure 3-17). The flower heads, 1 to 2 inches (2.5 to 5 cm) in diameter, have 25 to 50 ray florets, which are reflexed when the flower matures.

Figure 3-17 Desert baileya (*Baileya multiradiata*)

Principal Toxin

Sesquiterpene lactones, which are irritating to the nose, eyes, and gastrointestinal tract, are similar to those found in bitter sneezeweeds.

Other common species of plant that contain sesquiterpene lactones with the potential to cause poisoning in livestock that eat them include paper flower (*Psilostrophe* spp.), *Gaillardia* spp., and copper weed (*Oxytenia* spp.).[38]

Field Bindweed, Morning Glory
Convolvulus arvensis
Convolvulaceae (Morning glory family)

Description
Bindweed is an extremely persistent, invasive, perennial noxious weed. It is a twining or creeping weed with alternate leaves and white or pink funnel-shaped flowers (see Figure 1-24). The plant reproduces readily from seed and its extensive root system.

Figure 3-18A Field bindweed (*Convolvulus arvense*).

Principal Toxin

Tropane alkaloids (pseudotropine) with atropine-like action are present in all parts of the plant.[50,51] Calystegins present in bindweeds (*Calestegia* and *Convolvulus* spp.) inhibit glucosidase enzyme activity and therefore possibly play a role in poisoning animals that eat the plants (Figures 3-18A and 3-18B).[52] The glycosidase inhibitory activity of the calystegins is comparable to that of swainsonine, an indolizidine alkaloid found in locoweeds (*Astragalus* and *Oxytropis* spp.). Mice fed an exclusive diet of bindweed developed severe gastritis and liver necrosis possibly as a result of the combined effects of the toxins present in the plant.[51] Bindweed may also accumulate toxic levels of nitrate (see Chapter 1).

Bindweed is most likely to cause poisoning in animals when pastures are over grazed and the bindweed becomes the predominant plant available for the animals to eat. Hay contaminated with large amounts of bindweed seed may also cause colic especially in horses.

Clinical Signs

Horses develop colic as the result of intestinal stasis and flatulence induced by the tropane alkaloids. A slow heart rate and dilated pupils may result if toxic levels of the bindweed are consumed. Affected horses should be removed from the source of the bindweed, and given symptomatic therapy for colic.

Figure 3-18B *Calestegia sepium*.

Nightshades
Solanaceae (Nightshade family)

This large family of plants with some 88 genera and more than 2300 species has long been associated with poisoning of humans and animals. Deadly nightshade or belladona (*Atropa belladona*) was used in ancient times to dilate the pupils of women to enhance their beauty, and it has found use as a potent hallucinogen.[53] The black berries of belladona, the showy red berries of Jerusalem cherry (*Solanum pseudocapsicum*), and bittersweet (*S. dulcamara*) have caused poisoning in people who eat them. Animals are rarely poisoned by belladona and are more likely to be poisoned by various genera that include *Solanum* spp. (nightshades), *Datura stramonium* (jimson weed), *Hyoscyamus niger* (black henbane), *Lycopersicon* spp. (tomato), *Cestrum* spp. (jessamine), and *Physalis* (ground cherries or Chinese lanterns).[54,55] The more common members of the nightshade family associated with poisoning in animals are presented in Table 3-2. Livestock may be poisoned if they are fed potatoes (*Solanum tuberosum*) after they have sprouted and the skins turned green.[56] Similarly green tomato vines may cause poisoning if fed to livestock. Potato plants, however, can be effectively used as a source of food for livestock if ensiled, or fed with grass hay or cereal grains.[57]

Table 3-2 Common Nightshades with the Potential for Causing Poisoning	
COMMON NAME	SCIENTIFIC NAME
Black nightshade	*Solanum nigrum*
Huckleberry, wonderberry	*S. americanum*
Hairy nightshade	*S. sarrachoides*
Cutleaf nightshade	*S. triflorum*
Silverleaf nightshade	*S. elaeagnifolium*
Horse nettle, bull nettle	*S. carolinense*
Sodom apple	*S. sodomaeum*
Buffalo bur	*S. rostratum*
Tropical soda apple	*S. viarum*
Bittersweet, climbing nightshade	*S. dulcamara*

Principal Toxin

A variety of steroidal (tropane) glycoalkaloids are found in the Solanaceae, especially in the green parts of the plant and the unripe fruits.[58,59] The more common alkaloids including solanine, hyoscine (scopolamine), and hyoscyamine (atropine) act similarly on the autonomic nervous system by blocking the action of cholinesterase. This results in the accumulation of the neurotransmitter acetylcholine, and consequently inhibition of the parasympathetic nervous system, causing decreased salivation and intestinal motility, dilated pupils, and tachycardia. The alkaloids also have a direct irritant effect on the digestive system causing colic, constipation, or hemorrhagic diarrhea.

In addition to the effects of the tropane alkaloids, some members of the Solanaceae have other toxic compounds, not least of which is the highly toxic alkaloid nicotine found in the tobacco plant (*Nicotiana* spp.). Livestock that have access to either cultivated or wild tobacco are easily poisoned by nicotine.[54] Sudden death of cattle has been attributed to the consumption of the wild tree tobacco (*N. glauca*).[60,61] Also within the tobacco plant are alkaloids that cause fetal deformities if eaten by pigs and pregnant cows or sows[62] (see Chapter 8).

Some species of *Solanum* including *S. fastigiatum*, *S. kwebense*, *S. dimidiatum*, *S. cinerum*, *S. suriale*, and *S. viarum* (tropical soda apple) have been associated with a neurologic disease in cattle, sheep, and goats characterized by loss of equilibrium, falling down, tremors of the head, incoordination, opisthotonus, and seizures.[63-66] The lesions produced in the brain are confined to the cerebellum and include finely vacuolated cytoplasm of the Purkinje cells and neurons, with degeneration and necrosis[65-67] (see Chapter 6).

One species, *S. malacoxylon*, found in South America, contains vitamin D-like compounds toxic to cattle.[69] Affected animals absorb excessive amounts of calcium, which is deposited in tissues and results in severe lameness and weight loss (see Chapter 9).

Livestock will eat members of the nightshade family when other forages are scarce, or when crop residue products such as green-sprouted potatoes (*S. tuberosum*), potato vines and tomato plants are fed to them.[55] Signs of poisoning can be expected when 0.1 to 0.3 percent of an animal's body weight in green plant is eaten.[54] Grain contaminated with seeds of jimson weed (*Datura stramonium*) can be a significant source of poisoning.[70] Compared to other livestock, cattle may be more susceptible to the toxic effects of solanine alkaloids.[71]

Clinical Signs

Initially there may be central nervous system excitement, but depression follows with decreased heart and respiratory rate, muscle weakness, dilated pupils, colic, and watery diarrhea. Rupture of the stomach and paralysis of the digestive system in horses can be a sequel to the effects of the tropane alkaloids.[70] If large amounts of the tropane alkaloids are consumed over a short period of time, cardiac arrest may lead to death before digestive signs have time to develop.

Treatment

Animals showing severe anticholinergic signs consisting of muscle tremors, hyperesthesia, dilated pupils, intestinal stasis, and depressed respiratory rate may be treated with physostigmine.[72,73] Many animals, however, recover if treated symptomatically. Oral administration of activated charcoal as an adsorbent may be effective if given soon after the plants have been eaten.

Jimson Weed, Thornapple, Peru Apple, Stinkweed
Datura stramonium (D. tatula)
Solanaceae (Nightshade family)

Habitat

Introduced from the tropics, jimson weed has become a naturalized weed throughout North America, established in disturbed soils, disused corrals, roadsides, edges of cereal grain fields, and so on. Seeds are probably dispersed in cereal grains at harvest time. The name jimson weed or "Jamestown weed" was given to the plant in 1676, when a large number of soldiers were poisoned after eating the plant in Jamestown, Virginia.

Description

This erect, branching, glabrous, herbaceous annual often reaches heights of 3 to 6 feet (1 to 2 meters). Leaves are large, alternate, and simple with irregularly toothed edges. When crushed, the leaves have a strong, unpleasant odor. The showy, white, and sometimes purple fragrant flowers are carried singly in leaf axils (Figure 3-19A). The large 2 to 4 inches (6 to 10 cm), funnel-shaped flowers with a flared end have five sections with tapering points. The characteristic fruit is a spiny capsule that, when ripe, splits open into four sections exposing many dark brown, flattened, kidney-shaped seeds with a pitted surface. (Figure 3-19B).

There are several similar poisonous species of *Datura* including sacred Datura (*D. meteloides*) (*D. inoxia*), oak leaf thornapple (*D. quercifolia*), and *D. metel*.

Figure 3-19A Jimson weed flower and seed capsule (*Datura stramonium*).

Figure 3-19B Jimson weed seed capsule.

Sacred Datura, Tolguacha, Angel's Trumpet, Moonflower
Datura Wrightii (D. meteloides, D. metel)
Solanaceae (Nightshade family)

The sacred datura differs from jimson weed in that the leaves are grayish in color due to the covering of fine hairs, and the flowers are much larger 10 to 12 inches (20 to 30 cm). The fragrant white flowers open at night and last only a day (Figure 3-20A). The nodding fruit capsules are covered with long, fine spines (Figure 3-20B).

Figure 3-20A Sacred datura (*Datura wrightii*).

Figure 3-20B Sacred datura seed capsule

Angel's Trumpet
Brugmansia suaveolens (Datura suaveolens)
Solanaceae (Nightshade family)

Habitat and Description

Common in the southern states as an ornamental treelike shrub, angel's trumpet grows to 15 feet (5 meters) in height and produces numerous white, yellow, or pink, pendulous, large trumpet-like flowers 10 to 12 inches (20 to 30 cm) in length (Figure 3-21). Brugmansia hybrids do not produce fruits but are easily propagated from cuttings.

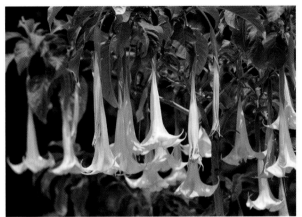

Figure 3-21 Angel's trumpet (*Brugmansia* spp.).

Principal Toxins

Datura species contain the tropane alkaloids hyoscyamine (atropine) and hyoscine (scopolamine).[74] Most poisoning of animals occurs when Datura seeds contaminate cereal grains that are fed to animals and birds.[54,70,75-77] The plant itself is rarely eaten because of its pungent odor and taste. Human poisoning most frequently occurs when the seeds are deliberately or unintentionally eaten or the leaves smoked or made into a tea.[78-80] The signs of jimson weed poisoning are essentially those attributable to atropine poisoning. In humans there is a period of hallucination before characteristic signs of decreased salivation, intense thirst, mydriasis, tachycardia, intestinal stasis, and eventual respiratory and cardiovascular collapse. In animals and birds the main effects seem to be on the digestive system causing intestinal stasis, decreased food consumption, and poor growth rates. Muscle tremors, disturbances in locomotion, hyperesthesia, rapid respiration, and decreased water consumption have been observed in sheep and goats experimentally poisoned with jimson weed.[81]

Cattle fed approximately 107 Datura seeds per kilogram body weight experimentally developed anorexia and rumen stasis and did not succumb to severe atropine intoxication because they stopped eating the seeds.[76] Similarly pigs find the seed quite unpalatable and are likely to reject the contaminated food before becoming seriously intoxicated.[77] Horses fed jimson weed seed in corn developed anorexia, weight loss, thirst, tachycardia, dilated pupils, diarrhea, and excessive urination.[70] Death resulted from rupture of the stomach and paralysis of the digestive system. Treatment of jimson weed poisoning is rarely necessary, and the use of physostigmine to counteract the atropine-like effects of the alkaloids is controversial.[82]

Conflicting reports exist as to the effects of jimson weed on the developing fetus. Piglets were reported to have developed bony deformities (arthrogryposis) after sows had consumed jimson weed in early gestation.[83] However, this teratogenic effect could not be reproduced experimentally.[84]

Deadly Nightshade, Belladonna
Atropa belladonna
Solanaceae (Nightshade family)

Habitat and Description

Introduced from Europe, deadly nightshade has escaped from cultivation on occasion. It is a robust herbaceous plant arising from a perennial root and grows up to 5 to 6 feet (1 to 2 meters) in height. The leaves are large, ovate, entire, and alternate and occur in pairs at each node, one member of the pair always being smaller. The flowers are solitary, nodding, five-parted, tubular and purple in color. The fruit, a berry, turns purple to black when ripe (Figure 3-22).

Figure 3-22 Belladona flowers and fruits (*Atropa belladona*).

Principal Toxin

The entire plant is poisonous because of the presence of the atropine-like alkaloid L-hyoscyamine, the isomeric form of atropine. The incidence of animal poisoning from deadly nightshade is rare. Because it is occasionally cultivated as an ornamental plant, it is a real hazard to children who are attracted to the black fruits.

Black Henbane
Hyoscyamus niger
Solanaceae (Nightshade family)

Habitat and Description

Black henbane is listed as a noxious weed in some areas, as it can be invasive in cultivated fields, roadsides and waste areas. It is a coarse, very hairy, sticky, branching annual or biennial, erect herb that grows to 5 feet (1 to 2 meters) in height. The leaves are alternate, simple, sessile, coarsely toothed, and 6 to 8 inches (10 to 18 cm) long. The flowers are greenish yellow or white with prominent purple veins. The flowers are produced in the leaf axils and are surrounded by a five-pointed calyx that enlarges to form the characteristic globular fruits (Figure 3-23). The top of the capsular fruit detaches to release many gray-brown hard-coated seeds.

Figure 3-23 Black henbane (*Hyoscyamus niger*).
Inset: Flower and partially dissected fruit.

Principal Toxin

The tropane alkaloids, hyoscyamine, hyoscine, and atropine, are found in all parts of the plant. Henbane has the potential to poison animals but because it is unpalatable, it is rarely eaten.

Black Nightshade
Solanum nigrum
Solanaceae (Nightshade family)

Habitat

Solanum nigrum is found throughout North America, preferring disturbed soils along roadsides, fences, and cultivated fields.

Description

Black nightshade is a hairless, spineless, erect, or trailing branched annual plant with simple ovate to lanceolate sinuate-toothed leaves. The flowers have five white petals arranged in a 0.5 to 1 cm (6 to 10 mm) star. The clusters of flowers arise from a stalk that is situated between leaf nodes. The smooth, round, 0.5 to 1 cm (5 to 10 mm) fruits are initially green, turning black when ripe (Figure 3-24). The green fruits containing solanine are toxic but the ripe black fruits are edible.

Distinguishing between eastern black nightshade (*S. ptycanthum*), plains black nightshade (*S. interius*), and black nightshade (*S. nigrum*) is difficult and some consider these species to be identical. *S. americanum* is very similar, with edible black berries that are often referred to as "garden huckleberry" or "wonderberry."[54]

Principal Toxin

A variety of steroidal (tropane) glycoalkaloids are found in black nightshade, especially in the green parts of the plant and the unripe fruits.[58,59] The more common alkaloids including solanine, hyoscine (scopolamine), and hyoscyamine (atropine) have similar effects on the autonomic nervous system by blocking the action of cholinesterase. This results in decreased salivation and intestinal motility, dilated pupils, and tachycardia. The alkaloids also have a direct irritant effect on the digestive system causing colic, constipation, or hemorrhagic diarrhea.

Figure 3-24 Black nightshade unripe and ripe berries (*Solanum nigrum*).

Bitter Nightshade, Climbing Bittersweet
Solanum dulcamara
Solanaceae (Nightshade family)

Habitat

Introduced from Europe, bitter nightshade has become established in most of North America, preferring moist soils of hedgerows, ditches, streams, and residential landscapes.

Description

Bitter nightshade is a trailing or climbing perennial with stems up to 10 feet (3 meters) in length. The leaves are heart-shaped with one to several basal leaflets. The flowers are star-shaped, with purple petals and prominent yellow or orange anthers. The fruits are berries that turn from green to yellow to red as they ripen (Figure 3-25).

Figure 3-25 Bittersweet (*Solanum dulcamara*).

Principal Toxin

Like other members of the nightshade family, glycoalkaloids are present in all parts of the plant. Livestock are not commonly poisoned by bittersweet unless they have access to a large quantity of the plants. Children are more likely to be affected after eating the attractive berries.

Silver Leaf Nightshade, White Horse Nettle, Tropillo
Solanum elaeagnifolium
Solanaceae (Nightshade family)

Habitat

Silver leaf nightshade is a weed of drier soils along roadsides, prairies, and unused areas, especially in the southern and western states.

Description

This nightshade is a perennial, branching, erect, spiny plant reaching 1 meter in height. It has an extensive root system. The leaves and stems are covered with white hairs. Leaves are simple, thick, linear to lanceolate, alternate, with spines along the main veins. Flowers are blue to purple and the fruits yellow to orange (Figure 3-26).

Figure 3-26 Silver leaf nightshade (*Solanum elaeagnifolium*).

Principal Toxin

Silver leaf nightshade contains the tropane alkaloid solanine that has similar properties to atropine, and acts on the gastrointestinal system.[71] Solanidine, a steroidal alkaloid, is also present and acts on the nervous system. All parts of the plant are toxic, the ripe yellow berries being most poisonous.

Horses that have recently been treated with the antiparasitic drug ivermectin and which then eat *S. elaeagnifolium* can develop a neurotoxicity due to the increased uptake of ivermectin in the brain.[72] Affected horses become severely depressed and ataxic. Drooling saliva, drooping lips and ears, and head pressing may develop and deaths have been reported.[72] High levels of ivermectin are detectable in the brains of fatally poisoned horses.

Horse Nettle, Bull Nettle
Solanum carolinense
Solanaceae (Nightshade family)

Habitat

This perennial weed is found in disturbed soils and unused areas along roads and field edges especially in the southern states.

Description

Horse nettle is an erect 16 feet (5 meters) high, branching plant with yellow spines on leaves and stems. Leaves are simple, alternate, oblong, and irregularly lobed. The flowers are simple, alternate, oblong, and irregularly lobed. The flowers are pale violet to white in cluster near the top of the plant (Figure 3-27). The five petals tend to be united. The fruits are yellow when ripe.

Figure 3-27 Horse nettle (*Solanum carolinense*).

Principal Toxin

The toxin is solanine, a tropane alkaloid with toxic properties similar to atropine.

Buffalo Bur, Kansas or Texas Thistle
Solanum rostratum
Solanaceae (Nightshade family)

Habitat

This is a common weed of drier, disturbed soils of the plains, roadsides, and so forth throughout the western and southern states.

Description

This annual weed grows 1 to 2 feet (0.5 to 1 meter) tall, with prominent long spines on the leaves and stems (Figure 3-28). The flowers are yellow with five partially united petals and a very spiny calyx, which encloses the berry. Numerous seeds are produced.

Figure 3-28 Buffalo bur (*Solanum rostratum*).

Principal Toxin

Solanine, a tropane alkaloid with toxic properties similar to atropine, is the main toxin. More often buffalo bur is a problem because the spiny burs cause mechanical injury, and devalue the fleece of sheep when it becomes entangled with burs.

Tropical Soda Apple
Solanum viarum
Solanaceae (Nightshade family)

Habitat

A common weed of South America, tropical soda apple has now invaded Florida and is spreading, being reported in states from Pennsylvania to Alabama and Mississippi. Cattle, goats, and wildlife that eat the fruits and pass the seeds through their feces to infect other areas spread this aggressive weed.

Description

Soda apple is a robust, erect, branching, perennial that grows to 4 to 6 feet (1 to 2 meters) in height. The stems and leaves are covered with sharp spines that deter animals from eating it. Fruits are about 1 inch (2.5 cm) in diameter, turning yellow when ripe (Figure 3-29).

Figure 3-29 Soda apple (*Solanum viarum*).

Principal Toxin

The toxin responsible for causing cerebellar degeneration and the resulting neurologic disease has not been identified. Similar neurologic signs and cerebellar lesions have been seen in goats that have eaten *S. dimidiatum* and *S. cinereum*.[65,67,68] To date, goats seem to be the only livestock to develop this neurologic syndrome from eating soda apple.

Clinical Signs

Goats develop a neurologic syndrome that consists of fine head and neck tremors and occasional general muscle spasms.[68] The tremors become pronounced during attempts to eat and drink. Affected goats maintain a base-wide stance and have a jerky, uncoordinated gait affecting the hind limbs most severely. General muscle weakness is common. Blindness is not a feature of the disease, but if blindfolded, the goats become disoriented and cannot maintain their balance.

There is no known specific treatment for soda apple poisoning. Affected animals should be denied further access to the plants and a nutritious diet provided. Recovery from the cerebellar lesions depends on the severity and duration of the disease. Histologic lesions in the brain consist of fine vacuolation of the Purkinje cells and neurons.[68]

Ground Cherries
Physalis spp.
Solanaceae (Nightshade family)

There are many species of *Physalis* throughout the United States that are potentially toxic to livestock and people owing to the presence of glycoalkoids including solanine. Symptoms of poisoning would therefore resemble those seen in poisoning caused by other plants of the genus *Solanum*. Only a few of the more common species will be described because they have characteristic features common to all species of *Physalis* that make them easily recognizable.

Ground Cherry
Physalis virginiana
Solanaceae (Nightshade family)

Habitat
The plant prefers the dry soils of the plains, roadsides, and waste ground.

Description
Ground cherry is an erect, 2 to 4 inches (5 to 10 cm) high, branching herbaceous, hairy plant. Leaves are alternate, ovate, and broadly toothed. The five-lobed, bell-shaped, dropping flowers are pale yellow with a dark center and are produced at the leaf axils. The characteristic fruit is covered by an enlarged podlike calyx that turns papery brown when the enclosed berry is mature (Figure 3-30A).

A closely related plant is the Chinese lantern *P. lobata* (*Quincula lobata*). It is a more prostrate plant and has showy blue flowers (Figure 3-30B).

Principal Toxin
The main toxin is solanine, a tropane alkaloid with toxic properties similar to atropine.

Figure 3-30A Ground cherry (*Physalis virginiana*).

Figure 3-30B Ground cherry (*Physalis lobata*).

Diarrhea

A variety of common plants may cause diarrhea when they are eaten by animals that do not have good quality forages available to them. Invasive pasture plants such as *Euphorbia esula* (leafy spurge), *Iris missouriensis* (wild iris), *Equisetum arvense* (horse tail, scouring rush), *Helenium* spp. (bitter weeds), pokeweed (*Phytolacca americana*), and a variety of *Brassica* spp. (mustards) may cause colic and diarrhea. English ivy, which can invade animal pastures and fence rows accessible to livestock, may cause gastroenteritis, colic, and diarrhea (Table 3-3).

Leafy Spurge
Euphorbia esula
Euphorbiaceae (Spurge family)

Habitat

Introduced from Eurasia, leafy spurge has become a troublesome weed of North America where it has infested millions of acres in Canada and the north-central United States. It is listed as a noxious weed in most states because of its ability to proliferate and displace normal forages.

Description

Leafy spurge is a prolific perennial, up to 3 feet (1 meter) tall, that reproduces by seeds and an extensive root system. Leaves are alternate, narrow 1 to 4 inches (2 to 10 cm) long. Multiple stems arise from root crowns. The plant contains a milky sap in the stems and leaves. The flowers are small and yellowish green in color and are arranged in terminal clusters. Conspicuous yellowish green heart-shaped bracts (often mistaken as the flower) surround each flower (Figure 3-31A). Seed capsules explode when dry, scattering the seeds widely, which aids in the plant's rapid proliferation.

Figure 3-31A Leafy spurge (*Euphorbia esula*).

Other spurges are potentially toxic to animals and include the tropical pencil tree (*Euphorbia tirucalli*), and snow on the mountain (*E. marginata*) (Figure 3-31B). Myrtle spurge (*E. myrsinites*) (Figure 3-31C) has a particularly irritating sap that causes dermatitis in some people who handle the plant.

Principal Toxin

The specific toxin in leafy spurge has not been defined. Cattle, through negative feedback, learn to avoid eating the plant once they have initially consumed small amounts of the plant.[85] Terpenes appear to be the aversive chemical in the plant.[86] Even ensiling leafy spurge does not improve its palatability.[87] Sheep and goats readily eat leafy spurge and appear to be unaffected by it.

Table 3-3
Miscellaneous Plants Infrequently Associated with
Gastrointestinal Poisoning

Scientific Name	Common Name	Symptoms
Achillea milleform	Yarrow	Colic, diarrhea
Baccharis halimifolia	Eastern baccharis	Colic, diarrhea, staggering, trembling
Brassica spp.	Mustards	Colic, hemorrhagic diarrhea
Cephalanthus occidentalis	Button bush	Vomiting, weakness, death
Datisca glomerata	Durango root	Anorexia, diarrhea, depression, death
Hedera helix	English ivy	Colic, diarrhea
Hydrangea spp.	Hydrangea	Hemorrhagic diarrhea
Iris spp.	Iris	Colic, diarrhea
Phoradendron spp.	Mistletoe	Severe colic and diarrhea
Tulipa spp.	Tulips	Colic, diarrhea

Clinical Signs

Spurges are usually not eaten by cattle when other forages are available. Sheep and goats, however, will eat the plants without apparent problem.[88] Spurges cause excessive salivation in some animals due to the irritant effects of the plant sap. Cattle frequently develop diarrhea if they are compelled to eat leafy spurge. Recovery is rapid once animals are provided more nutritious food.

Note: Leafy spurge is a noxious weed that should be vigorously controlled to prevent its rapid invasion of pastures and rangeland where it will

Figure 3-31B Snow on the mountain (*Euphorbia marginata*).

displace nutritious forbes and grasses. Sheep are effective biologic controls for leafy spurge and can be profitably used to graze rangeland heavily infested with leafy spurge.[89] Sheep can eat diets containing up to 40 to 50 percent leafy spurge without any evidence of dis-

Figure 3-31C Myrtle spurge (*Euphorbia myrsinites*).

ease or decrease in weight gain.[90] The sheep can reduce the plant's biomass and density but will not eradicate it. Approximately 5 percent of leafy spurge seeds eaten by sheep remain viable in the feces, and, therefore, sheep can help spread the plant if not appropriately managed. Ideally sheep should be used to graze the leafy spurge before it flowers and the seeds are produced. If sheep are grazing the plant when it has seeds, they should be kept confined for at least 5 days before they are moved to leafy spurge-free areas.[91]

Lectins

A group of unrelated plants including castor bean (*Ricinus communis*), rosary pea (*Abrus precatorius*), and black locust (*Robinia pseudoacacia*) contain highly poisonous compounds called lectins that are capable of causing severe poisoning in humans and animals. Lectins are some of the most toxic of known plant compounds, and as glycoproteins (toxalbumins) are capable of binding to receptor sites on cells causing inhibition of protein synthesis and cell death. Lectins are concentrated in the seeds, and it is the consumption of the seeds that is most commonly the cause of poisoning in humans and animals. The seeds have a tough outer coating and unless it is disrupted through chewing, the seeds will pass through the digestive system without causing problems. Castor beans and rosary peas have a long history of causing poisoning in children and adults.[92-95] Cattle, goats, horses, poultry, and dogs have been poisoned by castor beans, and less frequently by rosary peas when they contaminate animal foods.[95-101] Cattle fed "cake" made from castor beans following the extraction of castor oil may be poisoned unless the cake is first treated with heat to destroy the ricin.[95,102] Properly detoxified castor bean meal is a useful protein source for cattle.

Principal Toxins

The principal toxins in castor beans, rosary peas, and black locust are the glycoproteins ricin, abrin, and robinin, respectively. Castor oil derived from the beans contains ricinoleic acid, an irritant that can cause severe intestinal irritation with profound purgation. Various other alkaloids and proteins are present in the seeds but are generally not toxic and have been used for a multitude of medicinal purposes.[103] Some of the proteins act as agglutinins and are capable of causing red blood cell agglutination in vitro.[96] Agglutination is not a feature of orally ingested castor beans, and ricin, when injected systemically, does not cause hemolysis, indicating that compounds other than ricin are involved.[104]

Lectins, comprising two peptide chains joined by sulfide bonds, are capable of binding to certain cell receptor sites and inhibit cellular protein synthesis in the ribosomes.[105] Because this process takes time to occur, clinical signs of poisoning do not occur for several days after the ingestion of the lectins.[105] An additional property of lectins, being proteins, is their ability to induce antibody formation when injected into animals, a factor that has been explored in an attempt to develop antitoxins to ricin and abrin.[104] The allergic reaction encountered in humans exposed to dust from castor beans is not a reported problem in animals.[103,107] Ricin is also capable of inhibiting the growth of tumor cells that may have potential for treating some tumors.[108]

Ricin and abrin are some of the most poisonous plant compounds known, especially when administered by injection. As little as 1 mg ricin is lethal to humans.[93] Reports of castor bean poisoning have varied considerably as to the number of beans that will induce poisoning and death.[89] This may be accounted for by variations in the quantity of ricin present in some castor beans and the degree to which the beans had been chewed before they were swallowed. Human poisoning has occurred when 2.5 to 20 castor beans have been eaten, and animals may be fatally poisoned with 4 to 11 beans.[103] A dose of 2 g/kg body weight of ground castor beans is reported as lethal in cattle.[109] Horses are reportedly fatally poisoned by as few as 60 seeds; ruminants appear to be less susceptible because abrin is broken down in the rumen.[110] Goats experimentally fed 1 to 2 g/kg body weight of ground *A. precatorius* seeds died 2 to 5 days later.[96] Thousands of wild ducks have been fatally poisoned by eating castor beans, and experimentally domestic ducks could be poisoned by three to four seeds.[111] Studies in mice and dogs in which pure abrin and ricin were injected experimentally demonstrated that the minimum lethal dose was 0.7 and 2.7 µg/kg body weight, respectively.[104]

In reexamining the literature on castor bean poisoning in humans, Rauber and Heard propose a much less severe prognosis for poisoning in humans in light of modern medical practices.[93] Rapid removal of the castor beans from the stomach and aggressive fluid and electrolyte therapy to counteract the primary effects of diarrhea and dehydration greatly reduce the incidence of fatalities.

Plant parts of castor beans and rosary peas, other than the seeds, are rarely reported as a cause for poisoning in animals. Although the leaves of the castor bean plant are rarely eaten, they are reported to be toxic.[112] Cattle fed castor bean leaves develop signs distinct from those associated with ricin. Affected animals develop neuromuscular impairment characterized by muscle weakness, tremors, salivation, and excessive eructation.[112] Recovery may occur after a short period or the animal dies, presumably as a result of the quantity of leaves consumed.

Clinical Signs

The signs of castor bean and rosary pea poisoning are primarily associated with severe gastrointestinal irritation, and begin several days after the consumption of a toxic dose of lectins. [96,100,101,109,110] Affected animals stop eating and develop a severe hemorrhagic diarrhea. Lactation stops abruptly. Abdominal pain is often severe. In the case of black locust poisoning, horses, in particular, may also develop dilated pupils and cardiac arrhythmias.[114] Rapid loss of water and electrolytes through the diarrhea results in dehydration and hypovolemic shock. Increases in serum liver enzymes, creatinine, urea nitrogen, and sodium and potassium levels, and a decrease in serum total protein reflect the loss of fluid and electrolytes and the effects of the lectins on organ function.[96] Animals left untreated die from hypovolemic shock.

Postmortem findings include severe pulmonary congestion, ulceration of the stomach and intestines, and fatty degeneration and necrosis of the liver and kidneys.[96]

Diagnosis

The diagnosis of castor bean or rosary pea poisoning can be difficult to confirm unless animals are observed eating the seeds or the intact seeds or parts thereof are identifiable in the digestive tract at postmortem examination. Because it resembles sunflower and cotton seed cake, castor bean meal can be recognized in animal feeds by microscopic examination for the characteristic seed hull fragments.[110]

Treatment

Treatment is directed at removing the seeds from the stomach and digestive tract as quickly as possible. Vomiting can be induced in dogs and cats, or endoscopy can be used to remove the seeds from the stomach. Orally administered activated charcoal is of benefit. If cattle are known to have consumed feed contaminated with castor beans within the past day, purgatives such as magnesium hydroxide may be helpful in removing the toxin from the digestive system. Animals with diarrhea and resulting dehydration and hypovolemic shock should be given intravenous fluids and electrolytes. Such treatment has been the main reason that fatalities in humans who have consumed castor beans have been virtually eliminated.[93,115] The use of immune serum to treat ricin poisoning is rarely necessary.[115]

Castor Bean
Ricinus communis
Euphorbiaceae (Spurge family)

Description

Castor bean is an annual herb or short-lived perennial, and small tree in warmer areas. It is often cultivated as a garden annual. The leaves are alternate, large, and palmate, with 5 to 11 serrate lobes (Figure 3-32A). The petioles have conspicuous glands. New leaves are usually purple. The plants are monoecious. The flowers have no petals and are borne in terminal panicles with staminate (male) flowers below and pistillate (female) above. The calyx is of three to five parts, and the stamens are numerous, with many branched filaments. The ovary has three cells with one ovule in each cell; the styles are deep red with fine feather-like hairs. The fruit is a spiny capsule, which splits into three sections, each containing a shiny seed with gray and brown mottling (Figure 3-32B).

Principal Toxin

A lectin, ricin, is the principal toxin. All parts of the plant are toxic with the seeds containing the highest concentration of ricin, a heat-labile glycoprotein (toxalbumin). Other compounds in the seeds are responsible for agglutination and hemolysis. Ricinoleic acid present in castor oil is primarily responsible for its purgative action.

Note: Castor oil cake is poisonous to ruminants unless it has been heat treated. Castor bean plants should not be planted in or near livestock enclosures.

Figure 3-32A Castor bean plant with flower spike and spiny seed capsules (*Ricinus communis*).

Figure 3-32B Castor beans (*Ricinus communis*).

Precatory Bean, Rosary Pea, Jequirity Bean, Crab's Eye
Abrus precatorius
Fabaceae (Legume family)

Habitat

A weed of fence rows, roadsides and citrus groves, precatory bean was introduced from tropical countries. It has become established in Florida.

Description

Precatory bean is a twining, perennial vine, 10 to 20 feet (3 to 6 meters) long, using other plants for support. Lower, older portions of the stem become gray; the younger portions remain green. The leaves are alternate, opposite pinnately compound with 8 to 15 pairs of leaflets. The flowers are in axillary racemes and are red to purple in color. The fruit is a legume pod, 2 inches (4 cm) long, and produces ovoid seeds that are glossy red with a jet black eye (Figures 3-33A and 3-33B). Some varieties have seeds that are black with a white eye or are white with a black eye.

Figure 3-33A Precatory bean vine and pods with seeds (*Abrus precatorious*).

Figure 3-33B Precatory beans (*Abrus precatorious*).

Principal Toxin

Abrin, like ricin from castor beans, is a potent lectin found in highest concentration in the seeds. Other toxic compounds are also present in the seeds. Only if the seeds are chewed and swallowed is the lectin released. Abrin is one of the most toxic compounds known, requiring as little as 0.00015 percent of a person's body weight to be fatal. Animals are infrequently poisoned.

Black Locust
Robinia pseudoacacia
New Mexican Black Locust
Robinia neomexicana
Fabaceae (Legume family)

Habitat

Usually around dwellings and along fence rows, black locust is common in the southwestern states. It occasionally forms dense stands. Robinia neomexicana often grows along streams and in valleys.

Description

Black locust is a small tree up to 70 feet (21 meters) tall. The trunk is straight and slender; branches are spiny and glabrous when young. The leaves are alternate, pinnately compound with entire, elliptical leaflets in 3 to 10 pairs. The individual flowers are showy, white, and pealike, forming drooping clusters, 4 to 8 inches (10 to 20 cm) long (Figure 3-34A). The fruit is a straight, flat, many-seeded brown legume pod.

Figure 3-34A Black locust (*Robinia pseudoacacia*).

Robinia neomexicana differs from *R. pseudoacacia* in that it has rose-pink flowers, hairy leaflets, finely haired young twigs and glandular hairy pods (Figure 3-34B).

Principal Toxin

Robin, a lectin, is similar to but less toxic than ricin found in castor beans. The bark and seeds have the highest concentrations of lectins; the flowers are not toxic.

Note: Locust trees of the genus *Gleditsia*, commonly referred to as honey locusts, are unrelated to the black locust and are not poisonous.

Mistletoes (*Phoradendron* spp.) and *Viscum* spp. (English mistletoe) also contain toxic lectins that can cause severe gastrointestinal irritation resulting in vomiting and diarrhea. Cardiovascular collapse with bradycardia and hypotension may occur when a large dose of mistletoe has been eaten. Animals are rarely poisoned by mistletoe. Children are most likely to be poisoned after eating the white berries when mistletoe is brought into the house for festive occasions.

Figure 3-34B Black locust (*Robinia neomexicana*).

Mayapple, Mandrake
Podophylum peltatum
Berberidaceae (Barberry family)

Habitat

Mayapple is an indigenous plant of eastern North America extending westward to Minnesota and Texas. It prefers moist, fertile soils of woodlands and pastures.

Description

Mayapple is a perennial herb, 1.5 to 2 feet (0.5 to 0.60 meters) tall, with a simple stem bearing two large umbrella-shaped, five to nine lobed, hairless leaves. A single, white, nodding, flower, with six to nine petals is produced at the junction of the two leaf stems. The fruit is a 1 to 2 inch (2.5 to 5 cm) fleshy berry that turns yellow when ripe (Figure 3-35). The plant spreads by a fibrous creeping root system.

Figure 3-35 Mayapple, mandrake (*Podophyllum peltatum*).

Principal Toxin

Podophyllin, a bitter, resinous compound, is found in all parts of the plant The ripe yellow fruit, however, is edible. Podophyllin acts as an irritant and has strong laxative properties. It also interferes with cell division and may have anticancer properties.[116] Livestock generally will not eat the plant unless deprived of their normal forages. Human poisoning occurs more commonly when unripe may apples are eaten or when parts of the plant are inappropriately used as a medicinal herb.

Clinical Signs

Colic and diarrhea are the major signs of mayapple poisoning. Excessive salivation and swelling of the muzzle, intermandibular area, and eyelids may result from the irritant effects of the resinous toxin.[117] Excitement lasting about a day is a reported symptom of mayapple poisoning. Most animals will recover once they are prevented from eating the plant.

Description

Privet is a deciduous shrub, with opposite, lanceolate leaves 1 to 2 inches (2 to 5 cm) long. The leaves are dark green on the upper surface and paler underneath. Some varieties have yellow or white marbling. Numerous small, white flowers are produced in clusters at the ends of the branches (Figure 3-36). The fruits are drooping clusters of black berries containing one to four seeds.

Figure 3-36 Privet (*Ligustrum vulgare*).

Principal Toxin

The toxin has not been identified. However, it has irritant properties that cause gastroenteritis, hypotension, and kidney damage. The berries and leaves are toxic. Animals rarely eat the plant unless they are hungry and deprived of normal forage.

Privet hedges should be avoided around livestock corrals or pastures and pruned leaves and branches should not be given to livestock.

Clinical Signs

Vomiting, abdominal pain, and diarrhea are common signs of privet poisoning. Hypotension and kidney failure will occur in severe cases, and death may occur within a day of eating a lethal amount of the plant.[118]

Rhododendron
Ericaceae (Heath family)

Rhododendrons and closely related species including azaleas (*Rhododendron* spp.), laurel (*Kalmia* spp.), fetter-bush (*Leucothoe* spp.), mountain fetter-bush (*Pieris* spp.), maleberry (*Lyonia* spp.), mock azalea (*Monziesia* spp.), and Labrador tea (*Ledum* spp.) are both wild and cultivated plants of North America. All are poisonous to animals that eat them.[119-128] The honey produced by bees feeding on the nectar of rhododendrons is also poisonous to people eating it.[129,130]

Principal Toxin

Members of the *Ericaceae* (Heath) family contain grayanotoxins (andromedotoxin, deacetylandromedol, deacetylanhydroandromedol) that are water-soluble diterpenoid compounds. All parts of the plant, including the nectar of the flowers, contain the toxins.

Animals are most often poisoned during the winter because rhododendrons retain their green leaves year round in milder climates. As little as 0.2 percent of an animal's body weight of green leaves can cause poisoning. Cattle, sheep, goats, occasionally horses, and rarely other animals and birds have been poisoned by members of the *Ericaceae*.[121,123,125,131] Goats seem to be particularly susceptible to poisoning by rhododendrons

Grayanotoxins act by binding to cell membranes, thereby affecting sodium channels and causing prolonged depolarization of cells. The primary effects are on the heart, nervous system, and gastrointestinal tract.[132-134] A glycoside, arbutin, present in the plants may contribute to the toxicity of the Ericaceae.

Clinical Signs

Animals poisoned by rhododendrons initially have clinical signs of digestive disturbances characterized by anorexia, excessive salivation, vomiting, colic, and frequent defecation.[119-128] In severe cases, muscle weakness, bradycardia, cardiac arrhythmia, weakness, paralysis, and coma may precede death. Regurgitation of rumen contents may result in inhalation pneumonia. Fetal mummification has been reported in goats following severe Japanese pieris poisoning.[135] Depression, vomiting, slow erratic heart rate, painful neck, and weakness are reported in people who have consumed "mad honey" made by bees feeding on rhododendrons or who have consumed tea made from the leaves of rhododendrons.[124,130,135-137]

Diagnosis

A diagnosis of rhododendron and laurel poisoning is usually based on the clinical signs and evidence that the plant has been consumed. Postmortem findings are not specific and generally consist of multiple hemorrhages on internal organs. The detection of grayanotoxins in the rumen contents is also possible and is a means of confirming rhododendron poisoning.[137]

Treatment

Animals should be removed from the source of the toxic plants and given supportive therapy. Osmotic laxatives such as magnesium sulfate and activated charcoal may be useful early in the course of poisoning to reduce further intestinal absorption of the toxins. Oral and intravenous fluids should be given as necessary to counteract the effects of vomiting and diarrhea. If cattle, sheep, or goats are observed eating significant quantities of rhododendron or other related plants containing grayanotoxins, a rumenotomy may prove life-saving to remove the rhododendron leaves and prevent further absorption of the toxins. If severe bradycardia is present, atropine to increase the heart rate is indicated.

Rhododendron (Azalea)
Rhododendron spp.
Ericaceae (Heath family)

Habitat and Description

There are at least 250 species of rhododendrons found mostly in the acidic soils of western and eastern North America. Many hybrids have been developed for their showy flowers. Generally they are large shrubs or open trees growing to heights of 30 feet (10 meters) or more. The leaves are alternate, simple, leathery, lanceolate, and often evergreen. The flowers are produced in large, showy, terminal clusters, ranging in color from white to purple, to red (Figures 3-37A and 3-37B). The fruits are elongated capsules that split into five sections to release the small, scalelike seeds.

Azaleas are considered by some to be a subgenus of rhododendron (Figure 3-37C). Azaleas are generally deciduous and have been extensively hybridized to produce showy garden and house plants in a wide spectrum of colors.

Principal Toxin

All parts of the plant including the nectar contain grayanotoxins. Most poisoning occurs in the winter months because the leaves are generally evergreen and are attractive to animals when other forages are unavailable. Animals eating approximately 0.2 percent of their body weight of leaves are likely to develop signs of poisoning.

Figure 3-37B Great laurel (*Rhododendron maximum*).

Figure 3-37A Catawba rhododendron (*Rhododendron catawbiense*).

Figure 3-37C Flame azalea (*Rhododendron calandulaceum*).

Mountain Laurel
Kalmia latifolia
Ericaceae (Heath family)

Habitat and Description

Laurels are common to the eastern and southern areas of North America. They are common branching shrubs or small trees with glossy green, alternate, lanceolate leaves. The characteristic white to pink flowers are produced in showy clusters (Figures 3-38A and 3-38B).

Figure 3-38A Mountain laurel blooming shrub (*Kalmia latifolia*).

Principal Toxin

Grayanotoxins (andromedotoxin) and a glycoside arbutin are the principal toxins present in all parts of the plant. Similar toxins are also present in the genera *Rhododendron* (azalea), *Leucothoe* (fetter-bush), *Pieris* (mountain fetter-bush), and *Lyonia* (maleberry). The principal actions of the toxin are gastrointestinal irritation and disruption of myocardial activity.

Figure 3-38B Mountain laurel flowers (*Kalmia latifolia*) .

Clinical Signs

All animals are susceptible to laurel poisoning. Affected animals may show excessive green frothy salivation, vomiting, colic, frequent defecation, depression, weakness, and ataxia. Depending on the quantity of laurel that has been eaten, affected animals may become recumbent and comatose before death.

Treatment

Mineral oil via nasogastric tube and intravenous fluid therapy should be administered as necessary.

Japanese Pieris
Pieris japonica
Ericaceae (Heath family)

Habitat and Description

Introduced from Japan, *P. japonica* is grown as an ornamental flowering shrub in the acidic soils of eastern and western North America. Several species of Pieris are indigenous. Leaves are generally alternate, dark green, toothed, and evergreen. New foliage is bronze in color. The terminal flowers are in drooping pannicles and white to pink in color (Figure 3-39).

Principal Toxin

Grayanotoxins (andromedotoxin) are the principal toxins and are present in all parts of the plant.

Figure 3-39 Japanese pieris (*Pieris japonicus*).

Fetter-Bush, Black Laurel
Leucothoe spp.
Ericaceae (Heath family)

Habitat and Description

Several species are grown for their attractive foliage and flowers in the eastern and western regions of North America. Shrubs 4 to 6 feet (1 to 2 meters) in height, with leaves evergreen or deciduous, alternate, and carried on arching branches. Some cultivars have reddish purple leaves. Flowers are white to pink, borne along or at the tips of branches, and with five small teeth at the top of the flower (Figure 3-40).

Figure 3-40 Fetter-bush (*Leucothoe* spp.).

Principal Toxin

Grayanotoxins (andromedotoxin) are the principal toxins in fetter-bush and are present in all parts of the plant.

Maleberry
Lyonia ligustrina
Ericaceae (Heath family)

Description

These shrubs are 2 to 3 feet (0.5 to 1 meter) tall with alternate, deciduous, hairless, elliptical leaves, growing mostly in the southeastern region of North America. Flowers are white, 0.3 to 0.5 cm (3 to 5 mm) in length, in clusters at the ends of branches. Fruits are round capsules (Figure 3-41).

Principal Toxin

Grayanotoxins (andromedotoxin) are present in all parts of the plant.

Figure 3-41 Maleberry flowering branch (*Lyonia ligustrina*).

Common Box
Buxus sempervirens
Buxaceae (Box family)

Habitat

Originally from Europe and Asia, box is widely grown in North America as an ornamental shrub or hedge.

Description

Box is a heavily branched, perennial woody shrub, with dark green, opposite, leathery leaves up to 1.5 inches (4 cm) in length (Figure 3-42). The undersides of the leaves are lighter green or grayish in color. Small star-shaped yellow-green flowers are produced in the leaf axils.

Figure 3-42 Box (*Buxus sempervirens*).

Principal Toxin

All parts of the plant contain toxic alkaloids, the mode of action of which is yet to be defined. Horses, cattle, sheep, pigs, and camels are susceptible to poisoning from *Buxus* spp.[138] Approximately 1.5 lb of green leaves may be lethal to an adult horse. Most poisoning occurs when box clippings are carelessly fed to animals or when box hedges are placed around animal enclosures.

Clinical Signs

Severe gastroenteritis, colic, and hemorrhagic diarrhea can be expected in poisoned animals. In acute poisoning, death results from respiratory failure.[138]

REFERENCES

Traumatic plants

1. Bankowski RA, Wichmann RW, Stuart EE: Stomatitis of cattle and horses due to yellow bristle grass. *J Am Vet Med Assoc* 1956, 129:149–152.

Rhizoctonia

2. Crump MH: Slaframine (slobber factor) toxicosis. *J Am Vet Med Assoc* 1973, 163:1300–1302.

3. Sockett DC, Baker JC, Stowe CM: Slaframine (*Rhizoctonia leguminicola*) intoxication in horses. *J Am Vet Med Assoc* 1982, 181:606.

Buttercups (*Ranunculus* spp.)

4. Kingsbury JM: *Poisonous Plants of the United States and Canada.* Englewood Cliffs, NJ: Prentice Hall; 1964:140–145.

5. Turner NJ: Counter-irritant and other medicinal uses of plants in *Ranunculaceae* by native peoples in British Columbia and neighboring areas. *J Ethnopharmacol* 1984, 11:181–201.

6. Nachman NJ, Olsen JD: Ranunculin: a toxic constituent of the poisonous range plant bur buttercup (*Ceratocephalus testiculatus*). *J Am Agric Food Chem* 1983, 31:1358–1560.

7. Bai Y, Benn MH, Majak W, McDiarmid R: Extraction and HPLC determination of ranunculin in species of the buttercup family. *J Agric Food Chem* 1996, 44:2235–2238.

8. Winters JB: Severe urticarial reaction in a dog following ingestion of tall field buttercup. *Vet Med Small Anim Clinician* 1976, 71:307.

9. Kelch WJ, Kerr LA, Adair HS, Boyd JD: Suspected buttercup (*Ranunculus bulbosus*) toxicosis with secondary photosensitization in a Charolais heifer. *Vet Hum Toxicol* 1992, 34:238–239.

10. Olsen JD, Anderson TE, Murphy JC, Madsen G: Bur buttercup poisoning of sheep. *J Am Vet Med Assoc* 1983, 183:538–543.

11. Therrien HP, Hidiroglou, Charette LA: The toxicity of tall buttercup (*Ranunculus picris*) to cattle. *Can J Anim Sci* 1962, 42:123–124.

12. Hidiroglou M, Knutti HJ: The effects of green tall buttercup in roughage on the growth and health of beef cattle and sheep. *Can J Anim Sci* 1963, 43:68–71.

Tungnut (*Aleurities fordii*)

13. Kingsbury JM: *Poisonous Plants of the United States and Canada.* Englewood Cliffs, NJ: Prentice-Hall; 1964:182–184.

14. Emmel MW: The toxic principle of *Aleurities fordii* Hemsl. *J Am Vet Med Assoc* 1943, 103:1629.

15. Emmel MW: The toxic principle of the species Aleurities. *J Am Vet Med Assoc* 1947, 111:386.

16. Lin TJ, Hsu CI, Lee KH, *et al.*: Two outbreaks of acute tung nut (*Aleurities fordii*) poisoning. *J Toxicol* 1996, 34;87–92.

17. Emmel MW, Sanders DA, Swanson LE: The toxicity of the foliage of *Aleurities fordii* for cattle. *J Am Vet Med Assoc* 1943, 101:136.

Chinaberry (*Melia azadarach*)

18. Kingsbury JM: *Poisonous Plants of the United States and Canada.* Englewood Cliffs, NJ: Prentice-Hall; 1964:206–208.

19. Morton JF: *Plants Poisonous to People in Florida and Other Warm Places.* Stuart, FL:

Southeastern Printing; 1982:35–36.

20. Hare WR, Schutzman H, Lee BR, Knight MW: Chinaberry poisoning in two dogs. *J Am Vet Med Assoc* 1997, 210:1638-1640.

21. Oelrichs PB, Hill MW, Valley PJ. The chemistry and pathology of meliatoxins A and B constituents from the fruit of *Melia azedarach L.* var. *australasica. In* Plant Toxicology. Seawright AA, Hegarty MP, James LF *et al*, eds. Yeerongipilly, Australia 1985:387-394.

Pokeweed (*Phytolacca americana*)

22. Peixoto PV, Wouters F, Lemos RA, Loretti AP: *Phytolacca decandra* poisoning in sheep in southern Brazil. *Vet Hum Toxicol* 1997, 39:302–303.

23. Kingsbury JM: *Poisonous Plants of the United States and Canada.* Englewood Cliffs, NJ: Prentice-Hall; 1964:225-227.

24. Kingsbury JM, Hillman RB: Pokeweed (*Phytolacca*) poisoning in a dairy herd. *Cornell Vet* 1965, 55:534–538.

25. Storie GJ, McKenzie RA, Fraser IR: Suspected packalacca (*Phytolacca dioica*) poisoning in cattle and chickens. *Aust Vet J* 1992, 69:21–22.

26. Smith BP: *Large Animal Internal Medicine.* St. Louis: CV Mosby; 1990:1657–1658.

Corn cockle (*Agrostemma githago*)

27. Kingsbury JM: *Poisonous Plants of the United States and Canada.* Englewood Cliffs, NJ: Prentice-Hall; 1964:245–246.

Saponaria

28. Kingsbury JM: *Poisonous Plants of the United States and Canada.* Englewood Cliffs, NJ: Prentice Hall; 1964:249–250.

Sesbania

29. Kingsbury JM: *Poisonous Plants of the United States and Canada.* Englewood Cliffs, NJ: Prentice-Hall; 1964:353–357.

30. Kellerman TS, Coetzer JAW, Naude TW, Eds. *Plant Poisonings and Mycotoxicoses of Livestock in Southern Africa.* Cape Town, South Africa: Oxford University Press; 1988:151–152.

Mesquite (*Prosopsis* spp.)

31. Radostits OM, Blood DC, Gay CC, Eds: *Veterinary Medicine,* edn 8. Philadelphia: Bailliere Tindall; 1994:1533.

32. Tabosa IM, Souza IC, Graca DL, *et al*: Neuronal vacudation of the trigeminalnuclei in goats caused by ingestion of *Prosopis juliflora* pods (mesquite beans). *Vet Human Toxicol* 2000, 42:155-158.

Persimmon (*Diospyros* spp.)

33. Morgan SE, Bellamy J: Persimmon colic in a mare. *Equine Practice* 1994, 16:8–10.

34. Cummings CA, Copedge KJ, Confer AW: Equine gastric impaction, ulceration, and perforation due to persimmon (*Diospyros virginiana*) ingestion. *J Vet Diagn Invest* 1997, 9:311–313.

35. Kellam LL, Johnson PJ, Kramer J, Keegan KG: Gastric impaction and obstruction of the small intestine associated with persimmon phytobezoar in a horse. *J Am Vet Med Assoc* 2000, 216:1279-1281.

Sneeze and Bitterweeds

36. Rowe LD, Dollahite JW, Kim HL, Camp BJ: Hymenoxys odorata poisoning in

sheep. *Southwestern Vet* 1973, 26:287–293.

37. Aanes WA: Pingue (*Hymenoxys richardsonii*) poisoning in sheep. *Am J Vet Res* 1961, 22:47–52.

38. Herz W: Sesquiterpene lactones from livestock poisons. *In* Effects of Poisonous Plants on Livestock. Keeler RF, Van Kampen KR, James LF, Eds. Academic Press, New York 1078: 487-497.

39. Marsh CD, Clawson AB, Couch JF, Marsh H: Western sneezeweed (*Helenium hoopesii*) as a poisonous plant. *USDA Bulletin* 1921, 947:1–46.

40. Kingsbury JM: *Poisonous Plants of the United States and Canada.* Englewood Cliffs, NJ: Prentice-Hall; 1964:409–417.

41. Ivie GW, Witzel DA, Rushing DD: Toxicity and milk bittering properties of tenulin, the major sesquiterpene lactone of *Helenium amarum* (bitter sneezeweed). *J Agric Food Chem* 1975, 23:845-849.

42. Hill DW, Kim HL, Martin CL, Camp BJ: Identification of hymenoxon in *Baileya multiradiata* and *Helenium hoopesii. J Agric Food Chem* 1977, 25:1304–1307.

43. Ueckert DN, Calhoun MC: Ecology and toxicology of bitterweed (*Hymenoxys odorata*). *In* The Ecology and Economic Impact of Poisonous Plants on Livestock Production. James LF, Ralphs MH, Nielsen DB, Eds. Boulder, CO: Westview Press; 1988:131–134.

44. Witzel DA, Rowe LD, Clark DE: Physiopathologic studies in acute *Hymenoxys odorata* (bitterweed) poisoning in sheep. *Am J Vet Res* 1974, 35:931–934.

45. Calhoun MC, Ueckert DN, Livingston CW, Baldwin BC: Effects of bitterweed (*Hymenoxys odorata*) on voluntary feed intake and serum constituents of sheep. *Am J Vet Res* 1981, 42:1713–1717.

46. Kim HL, Anderson AC, Herrig BW, *et al.:* Protective effects of antioxidants on bitterweed (*Hymenoxys odorata*) toxicity in sheep. *Am J Vet Res* 1982, 43:1945–1950.

47. Rowe LD, Kim HL, Camp BJ: The antagonistic effect of L-cysteine in experimental hymenoxon intoxication in sheep. *Am J Vet Res* 1980, 41:484–486.

48. Witzel DA, Jones LP, Ivie GW: Pathology of acute bitterweed (*Hymenoxys odorata*) poisoning in sheep. *Vet Pathol* 1977, 14:73–78.

49. Taylor CA, Ralphs MH: Reducing livestock losses from poisonous plants through grazing management. *J Range Manage* 1992, 45:9–12.

Bindweed (*Convolvulus* spp.)

50. Todd FG, Stermitz FR, Schultheis P, *et al.:* Tropane alkaloids and toxicity of *Convolvulus arvensis. Phytochem* 1995, 39:301–303.

51. Schultheis PC, Knight AP, Traub-Dargatz JL, Stermitz FR: Toxicity of field bindweed (*Convolvulus arvensis*) to mice. *Vet Hum Toxicol* 1995, 37:452–454.

52. Molyneux RJ, Pan YT, Goldmann A, Tepfer DA, Elbien AD: Calystegins, a novel class of alkaloid glycosidase inhibitors. *Arch Biochem BioPhysics* 1993, 304:81-88.

Nightshades (*Solanaceae*)

53. Mercatante A: *The Magic Garden: The Myth and Folklore of Flowers, Plants, Trees, and Herbs.* New York: Harper & Row; 1976.

54. Kingsbury JM: *Poisonous Plants of the United States and Canada.* Englewood Cliffs, NJ: Prentice-Hall; 1964:275–294.

55. Case AA: Nightshade poisoning. *Southwestern Vet* 1956, 9:140–143.

56. Hansen AA: Potato poisoning. *North Am Vet* 1928, 9;31–34.

57. Nicholson JWG, Young DA, McQueen RE, *et al.:* The feeding value potential of potato vines. *Can J Anim Sci* 1978, 58:559–569.

58. Dalvi RR, Bowie WC: Toxicology of solanine: an overview. *Vet Hum Toxicol* 1983, 25:13–15.

59. Keeler RF, Baker DC, Gaffield W: Solanum alkaloids. *In* Mycotoxins and Phytoalexins. Sharma RP, Salunkhe DK, Eds. Boca Raton, FL: CRC Press; 1991:607–636.

60. Plumlee KH, Holstege DM, Blanchard PC, *et al.: Nicotiana glauca* toxicosis of cattle. *J Vet Diagn Invest* 1993, 5:498–499.

61. Bush LP, Crowe MW: Nicotiana alkaloids. *In* Toxicants of Plant Origin Volume 1. Cheeke PR, Ed. Boca Raton, FL: CRC Press; 1989:87–107.

62. Panter KP, Keeler RF, James LF, Bunch TD: Impact of plant toxins on fetal and neonatal development: a review. *J Range Manage* 1992, 45:52–57.

63. Riet-Correa F, Mendez MDC, Schield AL, *et al.:* Intoxication by *Solanum fastigiatum* var. *fastigiatum* as a cause of cerebella degeneration in cattle. *Cornell Vet* 1983, 73;240–256.

64. Kellerman TS, Coetzer JAW, Naude TW, Eds: *Plant Poisonings and Mycotoxicoses of Livestock in Southern Africa.* Cape Town, South Africa: Oxford University Press; 1988:64–67.

65. Menzies JS, Bridges CH, Bailey EM: A neurologic disease of cattle associated with *Solanum dimidiatum.* *Southwest Vet* 1979, 32:45–49.

66. Dunster PJ, McKenzie RA: Does *Solanum esuriale* cause humpy-back in sheep? *Aust Vet J* 1987, 64:119–120.

67. Bourke CA: Cerebella degeneration in goats grazing *Solanum cinereum* (*Narrawa burr*). *Aust Vet J* 1997, 75:363–365.

68. Porter MB, MacKay RJ, Uhl E, *et al.:* Neurologic syndrome in goats associated with tropical soda apple. *Proceedings of the 17th American College of Veterinary Internal Medicine.* 1999:222–223.

69. Wasserman RH: The nature and mechanism of action of the calcinogenic principle of *Solanum malacoxylon* and *Cestrum diurnum,* and a comment on *Trisetum flavescens.* *In* Effects of Poisonous Plants on Livestock. Keeler RF, Van Kampen KR, James L, Eds. New York: Academic Press; 1978:545–553.

70. Schulman ML, Bolton LA: Datura seed intoxication in two horses. *J S Afr Vet Assoc* 1998, 69:27-29.

71. Buck WB, Dollahite JW, Allen TJ: *Solanum elaeagnifolium,* silver-leafed nightshade, poisoning in livestock. *J Am Vet Med Assoc* 1960, 137:348–351.

72. Garland T, Bailey EM, Reagor JC, Binford E: Probable interaction between *Solanum elaeagnifolium* and ivermectin in horses. *In* Toxic Plants and Other Natural Toxicants. Garland T, Barr AC, Eds. CAB International, New York. 1998, 423-427.

73. Ceha LJ, Presperin C, Young E, *et al.:* Anticholinergic toxicity from nightshade berry poisoning responsive to physostigmine. *J Emerg Med* 1997, 15:65–69.

74. Friedman M, Levin CE: Composition of jimson weed (*Datura stramonium*) seeds. *J Agric Food Chem* 1989, 37:998–1005.

75. Day EJ, Dilworth BC: Toxicity of jimson weed seed and cocoa shell meal to broilers. *Poult Sci* 1984, 63:466–468.

76. Nelson PD, Mercer HD, Essig HW, Minyard JP: Jimson weed toxicity in cattle. *Vet Hum Toxicol* 1982, 24;321–325.

77. Worthington TR, Nelson EP, Bryant MJ: Toxicity of thorn apple (*Datura stramonium*) seeds to the pig. *Vet Rec* 1981, 108:208–211.

78. Guharoy SR, Barajas M: Atropine intoxication from the ingestion and smoking of jimson weed (*Datura stramonium*). *Vet Hum Toxicol* 1991, 33:588–589.

79. Gowdy JM: Stramonium intoxication: a review of symptomatology in 212 cases. *JAMA* 1972, 221:585–587.

80. Greene GS, Patterson SG, Warner E: Ingestion of angel's trumpet: an increasingly common source of toxicity. *South Med J* 1996, 89;365–369.

81. El Dirdiri NI, Wasfi IA, Adam SE, Edds GT: Toxicity of *Datura stramonium* to sheep and goats. *Vet Hum Toxicol* 1981, 23:241–246.

82. Rodgers GC, Von Kanel RL: Conservative treatment of jimson weed poisoning. *Vet Hum Toxicol* 1993,35:32–33.

83. Leipold HW, Oehme FW, Cook JE: Congenital arthrogryposis associated with the ingestion of jimson weed by pregnant sows. *J Am Vet Med Assoc* 1973, 162:1059–1060.

84. Keeler RF: Absence of arthrogryposis in newborn Hampshire pigs from sows ingesting toxic levels of jimson weed during gestation. *Vet Hum Toxicol* 1981, 23:413–415.

Spurges (*Esula* spp.)

85. Kronberg SL, Muntiferring RB, Ayers EL, Marlow CB: Cattle avoidance of leafy spurge: A case of conditioned aversion. *J Range Manage* 1993, 46:364–366.

86. Kronberg SL, Lynch WC, Cheney CD, Walker JW: Potential aversive compounds in leafy spurge for ruminants and rats. *J Chem Ecol* 1995, 21:1387–1399.

87. Heemstra JM, Kronberg SL, Neiger RD, Pruitt RJ: Behavioral, nutritional, and toxicological responses of cattle to ensiled leafy spurge. *J Anim Sci* 1999, 77:600–610.

88. Walker JW, Kronberg SL, Al-Rowaily SL, West NE: Comparison of sheep and goat preferences for leafy spurge. *J Agric Food Chem* 1994, 27: 429–434.

89. Williams KE, Lacey JR, Olson BE: Economic feasibility of grazing sheep on leafy spurge-infested rangeland in Montana. *J Range Manage* 1996, 49:372–374.

90. Landgraf BK, Fay PK, Havstad KM: Utilization of leafy spurge (*Euphorbia esula*) by sheep. *Weed Sci* 1984, 32:348–352.

91. Olson BE, Wallander RT, Kott RW: Recovery of leafy spurge seed from sheep. *J Range Manage* 1997, 50:10–15.

Lectins

92. Balint GA: Ricin: the toxic protein of castor oil seeds. *Toxicology* 1974, 2:77–102.

93. Rauber A, Heard J: Castor bean toxicity re-examined: a new perspective. *Vet Hum Toxicol* 1985, 27:498–502.

94. Wedin GP, Neal JS, Everson GW, Krenzolok EP: Castor bean poisoning. *Am J Emerg Med* 1986, 4:259–261.

95. Kinamore PA, Jaegrer RW, de Castro FJ: Abrus and ricinus ingestion: management of 3 cases. *Clin Toxicol* 1980, 17:401–405.

96. Barri M, Dirdiri NE, Damir HA, Idris OF: Toxicity of *Abrus precatorius* in Nubian goats. *Vet Hum Toxicol* 1990, 32:541–545.

97. Adam SEI: Toxicity of indigenous plants and agricultural chemicals in farm animals. *Clin Toxicol* 1979, 13:269–280.

98. Jacob DP, Peter CT: Toxicity of *Abrus precatorius* seeds in domestic fowl. *Kerala J Vet Sci* 1970, 1:125–127.

99. Rahman A, Mia AS: *Abrus precatorius* poisoning in cattle. *Ind Vet J* 1972, 49:1045–1049.

100. Albertson JC, Gwaltney-Brant SM, Khan SA: Evaluation of Castor bean toxico-

sis in dogs: 98 cases. *J Am Animal Hosp Assoc* 2000, 36:229-233

101.   Fox MW: Castor seed residue poisoning in dairy cattle. *Vet Rec 1961*, 73:885.

102.   Kingsbury JM: *Poisonous Plants of the United States and Canada.* Englewood Cliffs, NJ: Prentice-Hall; 1964:194–197.

103.   Duke JA: *Handbook of Medicinal Herbs.* Boca Raton, FL: CRC Press; 1988:408.

104.   Fodstad O, Johannessen JV, Schjerven L, Pihl A: Toxicity of abrin and ricin in mice and dogs. *J Toxicol Environ Health* 1979, 5:1073–1084.

105.   Olsnes S, Refsnes K, Pihl A: Mechanism of action of the toxic lectins abrin and ricin. *Nature* 1974, 249:627–631.

106.   Kanerva L, Estlander T, Jolanki R: Long lasting contact urticaria from castor bean. *J Am Acad Dermatol* 1990, 23:351–355.

107.   Mercier P, Panzani R: Human castor bean allergy and HLA-A,B,C,DR. *J Asthma* 1988, 25:153–161.

108.   Lin JY, Liu SY: Studies on the antitumor lectins isolated from the seeds of *Ricinus communis* (castor bean). *Toxicon* 1986, 24:757–765.

109.   Clarke EGC: Poisoning by castor seed. *Vet J* 1947, 103:273–278.

110.   Kellerman TS, Coetzer JAW, Naude TW, eds: *Plant Poisonings and Mycotoxicoses of Livestock in Southern Africa.* Cape Town, South Africa: Oxford University Press; 1988:144–146.

111.   Jensen WI, Allen JP: Naturally occurring and experimentally induced castor bean (*Ricinus communis*) poisoning in ducks. *Avian Dis* 1979, 25:184–194.

112.   Tokarnia CH, Dobereiner J, Canella CFC: Experimental poisoning by the leaves of *Ricinus communis* in cattle. *Pesq Agropec Bras Ser Vet* 1975, 10:1–7.

113.   Burrows GE, Edwards WC, Tyrl RJ: Toxic plants in Oklahoma–oak, black locust, Kentucky coffee tree, and buckeye. *Okla Vet Med Assoc* 1981, 33:37–39.

114.   Challoner KR, McCarron MM: Castor bean intoxication. *Ann Emerg Med* 1990, 19:1177–1183.

115.   Clarke EGC, Jackson JH: The use of immune serum in the treatment of ricin poisoning. *Br Vet J* 1956, 112:57–62.

Mayapple (*Podophyllum peltatum*)

116.   Duke JA, Ed. *Handbook of Medicinal Herbs.* Boca Raton, FL: CRC Press; 1988:387–388.

117.   Kingsbury JM: *Poisonous Plants of the United States and Canada.* Englewood Cliffs, NJ: Prentice Hall; 1964:145–146.

Privet (*Ligustrum vulgare*)

118.   Kingsbury JM: *Poisonous Plants of the United States and Canada.* Englewood Cliffs, NJ; Prentice-Hall; 1964:261–262.

Rhododendron

119.   Marsh CD, Clawson AB: Mountain laurel (*Kalmia latifolia*) and sheep laurel (*Kalmia angustifolia*) as stock poisoning plants. *USDA Bulletin* 1930, 219:1.

120.   Clawson AB: Alpine kalmia (*Kalmia microphylla*) as a stock-poisoning plant. *USDA Bulletin* 1933, 391:1–9.

121.   Kingsbury JM: *Poisonous Plants of the United States and Canada.* Englewood Cliffs, NJ: Prentice-Hall; 1964:258–259.

122.   Fowler ME. Plant poisoning in two pack llamas. *California Vet* 1985, 39:17–19.

123. Smith MC: Japanese pieris poisoning in the goat. *J Am Vet Med Assoc* 1978, 173:78–79.

124. Meier KH, Hemmick RS: Bradycardia and complete heart block after ingestion of rhododendron tea. *Vet Hum Toxicol* 1992, 34:351.

125. Higgins RJ, Hannam DAR, Humphreys DJ, Stodulski JBJ: Rhododendron poisoning in sheep. *Vet Rec* 1985, 116:294–295.

126. Miller RM: Azalea poisoning in a llama: a case report. *Vet Med/Small Anim Clinician* 1981, 76:104.

127. Frape D, Ward A: Suspected rhododendron poisoning in dogs. *Vet Rec* 1993, 132:515–516.

128. Hough I: Rhododendron poisoning in a western gray kangaroo. *Aust Vet J* 1997, 75:174–175.

129. Onat F, Yegen BC, Lawrence R, *et al.*: Site of action of grayanotoxins in mad honey in rats. *J Appl Toxicol* 1991, 11:119–201.

130. Lampe KF: Rhododendrons, mountain laurel, and mad honey. *JAMA* 1988, 259:2009

131. Labonde J: Avian toxicology. *Vet Clin North Am Small Anim Pract* 1991, 21:1329–1342.

132. Hikino H, Ohizumi Y, Konno C, *et al.*: Subchronic toxicity of ericaceous toxins and rhododendron leaves. *Chem Pharm Bull* 1979, 67:125–129.

133. Moran NC, Dresel PE, Perkins ME, Richardson AP: The pharmacological action of andromedotoxin, an active principle from *Rhododendron maximum*. *J Pharm Exp Therapeutics* 1954, 110:415.

134. Narahashi T: Toxins that modulate the sodium channel gating mechanism. *Ann N Y Acad Sci* 1986, 479:133–151.

135. Smith MC: Fetal mummification in a goat due to Japanese pieris (*Pieris japonicus*) poisoning. *Cornell Vet* 1979, 69:85–87.

136. Gossinger H, Hruby K, Haubenstock A, *et al.*: Cardiac arrhythmias in a patient with grayanotoxin-honey poisoning. *Vet Hum Toxicol* 1983, 25:328–329.

137. Humphreys DJ, Stodulski JB: Detection of andromedotoxins for the diagnosis of rhododendron poisoning in lambs. *J Appl Toxicol* 1986, 6:121–122.

138. Kingsbury JM: *Poisonous Plants of the United States and Canada*. Englewood Cliffs, NJ: Prentice-Hall; 1964:197–198.

CHAPTER 4

Plants Affecting the Skin and Liver

Plants that affect the skin and liver are considered together in this chapter because some plant toxins cause liver disease that results in secondary skin disease. One of the most noticeable signs indicative of an animal with liver disease is the appearance of dermatitis and hair loss affecting the white-skinned areas. Some plants contain compounds or pigments that once absorbed from the digestive system induce a direct effect on nonpigmented skin when it is exposed to light, without any effect on the liver. Other plants contain toxic alkaloids that cause irreversible liver disease resulting in a secondary photosensitization. Yet other plants have compounds that cause a skin reaction when they come in contact with skin, or they have spines or thorns that can cause mechanical injury to the skin.

Photosensitization

Photosensitization, resembling but distinct from sunburn, is a severe dermatitis of animals resulting from a complex reaction induced by plant pigments exposed to ultraviolet (UV) wave length sunlight in the skin of animals that have eaten certain plants[1-3]. This reaction is most severe in nonpigmented skin where these reactive compounds are most directly exposed to light in the UV spectrum.[2] The precise mechanism of this reaction is unknown, but it is thought to be a light-enhanced oxidation reaction.[2] The amino acids (histidine, tyrosine, tryptophan) are particularly susceptible to oxidation and once oxidized evoke an intense inflammatory response in the blood vessels and surrounding cells that results in tissue necrosis. In addition to plant pigments, fungal toxins, chemicals, and occasionally congenital diseases affecting porphyrin metabolism in the liver may induce photosensitization.[2] Quite frequently horses and cattle develop photosensitization while on pasture with no determinable cause.

Photosensitization may be conveniently classified into two basic types—primary and secondary. Primary photosensitization is associated with photodynamic compounds in certain plants, which once absorbed from the digestive tract, react in the nonpigmented with UV light to cause a severe dermatitis. Also in this category are the congenital photosensitivity diseases associated with defective pigment (porphyrins) metabolism in the liver of animals. Secondary or hepatogenous photosensitization, as the name implies, results when an animal's liver is sufficiently diseased to be unable to remove plant by-products that can react with UV light to cause photosensitization. Phylloerythrin, a bacterial breakdown product of chlorophyll, is the photosensitizing compound.[2] Normally phylloerythrin is removed by the liver and is excreted in the bile, but if the liver is severely diseased, it accumulates in the blood to cause photosensitization if a white skinned animal is exposed to UV light. Hepatogenous photosensitization can be further subdivided into that attributable to liver disease as opposed to that caused by biliary system disease that causes a back-up of bile. Secondary photosensitization is much more common in livestock than primary photosensitization, and because of the severity of the underlying liver disease, it always carries a poor prognosis.

Occasionally photoreactive pigments (porphyrins), produced in animals as a result of normal hemoglobin breakdown, accumulate and cause photosensitization.[4] Congenital porphyria is an inherited defect in various breeds of cattle as a result of a specific enzyme deficiency that normally regulates metabolism of porphyrins.[5] Southdown sheep may also develop photosensitivity due to a congenital defect in the liver's ability to excrete the photoreactive compound phylloerythrin.[6] As it accumulates in the skin, phylloerythrin causes photosensitivity when the animal is exposed to sunlight. Chemicals such as phenothiazine sulfoxide, a derivative of the anthelmintic phenothiazine, may also produce photosensitivity in ruminants if they are exposed to sunlight after treatment with phenothiazine.[2]

Primary Photosensitization

Primary photosensitization develops when animals eat plants containing polyphenolic pigments. These compounds are at highest concentration in the green plant and are readily absorbed from the gastrointestinal tract to circulate in the blood. In nonpigmented skin these compounds react with UV light to produce radiant energy that oxidizes essential amino acids in the skin's cells. The cells die in the photosensitization process, and the affected skin eventually sloughs off. Two plants associated historically with primary photosensitization are buckwheat (*Fagopyrum esculentum*),[1] and St. John's wort (*Hypericum perforatum*).[8-10] Both plants contain polyphenolic pigments capable of causing primary photosensitization. Several plant species including bishop's weed (*Ammi majus*), spring parsley (*Cymopterus watsonii*), and Dutchman's breeches (*Thamnosma texana*) contain photodynamic furanocoumarin compounds that have been associated with photosensitivity through ingestion and direct contact with the skin[11-13] (Table 4-1). In southeast Texas, a seasonal photosensitivity of cattle is associated with the consumption of the dead leaves of *Cooperia pedunculata*, a lily of the Amaryllis

Table 4-1
Primary Photosensitizing Plants

Scientific Name	Common Name
Ammi majus	Bishop's weed, greater ammi
Cooperia pedunculata	Rain lily
Cymopterus watsonii	Spring parsley
Fagopyrum esculentum	Buckwheat
Heracleum mentegazzianum	Giant hog weed
Hypericum perforatum	St. John's wort, Klamath weed
Thamnosma texana	Dutchman's britches

family.[14,15] Photosensitivity has also been reported in Europe as a result of exposure to giant hogweed (*Heracleum mantegazzianum*). Cow parsnip (*Heracleum* spp.), which occurs in North America, has the potential to cause photosensitivity.

The detection of primary photoreactive compounds in plants can be accomplished using a screening test that is based on the sensitivity of the fungus *Candida albicans* to irradiation.[16] The simple procedure involves exposing suspect plant material on agar plates seeded with *C. albicans* to UV light. Photoreactive plants will inhibit the growth of the *C. albicans*.

Secondary Photosensitization

Secondary or hepatogenous photosensitization in animals occurs more commonly than primary photosensitization. Liver disease, the underlying cause of secondary photosensitivity, results from ingestion of plants containing compounds toxic to the liver. A variety of compounds toxic to the liver are found in plants, the most important of which are the pyrrolizidine alkaloids (PAs). Once 80 percent or more of the liver is destroyed by these alkaloids, it is unable to eliminate phylloerythrin, a bacterial breakdown product of chlorophyll. Phylloerythrin then accumulates in the blood, and as it circulates through the skin and is exposed to UV light, it fluoresces and causes oxidative injury to the blood vessels and tissues of the skin.[2,7] The resulting intense inflammatory response is most severe in the nonpigmented skin. In severe cases of PA poisoning, acute liver failure and death may result before signs of photosensitization have time to develop.

Secondary photosensitization is also caused by a variety of plant toxins other than pyrrolizidine alkaloids (see Table 4-3). In the western range lands of North America, photosensitization in sheep (bighead) has for many years been attributed to the grazing of horsebrush (*Tetradymia glabrata* and *T. canescens*)[17,18] It has been shown, however, that sheep are much more susceptible to horsebrush photosensitivity if they concurrently browse on black sage (*Artemisia nigra*) or big sage (*A. tridentata*) or both.[17,18] Horsebrush and black and big sage frequently grow in the same locations in western rangelands, and when eaten together have a synergistic effect in causing photosensitivity. These plants including *Lantana camara*, *Agave lecheguilla*, *Tribulus terrestris*, and *Panicum* grass species cause secondary photosensitization through inflammation and obstruction of the biliary system.

Pyrrolizidine Alkaloid Poisoning

Pyrrolidine alkaloids are found in many species of the three major plant families Compositae, Fabaceae, and Boraginaceae[19] (Table 4-2). The important plant genera that cause liver disease and secondary photosensitization in North America are *Senecio*, *Crotolaria*, *Cynoglossum*, and *Amsinckia*. The most notorious PA-containing plants are those belonging to the genus *Senecio*, long known to cause liver disease in man and animals that eat them. Most PA poisoning of livestock in the western United States is attributable to three species of *Senecio*: tansy ragwort (*S. jacobaea*), threadleaf or wooly groundsel (*S. douglasii* var. *longilobus*),[20,21] and Riddell's groundsel (*S. riddellii*).[22,23] Throughout the world, *Senecio* species are also the most common cause of PA poisoning.[24-28] Severe economic losses to the livestock industry due to *Senecio* spp. poisoning have been estimated to exceed $1.2 million annually in the northwestern Pacific states alone.[29]

Other important plant species contain PAs including *Crotolaria spectablis*, a plant introduced to

Table 4-2 Plants Containing Pyrrolizidine Alkaloids	
SCIENTIFIC NAME	**COMMON NAME**
Amsinckia spp.	Fiddle neck, tarweed
Crotolaria spp.	Rattle box
Cynoglossum officinale	Hound' tongue
Echium vulgare	Blue weed, viper's bugloss
Heliotropium spp.	Heliotrope
Senecio spp.	Groundsels, Senecio
Symphytum officinale	Comfrey

the southeastern United States that is known for the effects of the PA, monocrotaline, found in the seeds. In man, monocrotaline induces severe veno-occlusive disease and pulmonary hypertension by causing hypertrophy of the smooth muscle of the pulmonary arteries.[30] In horses, cattle, pigs, and monkeys, *Crotolaria* causes severe liver and lung disease.[31-36] Fiddleneck (*Amsinckia intermedia*), and hound's tongue (*Cynoglossum officinale*), members of the Boraginaceae, have significant quantities of PA capable of causing secondary photosensitization in horses and cattle.[37-41] In other parts of the world, species of heliotrope (*Heliotropium* spp.), salvation Jane (*Echium plantagineum*), and *Trichodesma* species are important causes of PA poisoning.[42-47] Blue weed or viper's bugloss (*Echium vulgare*) is a common introduced weed of northeastern North America that contains PA and has the potential for causing livestock poisoning.

Human poisoning from PA occurs when people regularly drink teas and remedies containing groundsel (*Senecio* spp.).[48] Gordolobo yerba, a harmless herbal tea, is sometimes mistakenly made from *Senecio longilobus* that contains significant levels of

PA. Consequently people and especially children fed the tea over time develop veno-occlusive liver disease that can be fatal.[49,50] Teas made from comfrey (*Symphytum officinale*) are also potentially toxic because they contain at least three PAs that have been demonstrated to cause hepatogenic carcinoma in rats.[51,52]

A wide range of animal species including wild and domesticated ruminants, horses, and pigs are susceptible to PA poisoning. [27,37,38,53] Pyrrolizidine alkaloids are readily absorbed from the digestive tract and are bioactivated to toxic pyrroles and possibly other reactive metabolites by the liver's mono-oxygenase system.[54-57] It is the active pyrroles that affect the endoplasmic reticulum of the liver cells inhibiting mitosis and the replication of hepatocytes.[54,55] Animals experiencing poor nutrition, pregnancy, and other metabolic stress are more susceptible to PA poisoning.[58] At high doses, PAs cause hepatocellular necrosis, while at lower doses necrosis is less severe allowing time for the characteristic pathologic changes of megalocytosis, bile duct hyperplasia, and fibrosis to occur.[56,60,61] Similar changes may also be seen in the kidneys.[55] In low doses PAs cause endothelial changes in the capillaries of the lungs with resulting pulmonary hypertension and right heart failure.[30,61,62] In the case of monocrotaline found in *Crotolaria* species, the highly toxic metabolite dihydromonocrotaline not only affects the liver, but also the lung and possibly other tissues.[63] Horses develop an acute fatal fibrosing alveolitis after eating feed contaminated with crotolaria seeds.[64]

The PAs have been reported to be carcinogens, teratogens, and abortifactients.[65] Feeding toxic quantities of *S. jacobaea* to pregnant cows through the 15 to 30th days of gestation causes no detectable changes in the fetus, which may indicate PAs are not teratogenic in early gestation.[21] The PAs are secreted in the milk of cows and in very low quantities can cause mild liver changes in calves and kids consuming the milk.[66-70] The effects of PAs on people consuming milk containing PAs have not been established. Bees feeding on tansy ragwort (*S. jacobaea*), and Paterson's curse (*Echium plantagineum*) produce honey containing PAs that is potentially hazardous to those consuming it.[46]

Variation in the PA content of plants, the quantity of plant eaten, and individual animal species susceptibility affect the severity of poisoning seen in animals. The PA content of plants varies considerably, generally increasing with maturation of the plant and reaching a maximum just before the flower buds open.[71] Flowers tend to contain the greatest amount of the alkaloid, and seeds of *Crotolaria* and *Amsinckia* concentrate high levels of PA.[64,66] *Senecio redellii*, when near maturity, has been reported to contain exceptionally high levels of PA (10-18 percent dry weight).[72] The PA content of plants incorporated in hay remain stable for months, but appear to be largely degraded in properly prepared silage.[73] The stability of PAs in plants cured in hay can be a significant cause of poisoning in mid winter when plant toxicity may not be suspected.

Pigs are the most susceptible to PA poisoning, followed by poultry, cattle, horses, goats, and sheep.[59,74] Sheep can eat approximately 20 times the amount of *Senecio* it would take to poison a cow on an equivalent weight basis. This ability of sheep to tolerate greater amounts of PA is attributed to the presence of specialized rumen bacteria that can detoxify the alkaloids before they are absorbed.[76] It appears to be the smallest bacteria and possibly the protozoa in the rumen that are the primary detoxifiers of PAs.[76,77] Cattle and horses are highly susceptible to tansy ragwort (*S. jacobaea*), a lethal dose being 4 to 8 percent (0.05 to 0.2 kg/kg body weight) of their body of green plant.[41,78] Others have produced fatal poisoning in cattle by feeding as little as 0.2 lb dry weight of *S. jacobaea* per day for 4 weeks.[57,71,74,80] Sheep and goats, however, are quite tolerant of PA poisoning, requiring 200 to 300 percent of their body weight in green tansy ragwort to develop fatal poisoning.[10,78,80] Goats will abort, however, if they consume 1 percent of their body weight per day of dry *S. jacobaea*.[81] The chronic lethal dose of *S. jacobaea* in goats ranges from 1.2 to 4.04 kg/kg body weight. The toxicity of hound's tongue (*C. officinale*) varies considerably, ranging from 15 to 360 mg/kg body weight per day of dried plant.[38,41] The PA content of hound's tongue averages 0.8 percent in the early blossom stage.[41] This variation is possibly due to the stage of growth of the hound's tongue, and differences in the age of animals, with young animals being more susceptible to PA toxicity.

Animals will not readily eat plants containing PA unless they are forced to do so through lack of normal forage. Dried plants with only minimal reduction in the alkaloid content are, however, quite palatable, making them a particular risk when present in hay. The effects of the PA are cumulative so that symptoms of liver disease and photosensitization may not appear for many months after animals have eaten toxic quantities of PA. This makes identification of the suspected toxic plants difficult, because the plants will often not be present in the pasture or hay when liver disease and photosensitization become evident in the animal.

Biliary Occlussive Photosensitization

Photosensitivity in livestock has been attributed to ingestion of various plant species other than those containing primary photosensitizing compounds (see Table 4-1) or those causing liver disease as a result of pyrrolizidine alkaloids (see Table 4-2). The principal toxin(s) responsible for photosensitization in this diverse group of plants are not always known (Table 4-3). Some of these photosensitizing plants contain saponins that cause inflammation and obstruction of the biliary system. When bile cannot be excreted normally by the liver, photosensitizing compounds will accumulate in the animal's bloodstream and will result in photosensitization. A prime example of this is facial eczema, a disease of sheep principally in New Zealand and Australia. It is caused by the mycotoxin sporodesmin produced by the fungus *Pithomyces chartarum* growing on wet rye grass pastures. Sheep and cattle eating the affected rye grass excrete the sporodesmin in the bile, which causes a cholangitis and biliary occlusion. This in turn results in the accumulation of phylloerythrin and subsequently photosensitization.[81] Sheep eating puncture vine (*Tribulus terrestris*) develop photosensitivity secondarily to biliary obstruction that was initially thought to be caused by a mycotoxin, but it is now shown to be the result of steroidal

saponins in the plant.[82-84] A similar biliary occlusive photosensitivity occurs in sheep that graze on Kleingrass (*Panicum coloratum*) during hot humid weather.[85,86] Agave poisoning (*Agave lecheguilla*), bear grass (*Nolina texana*) in North America, and alveld, a photosensitivity of sheep in Great Britain and Western Europe due to bog asphodel (*Narthecium ossifragum*) are induced by plant saponins that form crystals in the bile ducts and occlude the biliary system.[88-91] Lantana (*Lantana camara*) also causes photosensitization through cholestasis, effectively preventing the elimination of phylloerythrin through the bile, and thereby causing photosensitization.[90-93]

Alsike Clover Poisoning

Alsike clover (*Trifolium hybridum*) (Figure 4-1) is a perennial legume that is commonly grown for livestock consumption in northern regions of North America. Two disease syndromes encountered in horses have been associated with the grazing of alsike clover. The first is an irre-

Figure 4-1 Alsike clover (*Trifolium hybridum*)

Table 4-3
Plants Associated with Photosensitization[3]

Scientific Name	Common Name
Agave lecheguilla	Agave
Avena sativa	Oats
Brachiaria decumbens	Signal grass
Bassia hysopifolia	Basssia
Brassica spp.	Rape, kale
Cenchrus spp.	Sandbur
Cynodon dactylon	Bermuda grass
Descurainia pinnata	Tansy mustard
Daucus carota	Wild carrot
Euphorbia maculata	Milk purslane
Hordeum spp.	Barley
Kalstroemia	Caltrops
Kochia scoparia	Kochia, Mexican fire weed
Lantana camara	Lantana
Lolium perenne	Perennial rye grass
Medicago sativa	Alfalfa
Microcystis spp.	Blue-green algae, water bloom
Narthecium ossifragum	Bog asphodel
Nolina texana	Sacahuiste
Panicum coloratum	Klein grass
Panicum spp.	Panic grasses
Pastinaca spp.	Parsnip
Psoralea spp.	Scurf pea
Polygonum spp.	Knottweed
Ranunculus bulbosus	Buttercup
Sorghum vulgare sudanensis	Sudan grass
Tetradymia spp.	Horsebrush
Thamnosma texana	Dutchman's breeches
Tribulus terrestris	Puncture vine, caltrop
Trifolium spp.	Clovers

versible liver disease often accompanied by neurologic disturbances and referred to as alsike clover poisoning.[94-97] The second syndrome is one of photosensitivity without apparent concurrent liver disease and is referred to as trifoliosis (dew poisoning).[94,96,98,99] The specific toxins responsible for either of these syndromes have not been determined. A fungal toxin may be involved because alsike clover poisoning is often a problem in years when there is high rainfall and humidity.[100]

Although evidence is insufficient to conclude that alsike clover (*T. hybridum*) is the primary cause of poisoning, a close association exists between the liver disease and photosensitivity syndrome seen in horses and the grazing of the plant.[94] Horses with alsike clover poisoning typically develop signs of liver failure including weight loss, jaundice, depression. Other neurologic abnormalities may develop. Significant elevations of specific liver enzymes are usually present and are indicative of severe liver disease.[96,97] The abnormal neurologic findings in affected horses are possibly due to hepatic encephalopathy caused by the failing liver. This can be substantiated by the elevation of blood ammonia levels that typically occur in advanced liver disease.[96,97] The appearance of photosensitivity can be attributed to the accumulation of phylloerythrin in the horse's circulation as a result of liver failure.[95-97] The primary lesion found in horses with alsike clover poisoning is a grossly enlarged, fibrotic liver, with marked bile duct proliferation and perilobular fibrosis.[95,101]

Mycotoxin Photosensitivity

Photosensitization can develop in animals that eat moldy grains containing hepatotoxic mycotoxins (aflatoxins) produced by fungi belonging to the genera of *Aspergillus* and *Penicillium*.[102] Water-damaged alfalfa hay and moldy straw may also be implicated in photosensitivity of cattle probably as a result of mycotoxins that induce hepatitis and cholangitis.[103-106] Feeding moldy alfalfa hay and silage to cattle has been associated with a hepatogenous photosensitivity thought to be due to liver toxins produced by a variety of fungi cultured from the hay.[107] The unidentified toxins, however, appear to specifically affect cattle and not sheep, goats, horses, or mice fed the same hay, suggesting the toxin may be a rumen or liver metabolite unique to cattle.[108]

Blue-Green Algae Poisoning

Blue-green algae or cyanobacteria may develop in large numbers in stagnant ponds and dams producing a "bloom" or green-blue discoloration of the water. Heavy growth of cyanobacteria usually occurs in late summer when water temperatures are high and the mineral or organic matter present in the water is also elevated. The primary cyanobacteria reported to cause poisoning in North America include members of the genera *Microcystis*, *Anabaena*, and *Aphanizomenon*.[109-111] As these organisms die they release potent toxins into the water that can cause severe poisoning in animals that drink the water.[109-115] The toxins have either primary neurotoxic effects with acute deaths, or they act on the liver to cause liver failure and photosensitization.[110,111]

Clinical Signs of Photosensitization

Photophobia, excessive tearing, and swelling, redness, and increased sensitivity of nonpigmented skin initially characterize photosensitization in animals.[116] Frequently the skin around the lips, eyes, and coronary band of the hooves is most severely affected because it is heavily vascularized and poorly protected by hair or wool. White skin on the face, back, and legs of animals is particularly prone to photosensitization. White breeds of sheep often only develop lesions on the ears and face because of the protective fleece covering unless they have been recently sheered. Cows with nonpigmented udders may develop photosensitization of the teats that may be especially severe in wintertime when light is reflected off the snow. Affected skin rapidly becomes reddened, painful, and raised above areas of adjacent pigmented skin. Serum often oozes through the affected skin to form crusts in the hair. After 2 to 3 weeks, the necrotic skin becomes dry and parchment-like, and the hair and white skin slough leaving ulcerated areas that may develop secondary bacterial infections (Figure 4-2A). Lameness is common in horses, which have photosensitization involving the skin over joints and the coronet (Figure 4-2B).

In cases of secondary photosensitization from PA poisoning, weight loss, abnormal neurologic signs, and typical photosensitization signs may be the first signs of disease noted. Other signs of liver disease including jaundice, abdominal distention due to ascites, diarrhea, tenesmus, and rectal prolapse may be evident.[23,69,117-119] Neurologic signs develop when the severity of the liver disease precludes the liver's ability to remove toxins from the animal's system. The accumulating toxins act on the brain interfering with normal function and causing its degeneration (hepatic encephalopathy). Horses and cattle with hepatic encephalopathy show abnormal behavior that may include yawning, drowsiness, aimless wandering, head press-

Figure 4-2A Generalized photosensitization in a cow.

ing, incessant licking of objects, and terminal coma.[26,116,117] Hemolysis and hemoglobinuria from copper poisoning can occur in sheep and occasionally horses, associated with severe PA-induced liver disease that results in the release of copper stored in the liver.[121] Sheep, however, fed tansy ragwort until they developed severe PA poisoning did not develop copper poisoning, suggesting other mechanisms may be active in sheep.[122] In one instance horses died from an acute hemolytic crisis associated with the ingestion of tansy ragwort and the chewing of wooden fencing impregnated with a copper preservative.[123] Gastric impaction in ponies has been reported as an unusual finding in PA poisoning.[124] The prognosis for animals with secondary photosensitization is always poor because the liver disease is generally irreversible and affected animals eventually die from liver failure.

Figure 4-2B Secondary photosensitization involving the white skinned areas of a horse's legs.

Diagnosis

Dermatitis limited to areas of nonpigmented skin is indicative of photosensitization. Identification of photosensitizing plants in the animal's environment and food substantiates the diagnosis. Removal of the suspected plants from the diet with subsequent recovery of the animal suggests a primary photosensitization. Recovery is not likely to occur in cases of secondary photosensitization especially if due to PA toxicity because the underlying liver disease is usually irreversible, and signs of photosensitivity return as soon as the animal is exposed to sunlight.

The optimal test for the diagnosis of PA hepatotoxicity is a liver biopsy to confirm the presence of megalocytosis, fibrosis, and biliary hyperplasia, the classic signs of PA poisoning.[125] In early cases of poisoning, intranuclear vacuoles, biliary hyperplasia, and fibrosis are more prevalent, megalocytosis being more evident in chronic cases.[126-128] Laboratory tests to determine the presence of liver disease should be performed to confirm a diagnosis of secondary photosensitization. There are, however, no commonly available specific liver tests to confirm PA poisoning, and a profile of tests is necessary to help define the severity and duration of liver disease.[128,129] Elevation of serum enzymes glutamate dehydrogenase and sorbitol dehydrogenase (horses, cattle) are indicative of active liver necrosis due to PA poisoning.[130] Serum glutamate dehydrogenase is the first enzyme to increase and also the first to return to normal following PA poisoning.[126] Serum γ-glutamyltransferase (GGT) has both sensitivity and specificity to serve as a useful screening test for PA-induced liver disease because it remains elevated especially in the presence of bile duct disease.[130,131] Alkaline phosphatase is less specific but its elevation in conjunction with GGT adds to its usefulness as part of a screening profile.[131] Other enzymes and liver function tests are often elevated with PA toxicosis but are nonspecific and vary considerably depending on the stage of liver disease.[132] A decline in the ratio of the sum of serum branched-chain amino acids (leucine, isoleucine, valine) to the sum of phenylalanine and tyrosine is indicative of the severity of liver disease and can be used in the prognosis and treatment of PA poisoning in horses.[133,134] Measurement of serum bile acid levels in horses has been also shown to have predictive value because horses with serum bile acid levels of 50 μmol/L or greater are unlikely to survive.[127]

Detection of metabolites of toxic pyrroles formed in the liver have been shown to alkylate tissue-bound thiol groups forming pyrrole thioethers.[135] These compounds can be extracted and identified by various chromotography methods after the PAs are undetectable, thus providing a means of positively identifying the residual metabolites of pyrrolizidine alkaloids even in formalin-fixed tissues.[27,135-137]

Treatment

Animals showing signs of photosensitization should be provided with shelter from the sun and preferably kept stalled out of the sunlight. Further access to the photosensitizing plant should be prevented. Gentle daily cleaning of the affected skin with a mild antiseptic solution will aid in the healing process. Antibiotics may be indicated in animals with severe secondary bacterial dermatitis.

A variety of different ways have been proposed for treating and preventing PA toxicosis since the toxicity of the alkaloids was first recognized. Few if any have been proven to be effective in cattle and horses.[138] Antioxidants have shown some promise in research studies because they increase liver thiol levels.[138] Administration of cysteine, B vitamins, and synthetic antioxidants may have some protective effect against PA toxicosis, although they do not prevent liver damage due to the alkaloids.[76,133] However, feeding butylated hydroxyanisole, cysteine, and vitamin B had no detectable beneficial effect on PA poisoning in ponies.[139] Mineral and vitamin supplements have similarly failed to prevent tansy ragwort poisoning in cattle.[140]

Symptomatic treatment should include a balanced high-energy, low-protein diet to avoid overloading the liver with nitrogenous compounds, and intravenous polyionic fluids with glucose in severely affected animals. In horses, the dietary supplementation of branched-chain amino acids leucine, isoleucine, and valine may result in clinical improvement by ensuring an adequate level of these essential amino acids.[133,134]

Complete recovery from primary photosensitization usually occurs provided the animal does not return to eating the photosensitizing plant. However, secondary photosensitization due to PA poisoning has a poor prognosis because of the irreversible underlying liver disease. Symptoms of photosensitization usually do not appear until at least 80 percent of the liver is affected by the disease process. Affected animals, therefore, usually die from the effects of liver failure or are euthanized to prevent needless suffering. Plant-induced biliary occlussive photosensitivity cases have a better prognosis and can recover if the toxicity has not persisted for a long time.

Control of Photosensitizing Plants

Eradication of plants containing PAs is not practical on a large scale. Selective cultivation and removal of the plants works well in a limited area. Herbicides are effective when applied to the immature stages of *Senecio* and hound's tongue. Regardless of the control method implemented it is important to do so before heads are produced. Biologic control methods offer possibilities for control in large geographic areas. A beetle (*Chrysolina quadrigemina*) has been successfully used in the past to control St. John's wort. In limited areas of northern California, the larvae of the Cinnabar moth (*Tyria jacobaeae*) in conjunction with a root-feeding flea beetle (*Longitarsus jacobaeae*) have been used successfully to control *S. jacobaea*.[141] It is also feasible to use sheep to graze *Senecio* spp. because they are relatively resistant to pyrrolizidine poisoning. In due course new biological methods may be developed to aid in the control of these poisonous plants.

Bishop's Weed, Greater Ammi
Ammi majus
Apiaceae (Parsley family)

Habitat

Introduced from Asia, bishop's weed has become established in the wet, brackish meadows of the coastal region of the southern United States. It is also commonly grown as an ornamental annual.

Description

Bishop's weed is an annual, 1 to 2 feet (0.30 to 0.60 meters) tall with ascending branches and leaves that are finely dissected into filiform segments. The inflorescence is umbellate with small white flowers (Figure 4-3). The involucral bracts are pinnately parted, the involucels are of several linear bracts.

Figure 4-3 Bishop's weed (*Ammi majus*).

Principal Toxin

The principal toxins are the furocoumarins. All parts of the plants, but especially the seeds, may be phototoxic to cattle, sheep, fowl, and humans as a result of ingestion or skin contact with the plant and subsequent exposure to sunlight.[142-145]

Buckwheat
Fagopyrum esculentum
Polygonaceae (Buckwheat family)

Habitat

Cultivated as a grain or cover crop, buckwheat has since escaped and become a plant of disturbed soils and roadsides. It is commonly used as a cover crop for soil enrichment.

Description

Buckwheat is a glabrous, herbaceous annual plant with an erect stem. The leaves are alternate and have hastate or cordate blades. The stipules are united as a sheath (ochrea) around the stem at the nodes. The greenish white flowers occur at terminal or axial panicles (Figure 4-4). The stamens number eight, styles three-parted, and the achenes three-angled and brown in color.

Figure 4-4 Buckwheat (*Fagopyrum esculentum*).

Principal Toxin

Fagopyrin, a plant pigment present in the green and to a lesser extent the dry plant, is capable of producing photosensitization in domestic livestock and people.[146] Buckwheat flour is not toxic.

St. John's Wort, Klamath Weed
Hypericum perforatum
Hypericaceae (St John's wort family)

Habitat

St. John's wort is an introduced weed of dry soils throughout most of North America, especially in the northwestern states. There are numerous cultivated hybrids in existence.

Description

St. John's wort is an erect perennial herb up to 3 feet (1 meter) tall with woody lower stems. The branches are opposite and sterile. Usually both stems and branches are two-edged or winged. The leaves are opposite, sessile, and linear-oblong, 1 inch long and spotted with tiny dots that are translucent when held against the light. The inflorescence is a cyme with numerous flowers, two-thirds to an inch in diameter having five bright yellow petals, five green sepals, many stamens in three to five clusters and an ovary with three widely spreading styles (Figure 4-5). The petals have finger-like margins and may have black glandular dots on the margins.

Figure 4-5 St. John's wort (*Hypericum perforatum*).

Principal Toxin

Hypericin, a photoreactive pigment, is readily absorbed from the digestive tract and remains chemically intact through the digestion process. Hypericin has no effect on the liver and causes primary photosensitization after ingestion.[147] Hypericin is present in the glandular dots on the leaves suggesting that all *Hypericum* spp. with similar glands are potentially toxic. Hypericin is stable to drying and therefore hay containing St John's wort may cause poisoning. The young plants are as toxic as the mature plant and more palatable to livestock although the content of hypericin varies with growing conditions of the plant.[148]

Hypericin has found recent popular use in people as an herbal stimulant and will induce photosensitivity in some individuals, especially if overdosed. Hypericin has also shown some antiviral properties and has been investigated for its effects on human immunodeficiency virus.[149,150]

Lechuguilla
Agave lecheguilla
Agavaceae (Agave family)

Habitat

Lechuguilla plants are found in low limestone hills, dry valleys, and bordering canyons especially in the southwestern United States and Mexico.

Description

The plant is a stemless perennial with leaves 10 to 30 in number, fleshy, bayonet-like, erect and attached to a short, broad crown at ground level. Each leaf is 1 to 1.5 inches (2.5 to 4 cm) wide at the base and tapering to a very sharp point (Figure 4-6). They range from 12 to 20 inches (30 to 51 cm) long with recurring marginal teeth. The plant flowers once after 10 to 15 years of vegetative growth, producing a large terminal panicle on a thick stalk ranging from 6 to 12 feet (2 to 3.5 meters) tall. The flowers are tubular with three sepals, three petals, six stamens, and a three-carpellate ovary that matures into a leathery capsule producing many flat black seeds. After flowering, the plant dies. Reproduction may continue vegetatively by offshoots from the parent plant.

Figure 4-6 Agave (*Agave lecheguilla*).

Principal Toxin

Hepatotoxic saponins in lechuguilla appear to be the primary cause of poisoning in sheep and goats and less frequently cattle.[151] Photosensitivity results from the accumulation of phylloerythrin as a result of bile duct obstruction similar to that which occurs in *Tribulis terrestris* photosensitivity.[82,83] Animals seldom eat the plant except in times of drought. As little as 1 percent of the animal's body weight of the leaves has caused poisoning and death. Liver disease with icterus, weight loss, and secondary photosensitization are common features of lecheguilla poisoning.

Fiddleneck, Tarweed
Amsinckia intermedia
Boraginaceae (Forget-me-not family)

Habitat

Fiddleneck is a weed of dry cultivated soils, waste ground, and wheat fields. It continues to spread eastward from western North America.

Description

Fiddleneck is an erect, sparsely branching, 2 to 3 feet (1 meter) tall annual weed, covered with numerous white hairs. Leaves are hairy, lanceolate, and alternate (Figure 4-7A). The perfect five-parted small orange to yellow flowers are born terminally on a characteristic fiddleneck-shaped raceme, the flowers all inserted on one side of the axis (Figure 4-7B). Mature fruits separate into two to four black-ridged nutlets.

Figure 4-7A Fiddleneck (*Amsinckia intermadia*).

Figure 4-7B Fiddleneck showing the typical "fiddleneck" (*Amsinckia intermedia*).

Principal Toxin

Pyrrolizidine alkaloids are the principle toxins responsible for poisoning in horses, cattle, and pigs that eat the plants, especially the seeds.[1,31] The symptoms and lesions of Amsinckia poisoning in all species of animals consist of liver necrosis and fibrosis characteristic of PA poisoning. *Amsinckia* spp. may also accumulate potentially toxic levels of nitrate as rapidly growing annual weeds.

Rattlebox
Crotalaria spp.
Fabaceae (Legume family)

Habitat

Some species of *Crotolaria* were introduced as a soil-building cover crop for the sandy soils in the southeastern United States and have since become established in disturbed soils along fences and roadsides in Florida and Georgia. An indigenous species, *C. sagittalis* is common along river bottomland.

Description

Crotalaria are erect, herbaceous, variably hairy plants, and may be annual or perennial. The leaves are simple, alternate, lanceolate to obovate, with a finely haired under surface (Figure 4-8A). The flowers are yellow, with the leguminous calyx longer than the corolla (Figure 4-8B). The fruit is a leguminous pod, inflated, hairless, becoming black with maturity, and contains 10 to 20 glossy black, heart-shaped seeds, which often detach and rattle with the pod. Several species of *Crotalaria* have been associated with livestock poisoning including *C. sagittalis*, *C. spectabilis*, and *C. retusa*.

Principal Toxin

The principal toxins in *Crotalaria* spp. are the pyrrolizidine alkaloids (PA), the most notable of which is monocrotalamine. The alkaloid is present in greatest quantity in the seeds, with lesser amounts in the leaves and stems. All livestock, including domestic fowl are susceptible to poisoning. Although acute deaths will occur from eating large quantities of the crotolaria seeds or plant, more typically animals will develop signs of liver disease and photosensitization from a few days up to 6 months later. Monocrotaline also causes severe pulmonary changes, and horses have been reported to die after developing an acute fibrosing alveolitis from eating a feed containing 40 percent crotolaria seeds.[63,64]

Figure 4-8B Rattlebox flower and seed pods (*Crotolaria spectabilis*).

Figure 4-8A *Crotolaria* spp.

Senecio
Groundsel, tansy ragwort, butterweed
Senecio spp.
Asteraceae (Sunflower family)

Habitat

More than 1200 species of *Senecio* are distributed throughout the world, with about 70 species occurring in North America. Approximately 25 species have been proven poisonous to animals, but there is high probability that other species of *Senecio* are toxic. *Senecio* spp. have a wide overlapping geographic range, but many are, however, selective in their habitat, some preferring high altitude, subalpine, moist conditions, whereas others prefer dry, rocky, or sandy soils at lower elevations. The more important toxic species of *Senecio* in North America are listed in Table 4-4.

Table 4-4 Species Associated with Livestock Poisoning	
SCIENTIFIC NAME	**COMMON NAME**
Senecio jacobaea	Tansy ragwort, stinking willie
S. intergerrium	Lamb's tongue groundsel
S. douglasii	Woody or threadleaf groundsel
S. riddellii	Ridell's ragwort
S. plattensis	Prairie ragwort
S. spartioides	Broom groundsel
S. glabellus	Butterweed
S. vulgaris	Common groundsel

Figure 4-9A Senecio showing typical bract formation (*Senecio* spp.).

Description

Identification of individual *Senecio* spp. is difficult without being an experienced taxonomist. However, recognition of a plant as a member of the genus *Senecio* can be based on the presence of a single layer of touching, but not overlapping, greenish bracts surrounding the flower (Figure 4-9A). *Senecio* spp. have alternate leaves, generally lanceolate to ovate, dentate and often irregularly and deeply pinnately divided (Figure 4-9B). The composite flower heads are flattened terminal clusters with showy yellow ray flowers (Figure 4-9C). Seed is produced in both disc and ray florets, each seed with a tuft of white hairs that aid in wind dissemination (Figure 4-9D).

Figure 4-9B Tansy ragwort (*S. jacobaea*).

Figure 4-9C Broom groundsel (*S. spartioides*).

Principal Toxin

Pyrrolizidine alkaloids (PA) are the principal toxins in senecios.[19,59,63] The quantity of alkaloid in *Senecio* spp. varies with the species and stage of growth, the young pre-flowering plant being most toxic. Acute liver necrosis and death in 1 to 2 days has been associated with feeding cattle 4 to 8 percent of an animal's body weight in green plant over a few days.[78]

Chronic poisoning, which is the more natural form of poisoning encountered in cattle and horses, is usually associated with ingestion of smaller quantities of *Senecio* over a period of 3 weeks or more. Cattle fed 1.5 g of *S. jacobaea*/kg of body weight daily for a minimum of 20 days had 100 percent mortality. This equates to a 20-day cumulative dose of 2 percent of an animal's body weight of dry plants. The cumulative effects of PAs on the liver may take 6 or more months to reach the point where 80 percent or more of the liver is affected and clinical signs of liver disease become evident.

Figure 4-9D Common groundsel showing coarsely toothed leaves, flowers with no ray florets, and the seed heads with white pappus that aids in wind dispersal (*S. vulgaris*).

Hound's Tongue
Cynoglossum officinale
Boraginaceae (Forget-me-not family)

Habitat

Imported in wheat seed from Russia, hound's tongue has become widespread in most regions of North America and especially in the Rocky Mountain region where it is considered a noxious weed. It prefers waste areas, roadsides, meadows, and hay fields.

Description

Hound's tongue is a biennial, forming a rosette the first year, with basal leaves up to 18 inches (45.5 cm) long, densely pilose, oblong or oblong-lanceolate in shape (Figure 4-10A). The second year the plant produces stems that are 2 to 4 feet (0.5 to 1 meter) tall, erect, stout, softly pilose, and flowering. The upper leaves are lanceolate and sessile. The flowers are regular, reddish purple, in scorpioid racemes (Figure 4-10B). The fruit is pyramidal, separating into four nutlets at maturity, which are covered with hooked barbs ("velcro") facilitating their adherence to clothing and animal hair.

Figure 4-10A Hound's tongue first-year rosette (*Cynoglossum officinale*).

Figure 4-10B Hound's tongue flowers and fruits (*C. officinale*).

Principal Toxin

Hound's tongue contains significant quantities of pyrrolizidine alkaloids (PA) comparable to the levels found in the most toxic *Senecio* spp. The plant is rarely eaten in the green state, but livestock find it palatable when it is dried in hay. As little as 15 mg of dried plant/kg body weight fed to horses over a 2-week period induced fatal liver disease.[137] This equates to about 6 percent of a horse's daily intake of food if the hound's tongue is dried in the early blossom stage when the PA content averages 0.08 percent.[38]

Note: Viper's bugloss or blue weed (*Echium vulgare*) is an introduced European plant that contains PAs. It is an invasive weed in some areas and is often cultivated as an ornamental flower. In the same family as hound's tongue (Boraginaceae), it is a very hairy plant with showy blue, pink, or white, tubular flowers, produced in one-sided clusters that uncoil as the flowers bloom sequentially (Figure 4-10C). The leaves are hairy, 2 to 6 inches (5 to 15 cm) in length and lanceolate. Numerous rough nutlets are produced that promote its proliferation.

Figure 4-10C Viper's bugloss, blue weed (*Echium vulgare*).

Lantana
Lantana camara
Verbenaceae (vervain family)

Habitat

Lantana is an ornamental shrub in the southern warmer areas of North America, often escaping along stream bottoms in the southwest to become an invasive noxious weed.

Description

The plant is a shrub with square stems and a few scattered spines. The leaves are simple, opposite or whorled, and oval-shaped. The margins are serrate. The flowers are born in flat-topped clusters and are small, tubular, and white, yellow, orange, red, or purple (Figure 4-11). The fruits are produced in clusters and turn black when ripe.

Figure 4-11 Lantana (*Lantana camara*)

Principal Toxin

At least 15 of the 29 described taxa of *L. camara* are known to be toxic to livestock.[91,92] Lantana poisoning is a disease of ruminants characterized by an intrahepatic cholestasis induced by the triterpene acids lantadene A and B. Horses are apparently not affected. Acute poisoning frequently results in death in 7 to 10 days. In more chronic poisoning, sheep in particular stop eating and develop a metabolic acidosis with resulting increases in serum potassium. Death probably results from a combination of liver disease, anorexia, and accumulation of triterpene acids. Photosensitization may be evident in white skinned areas of affected animals.

Sacahuista, Beargrass
Nolina texana
Agavaceae (Agave family)

Habitat

Sacahuista grows on dry hillsides at elevations from 3000 to 6500 feet (914 to 1,981 meters) in milder climates of the southwestern states.

Description

Sacahuista is an evergreen perennial grasslike plant with a thick woody, mostly haired basal stem growing 2 to 3 feet (0.5 to 1 meter) tall (Figure 4-12). Many linear leaves, 0.5 inch (1 cm) wide, 5 feet (1.5 meters) long, with finely toothed margins arise from the basal stem. Small white to green flowers are born on stalks in elongated clusters. Fruits are three-parted capsules containing many seeds. *Nolina microcarpa* is very similar in appearance and distribution.

Figure 4-12 Beargrass, sacahuista (*Nolina texana*).

Principal Toxin

An undefined hepatotoxin saponin concentrated in the flower buds, flowers, and fruits appears to be responsible for a crystalloid material that obstructs the biliary system, causing phylloerythrin accumulation and photosensitivity.[83,152] The leaves are not toxic and are eaten by livestock at times when other forages are unavailable. All ruminants are susceptible to poisoning, but sheep and goats are most commonly affected.[152] About 1.1 percent of an animal's body weight in flowers or fruits will induce poisoning. In severe cases animals die from liver failure.

Note: Beargrass is the common name given to *Xerophyllum tenax* that grows in British Columbia, Montana, and from Wyoming to northern California, and is not related to *Nolina* species. It is not known to be toxic.

Horsebrush
Tetradymia spp.
Asteraceae (Sunflower family)
T. glabrata Coal oil brush, little horsebrush, spring rabbit brush
T. canescens Spineless or gray horsebrush

Habitat
Horsebrush is generally confined to southwest North America and is found growing in sagebrush (*Artemisia* spp.) habitats. *Tetradymia glabrata* is found in semidesert foothills, whereas *T. canescens* grows at higher elevations in the Rocky Mountains.

Description
Horsebrush is a low shrub with stiffly, much-branched stems that are usually covered with white, wooly hairs, especially when young. (Figure 4-13A). The leaves are alternate, narrow, and entire. The inflorescence is a head that is solitary from the upper axils or clustered at the tips of the branches (Figure 4-13B).

Figure 4-13A Horsebrush (*Tetradymia canescens*).

Figure 4-13B Horsebrush flowers (*T. canescens*).

There are four bracts present. The florets are yellow with a large number of straw-colored bristles or pappus. *Tetradymia spinosa* is a shrub between 1.5 to 4 feet (0.45 to 1 meter) tall and varies from *T. canescens* by the fact that the flowers number five to nine to a head and the bracts five to six. The primary leaves are converted to spines.

Principal Toxin
Tetradymol, a furanoeremophilane, is the main toxic compound that causes liver toxicity. Sheep are primarily affected by horsebrush, but must eat black sagebrush (*Artemisia nova*) before or at the same time they eat horsebrush if poisoning is to occur.[153,154] The sagebrush preconditions the sheep to the toxic effects of the horsebrush indicating strong synergistic action of the two plants on the liver. Sheep vary in their susceptibility to horsebrush but 0.5 to 1.0 lb of little leaf horsebrush (*T. glabrata*), which is about twice as toxic as *T. canescens*, will produce photosensitization (big head). Consumption of larger amounts of the plant causes death in a few days. Abortions have also been attributed horsebrush when the plant is grazed by pregnant ewes.

Puncture Vine, Caltrop, Goat's head
Tribulus terrestris
Zygophyllaceae (Caltrop family)

Habitat

Puncture vine is found in the dry soils of wastelands, roadsides, and deserts in most eastern, southern, and western states, and other regions of the world.

Description

Puncture vine is a prostrate, annual weed with recumbent stems that are pubescent and branching occasionally up to 3 feet (1 meter) in length. Leaves are opposite and pinnately compound with leaflets occurring in four to seven pairs, oblong, elliptical, and about 0.5 inch (1 cm) long. Flowers are solitary, occurring in the axils of the leaves with a corolla composed of five yellow petals (Figure 4-14). The fruit is a small hard capsule that breaks apart into five spiny sections, each having two prominent, sharp, woody spines, that resemble a goat head in appearance.

Puncture vine is similar in appearance to hairy caltrop (*Kallstroemia hirsutissima*), differing in the very hairy leaves and stems of the latter.

Principal Toxin

A number of steroidal sapogenins have been found in *Tribulus terrestris* that form an insoluble crystalloid substance in the bile ducts.[84,155] The development of photosensitivity appears to be secondary to obstruction of the biliary system with this calcium rich crystalloid substance that forms microliths (caleuli or stones). These microliths in the bile ducts increase the retention of phylloerythrin and the development of secondary photosensitivity.[83,84,155] A mycotoxin present in the plant has also been suspected of being associated with hepatogenic photosensitivity in livestock.[82]

Figure 4-14 Puncture vine (*Tribulus terrestris*). Inset: seed head

An entirely different syndrome of sheep characterized by hindleg weakness and paresis has been associated with the chronic consumption of *T. terrestris* in Australia.[156] Caltrops (*Kalstroemia* spp.) have produced a similar neurologic syndrome in sheep, goats, and cattle in Texas.[157] The plant toxins causing the neurologic signs are the β-carboline indoleamines harmane and norharmane.[158]

The spiny burrs are mechanically injurious to animals and people.

Panic grasses
Panicum spp.
Kleingrass *(Panicum coloratum)*
Blue panicum *(P. antidotale)*
Poaceae (Grass family)

Habitat

Panic grasses are mostly confined to southwest North America in dry woodlands, plains, and valleys.

Description

Panic grass is a warm-season bunch grass growing to 5 feet (1.5 meters) tall with blue-green leaf blades to 24 inches (70 cm) long. There are often soft straight hairs on the upper side of the leaf next to the culm. The inflorescence is an open, widespreading panicle 6 to 18 inches (15 to 45 cm) long.

Principal Toxin

Sheep and occasionally cattle grazing on panicum pastures in some years under hot humid conditions can develop photosensitization. Saponins in the grass are assumed to be the cause of the formation of crystaloid substances in the biliary system that subsequently obstruct the bile duct and cause secondary photosensitization.[159]

Tansy Mustard
Descurainia pinnata
Flixweed
Descurainia sophia
Brassicaceae (Mustard or crucifer family)

Habitat

These plants are commonly found throughout North America, being opportunistic weeds of cultivated fields, disturbed soils, waste areas, roadsides, and open dry areas.

Description

Flixweed is an erect, terminally branched, annual, herbaceous plant growing 1 to 2 feet (0.5 meters) in height. The leaves are alternate, two to three times pinnately compound, each segment narrow or linear, the basal leaves being largest. The plant is pubescent or nearly hairless, with hairs being branched. The small greenish yellow flowers are produced on a raceme (Figure 4-15A). The seed pod (silique) is cylindrical, 0.5 to 1.25 inches (1 to 3 cm) in length, the silique septum having two to three longitudinal nerves.

Tansy mustard is a very similar annual that is differentiated from flixweed on the basis of its shorter, fatter, siliques (less than 0.75 inches (2 cm) in length), that contain two rows of seeds in each seed pod (Figure 4-15B). Both species spread readily by seed.

Principal Toxin

No specific toxin has been identified in either tansy mustard or flixweed that has been associated with the clinical signs of liver disease and photosensitization reported in cattle grazing the plants.[160,161]

Clinical Signs

Two apparently different syndromes in cattle have been attributed to tansy mustard. In cattle, a neurologic syndrome of partial or complete blindness, accompanied by the inability to use the tongue or swallow, aimless wandering, head pressing, and eventual death from dehydration and starvation has been sporadically reported.[1] In some areas, and in some years, the principal toxicity associated with tansy mustard is one of liver disease and photosensitization without neurologic signs.[160,161] Attempts at reproducing the disease by feeding tansy mustard have not been successful.[161]

Figure 4-15A Flixweed (*Descurainia sophia*).

Figure 4-15B Seed pods of Tansy mustard (*D. pinnata*) (left) and Flix weed (*D. sophia*) (right).

Kochia, Summer cypress, Mexican Fireweed, Belvedere, Morenita
Kochia scoparia
Chenopodiaceae (Goosefoot family)

Habitat

Originally from Asia, and introduced from Europe, kochia has become an established weed of North and South America because of its wind-distributed seeds and adaptation to dry conditions. It is commonly found in disturbed soils, croplands, pastures, and dry rangeland. In many areas of southwestern North America, kochia weed is favored as a useful forage for livestock and is often cut and baled as a winter feed.

Description

Kochia is an annual, herbaceous, rapidly growing plant with branching stem and simple, entire, hairy leaves (see Figures 1-17A and 1-17B). The leaves are alternate and vary in size from 0.5 to 3.0 inches (1 to 7.5 cm) in length on the mature plant. Stems are either green or red-tinged, the stems becoming particularly red in the fall. The size of the plant varies considerably, attaining heights of 6 feet (2 meters) where the moisture and organic content of the soil are high. The flowers are green and produced in dense spikes in the upper leaf axils (see Figure 1-17C). The seeds are small, brown, and slightly ribbed.

Kochia weed is often erroneously called tumbleweed because it breaks off at ground level in the winter and gets blown about by the wind. True tumbleweed is another common weed called Russian thistle (*Salsola kali*).

Principal Toxin

Kochia has a mixed reputation as a nutritious forage and under some circumstances as a poisonous plant to cattle and sheep. A specific toxin has not been identified in kochia that will account for the various syndromes of toxicity encountered in cattle and sheep whose diet consists predominantly of kochia. Nitrates, oxalates, sulfates saponins, and alkaloids have been identified in the plant depending on the stage of growth, growing conditions, and geographic region.[162-166]

Kochia is used as a forage and is made into hay for livestock in many arid regions of North America. If grazed or cut before it reaches maturity, kochia does not appear to be toxic, although if high in nitrates it is hazardous to ruminants.[162] Its nutritive value is similar to that of alfalfa.[165] However, livestock grazing kochia when it is mature or stressed by drought are prone to poisoning at least in some regions.[162,163] Poisoning is more likely to occur if the animal's diet consists predominantly of kochia. Experimental feeding trials in which cattle were allowed to graze pure stands of irrigated and fertilized kochia from 2 to 15 weeks resulted in poor weight gain when compared to cattle grazing native grasses.[167] Furthermore, those cattle eating kochia alone developed degenerative liver and kidney disease. Sheep may be fed kochia up to 50 percent of their total dry matter intake without affecting weight gain or causing toxicity.[168] Other studies have shown that sheep can graze greater than 95 percent kochia with only slight elevations of liver enzymes indicative of mild liver degeneration without apparent clinical disease or effect on growth weights.[166] The variable toxicity of kochia would appear to be related to the stage of maturity of the plant, and perhaps a combination of the effects of the various toxic components of the plant. The general toxic effect appears to be that of degenerative liver disease with secondary photosensitization and liver failure.

Clinical Signs

Cattle, sheep and occasionally horses appear to be the most commonly affected when mature kochia is the predominant constituent of their diet.[163,167-170] The most common clinical signs observed in cattle include poor weight gains, depression, weakness, excessive tearing, incoordination, and photosensitization.[167,169] Other clinical signs associated with kochia poisoning include loss of appetite, diarrhea, icterus, and mouth ulcerations.[162] Affected animals frequently have elevated serum liver enzymes and bilirubin levels.[166,167] Some animals will be blind and walk in circles or follow fences endlessly.[162,169] These neurologic signs may be due to the high levels of sulfate that once reduced in the rumen to hydrogen sulfide result in degenerative changes in the brain leading to depression and blindness.[171] Long term consumption of mature kochia results in eventual death of the animal. However, cattle and sheep will recover from kochia poisoning provided they are given a balanced ration where kochia weed is not the only food available.

Common Cocklebur
Xanthium strumarium
Asteraceae (Sunflower family)

Habitat and Description

Found throughout North American, cockleburs are annual, bushy weeds 2 to 5 feet (0.5 to 1.5 meters) tall, with stout stems, often with dark spots. The leaves are large, rough, glandular, and triangular from 2 to 14 inches (5 to 35.5 cm) long and 1 to 8 inches (2.5 to 20 cm) wide. Flowers are produced in the leaf axils, the inconspicuous male flowers being clustered at the top, with the larger female flowers that form the burs toward the base. (Figure 4-16A). Hooked spines cover the characteristic oval burs. Each bur contains two seeds, which can remain dormant in the soil for years. Consequently cockleburs can reappear many years after the parent plant disappeared.

Principal Toxin

Carboxyactractyloside, a sulfated glycoside, is the primary liver toxin present in cockleburs.[172,173] The glycoside is present in high concentration in the seeds and the two-leafed cotyledon stage (Figure 4-16B), but disappears by the four-leaf stage and is not present in the mature plant. The liver appears to be the primary target organ with pigs, ruminants, and horses being suscepti-

Figure 4-16A Cocklebur (*Xanthium strumarium*).

Figure 4-16B Two-leafed stage of cocklebur.

ble to poisoning. Fatalities occur when 0.75 to 3.0 percent body weight of cotyledons are consumed.[174] Acute deaths, convulsions, blindness, and recumbency have been encountered in cattle eating hay contaminated with mature cockleburs.[175] The most consistent effect of the toxin on the liver is acute diffuse central-lobular and paracentral coagulative necrosis.[173] Similar renal tubular necrosis may occur. Severe hypoglycemia and marked elevations of liver enzymes (sorbitol dehydrogenase) are encountered in experimental poisoning of pigs.[172,177]

The spiny burs are a source of mechanical injury to the mouth of dogs, especially long haired breeds, that pull them from their hair coat with their teeth when self grooming. If dogs swallow the burs they vomit subsequently, but it is unknown whether this is due to mechanical irritation or the toxic glycoside. Significant economic losses can also occur to wool producers when the burs become entangled in the fleece reducing its market value.

Clinical Signs

Poisoning usually occurs when animals eat large numbers of the two-leafed stage of the cocklebur or eat the seeds that may contaminate cottonseed or other food sources. Pigs and cattle show similar signs of poisoning including depression, reluctance to move, hunched back, ataxia, and recumbency.[174-176] Hypoglycemia may be present in severe cases. Paddling of the limbs, convulsions succeeded by coma, and death in 24 hours are common. Serosanguinous ascites, edema of the gallbladder, pericardial and pleural effusion, and hepatic congestion are the usual postmortem findings.

There is no known effective treatment for cocklebur poisoning. Attempts to eliminate the toxin from the gastrointestinal tract may be made with neostigmine and mineral oil. Activated charcoal may also be effective as an adsorbent. Intravenous glucose will help counteract the severe hypoglycemia induced by the toxin.

Cockle burs are best controlled by mowing, cultivation, or herbicides, provided the plants are destroyed before they produce burs. A single mature plant that goes to seed will effectively repopulate the area becoming a troublesome and aggressive weed.

Hairy Vetch
Vicia villosa
Fabaceae (Legume or pea family)

Habitat

Introduced from Europe as a legume for pasture improvement and as a cover crop, hairy or woolly vetch has become an established weed in many areas of North America, especially along roadsides, waste areas, and in croplands.

Description

Hairy vetch is a climbing or sprawling annual 4 to 6 feet (1 to 2 meters) in length, with hairy stems and leaves. The leaves have 10 to 20 leaflets up to 1 inch (2.5 cm) in length that are narrow and lance-shaped. Tendrils at the end of the leaves are well developed. Flowers are purple to red in color, 20 to 60 per spike, all on one side of the flower stalk.(Figure 4-17A) Pea-like seed pods are about 1 inch in length containing several hard seeds.

Figure 4-17A Hairy vetch (*Vicia villosa*)

Common vetch (*Vicia sativa*) that can be mistaken for hairy vetch, is similar but is not hairy and the flowers are not arranged on one side of the flower stalk. (Figure 4-17B)

Principal Toxin

The specific toxin(s) in hairy vetch responsible for the symptoms encountered in hairy vetch poisoning has not been determined. The generalized granulomatous disease that is characteristic of hairy vetch poisoning in cattle and horses is suggestive of a hypersensitivity (type IV) reaction induced by a foreign substance that activates the immune system response.[178-184] Vetch lectins have been proposed as the instigators of the immune mediated response that results in the granulomatous inflammatory response seen in many different tissues of the affected animal.[185] Not all animals are susceptible to vetch hypersensitivity, the disease being more prevalent in cattle over 3 years of age.[178,183] Although reported in many breeds of cattle, hairy vetch poisoning appears to be more common in the Angus and Holstein breeds.[181,183]

Hairy vetch poisoning occurs most often when it forms a major part of the diet of cattle and horses, and when the plant is nearing maturity. The plant is less likely to cause a problem in hay or when ensiled.[183] However, the mechanism by which hairy vetch induces poisoning remains unclear as the plant is frequently consumed by cattle without apparent problem.

Clinical Signs

Hairy vetch poisoning sporadically affects adult cattle and occasionally horses and is characterized by signs of pruritic dermatitis, weight loss, conjunctivitis, and diarrhea.[179,181,183] The dermatitis is not confined to white skinned areas. Initially the hair

Figure 4-17B Common vetch (*Vicia sativa*)

coat is rough and stands erect where lesions develop. Papules then develop that exude serum and result in superficial crusts. Continual rubbing and scratching of the lesions causes hair loss and thickening of the skin.[183] Abortions and red urine have also been associated with hairy vetch poisoning.[183] Lymphocytosis and hyperproteinemia are a feature of hairy vetch poisoning. A more acute form of hairy vetch poisoning characterized by subcutaneous swellings, ulcers of the oral mucous membranes, purulent nasal discharge, and coughing with significant mortality may also occur. Ingestion of large amounts of hairy vetch seed may induce neurologic signs and death in cattle and horses.[186]

At post mortem examination, granulomatous lesions can be found in multiple organs including the skin, liver, kidneys, heart, spleen, lymph nodes, and digestive system. Histologically, the lesions consist of monocytes, eosinophils, multinucleated giant cells and lymphocytes.[180,181,183]

Leucaena, lead tree
Leucaena leucocephala
Mimosaceae (Legume family)

Habitat

Leucaena is a tropical legume found in many parts of the world that is well established in Hawaii and has been introduced into Florida, Texas, and coastal areas in between. As a legume it has potential as a protein source for livestock in tropical areas of the world.

Figure 4-18 Leucaena showing typical leaves, flowers, and seed pods. (Courtesy of Gerald D. Carr. University of Hawaii).

Description

Rapidly growing shrub or tree attaining heights of 30 feet (10 meters), with bipinnately compound leaves, 6 to10 inches (15 to 25 cm) long, with 4 to 8 pairs of pinnae, each with 10 to 15 pairs of lanceolate leaflets. Numerous white to pink tightly clustered flowers are produced in spherical heads on long pedicels from the leaf axils. Fruits are many seeded legume-like pods. (Figure 4-18)

Principal Toxin

Leucaena contains 25 to 35 percent crude protein in its leaves that has potential as a livestock protein source especially in those parts of the world deficient in available sources of protein rich forages for animals. However, leucaena contains the amino acid mimosine that is poisonous to animals that are not adapted to eating the plant.

The precise mode of action of mimosine is unclear, but it and its metabolites are presumed to act as amino acid antagonists and/or will complex with metals such as zinc, copper, and magnesium thereby inhibiting a variety of enzymes important to protein metabolism.[187-189] Mimosine may also act as goitrogen, causing goiter and signs of thyroid deficiency.[187]

In some parts of the world cattle and goats thrive on eating leucaena, while in other areas animals lose weight, lose their hair and grow poorly when eating the plant. The ability of some animals to thrive on leucaena and others not is related to the ability of the animal's rumen microflora to detoxify mimosine. Specifically it is the presence of the anaerobic rumen bacterium *Synergistes jonesii* that detoxifies mimosine and enables the animal to make use of the plant's proteins without ill effect.[190-192]

It is possible and practical to transfer this bacterium from one animal to the next through transferring rumen contents thereby populating the animal's rumen with the bacteria and enabling it to eat leucaena without problem. The organism once established in the rumen will survive from season to season even if the animal does not have daily access leucaena.[193] Nonruminants do not have the capability of adapting to leucaena and should not be fed the plant.

Clinical Signs

Cattle, sheep, and goats eating leucaena exhibit signs of poor growth, weight loss, hair loss, lameness due to swollen and ulcerated coronary bands, and mouth and esophageal ulcers.[193,194] Goiter with decreased thyroxine levels, cataracts, atrophy of the gums, decreased fertility due to embryonic death, and fetal abnormalities may also be seen in animals consuming leucaena.[194,195]

Once ruminants establish mimosine degrading bacteria in their rumen they are able to utilize leucaena as a good protein source. Nonruminants should not be fed leucaena.

Habitat

Scorpion weed is a common plant of drier climates and desert areas of southwestern North America, preferring sandy soils of roadsides and waste areas. Other species of Phacelia are found throughout the continent and are considered a desirable wild flower.

Description

Often forming dense stands, scorpion weed is an annual or biennial, growing to 18 inches (45 cm) in height, with pinnately lobed hairy leaves. Flowers are small, blue or white, and open sequentially on one-sided terminal coils (Figure 4-19). The name scorpion weed refers to the coiled flower head that resembles a scorpion's tail.

Figure 4-19 Scorpion weed (*Phacelia crenulata*).

Principal Toxin

The hairs of the plants contain phenolic compounds that are contact allergens.[196] These compounds have an allergic equivalent to the compound urushiol found in poison ivy. Some dogs and people coming in contact with the hairs, especially on the dry plant, develop an acute skin reaction.

Clinical Signs

Dogs especially develop swollen feet and may also have swelling and inflammation of the lips, eyelids, and mouth if they are running through dense stands of the scorpion weed. Affected animals should have the affected skin areas washed with a mild soap and water to remove the residual plant hairs. Antihistamine and anti-inflammatory medications may be necessary in severe cases.

Poison Ivy
Toxicodendron radicans
(Synonym Rhus radican)
Anacardiaceae (Cashew family)

Figure 4-20A Poison ivy with fruits (*Toxicodenron radicans*).

Habitat

Poison ivy is found throughout North America except on the West Coast where poison oak (*Toxicodenron quercifolium*) prevails. Varieties of the plant may be found in dry sandy soils, or moist loamy soils of woodlands and forests.

Description

Poison ivy is a perennial, woody plant that is typically a trailing or climbing vinelike plant in eastern North America, in the drier western areas of North America it is a low growing, bushy shrub. The leaves are glossy, hairless, alternate, always three-parted, and may be toothed, or lobed (Figure 4-20A). The climbing stems produce aerial roots that anchor the pant to the bark of trees on which it is climbing. The flowers are small, whitish, hanging clusters, and the fruits are white and berry-like with longitudinal grooves. The leaves turn a bright red in the fall.

Poison oak (*T. quercifolium*) is a woody, perennial shrub, with alternate, deeply lobed, oaklike leaves found in western North America (Figure 4-20B). The flowers are small, greenish, and are produced in hanging clusters. The fruits are yellowish, hairy berries. Also found in the same region is western poison oak (*T. diversilobum*), which is similar to poison oak but is a vine that can reach 50 feet (15 meters) in length. Its fruits are white berries.

Poison sumac (*T. vernix*) is a woody perennial, branching shrub or small tree with gray bark and leaves to 12 inches (30 cm) long, each with a distinctive red-purple rachis. Leaflets are smooth, elliptic and entire (Figure 4-20C). The flowers are small and yellowish white and produced in panicles. The fruits are white berries. Poison sumac is usually confined to wet areas such as pineland bogs in southern North America.

Principal Toxin

Poison ivy, poison oak, and poison sumac all contain a highly irritating allergenic phenolic compound urushiol (oleoresin). The most toxic being the resin 3-n-pentadecyl-catechol.[197] All parts of the plant, green or dried, contain these compounds. The oily resin is not volatile or soluble in water but is soluble in alcohol. Smoke from burning

Figure 4-20B Poison oak (*T. quercifolium*)

the plants can contain droplets of the toxin and will affect people who are highly allergic to the toxin. The hair coat of animals may become contaminated with the resinous toxin if they are in close contact with the plants, and therefore can transmit the toxin to people who handle the animals. The animals themselves are rarely if ever affected by the urushiol, and goats and sheep have been observed eating poison ivy without apparent problems.

Figure 4-20C Poison sumac (*T. vernix*) (Courtesy of Drs. John and Emily Smith, Baldwin, Georgia).

REFERENCES

1. Kingsbury JM: *Poisonous Plants of the United States and Canada.* Englewood Cliffs, NJ: Prentice-Hall, 1964:52–57.

2. Clare NT: Photosensitization in animals. *Adv Vet Sci* 1955, 2:182–211.

3. Rowe LD: Photosensitization problems in livestock. *Vet Clin North Am: Food Anim Pract* 1989, 5:301–323.

4. Franco DA, Tsang-Long L, Leder JA: Bovine congenital erythropoietic porphyria. *Compendium of Continuing Education* 1992, 14:822–826.

5. Ruth GR, Schwartz S, Stephenson B, *et al.*: A new disease of cattle: bovine protoporhyria, clinical and diagnostic features. *American Association of Veterinary Laboratory Diagnosticians 21st Annual Proceedings.* 1978:91–96.

6. Hancock, JJ: Congenital photosensitivity in Southdown sheep. *N Z J Sci Tech Net* 1950, 32:16–24.

7. Johnson AE: An assessment of toxic pyrrolizidine alkaloids in US plants. *J Anim Sci Suppl 1* 1982, 230.

8. Marsh CD: Toxic effect of St John's wort (*Hypericum perforatum*) on cattle and sheep. *USDA Bulletin* 1930, 202:1–23.

9. Sampson AW, Parker KW: St Johnswort on range lands of California. University of California Experimental Station Bulletin 1930, 503:1–48.

10. Araya OS, Ford EJH: An investigation of the type of photosensitization caused by the ingestion of St John's wort (*Hypericum perforatum*) by calves. *J Comp Pathol* 1981, 91:135–141.

11. Dollahite JW, Younger RL, Hoffman GO: Photosensitization in cattle and sheep caused by feeding *Ammi majus* (Greater Ammi, Bishop's weed). *Am J Vet Res* 1978, 39:193–197.

12. Stermitz FR, Thomas RD: Furocoumarins of *Cymopterus watsonii*. *Phytochemistry* 1975, 14:1681.

13. Oertli EH, Rowe LD, Lovering SL, *et al.*: Phototoxic effect of *Thamnosa texana* (Dutchman's breeches) in sheep. *Am J Vet Res* 1983, 44:1126–1129.

14. Rowe LD, Norman JO, Corrier DE *et al.*: Photosensitization of cattle in southeast Texas: identification of phototoxic activity associated with *Cooperia pedunculata*. *Am J Vet Res* 1987, 48:1658–1661.

15. Casteel SW, Rowe LD, Bailey EM, *et al.*: Experimentally induced photosensitization in cattle with *Cooperia pedunculata*. *Vet Hum Toxicol* 1988, 30:101–104.

16. Rowe LD, Norman JO: Detection of phototoxic activity in plant specimens associated with primary photosensitization in livestock using a simple microbiological test. *J Vet Diagn Invest* 1989, 1:269–270.

17. Flemming CE, Miller MR, Vawter LR: The spring rabbit-brush. *Nevada Agricultural Experimental Station Bulletin* 1922,104:1–29.

18. Johnson AE: Predisposing influence of range plants on Tetradymia-related photosensitization in sheep: work of Drs AB Clawson and WT Huffman. *Am J Vet Res* 1974, 35:1583–1585.

19. Bull LB, Culvenor CCK, Dick AT: *The Pyrrolizidine Alkaloids: Their Chemistry, Pathogenicity and Other Biological Properties.* Amsterdam: North Holland Publishing, 1968.

20. Johnson AE, Molyneux RJ, Ralphs MH: Senecio: A dangerous plant for man and beast. *Rangelands* 1989, 11:261–264.

A GUIDE TO PLANT POISONING

21. Johnson AE, Molyneux RJ: Toxicity of threadleaf groundsel (*Senecio douglasii var longilobus*) to cattle. *Am J Vet Res* 1984, 45:26–31.

22. Johnson AE, Molyneux RJ, Stuart LD: Toxicity of Riddell's groundsel (*Senecio riddellii*) to cattle. *Am J Vet Res* 1985, 46:577–582.

23. Molyneux RJ, Johnson AE, Olsen JD, Baker DC: Toxicity of pyrrolizidine alkaloids from riddell's groundsel (*Senecio riddellii*) to cattle. *Am J Vet Res* 1991, 52:146–151.

24. Lombardo de Barros CS, Driemeier D, Pilati C, *et al.*: *Senecio* spp. poisoning in cattle in southern Brazil. *Vet Hum Toxicol* 1992, 34:241–245.

25. Walker KH, Kirkland PD. *Senecio lautus* toxicity in cattle. *Aust Vet J* 1981, 57:1–7.

26. Noble JW, Crossley JdeB, Hill BD, *et al.*: Pyrrolizidine alkaloidosis of cattle associated with *Senecio lautus*. *Aust Vet J* 1994, 71:196–200.

27. Winter H, Seawright AA, Noltie HJ, *et al.*: Pyrrolizidine alkaloid poisoning of yaks: identification of the plants involved. *Vet Rec* 1994, 134:135–139.

28. Odriozola E, Campero C, Casaro A, *et al.*: Pyrrolizidine alkaloidosis in Argentinian cattle caused by *Senecio selloi*. *Vet Hum Toxicol* 1994, 36:205–208.

29. Snyder SP: Livestock losses due to tansy ragwort poisoning. *Oregon Agric Rec* 1972, 255:2–4.

30. Lafranconi M, Huxtable RJ: Pyrrolizidines and the pulmonary vasculature. *Rev Drug Metab Drug Interactions* 1981, III:271–315.

31. McCulloch EC: Hepatic cirrhosis of horses, swine, and cattle due to the ingestion of the seeds of tarweed, *Amsinckia intermedia*. *J Am Vet Med Assoc* 1940, 96:5–17.

32. Piercy PL, Rusoff LL: *Crotolaria spectabilis* poisoning in Louisiana livestock. *J Am Vet Med Assoc* 1946, 108:69–73.

33. Gibbons WJ, Hoakson JF, Wiggins AM, Schmitz MB: Cirrhosis of the liver in horses. *North Am Vet* 1950, 31:229–233.

34. Sanders DA, Shealy AL, Emmel MW: The pathology of *Crotolaria spectabilis* Roth poisoning in cattle. *J Am Vet Med Assoc* 1936, 89:150–162.

35. McGrath JPM, Duncan JR, Munnell JF: *Crotolaria spectabilis* toxicity in swine: characterization of the renal glomerular lesion. *J Comp Pathol* 1975, 85:185–193.

36. Allen JR, Carstens LA, Knezevic AL: *Crotolaria spectabilis* intoxication in monkeys. *Am J Vet Res* 1965, 26:753–757.

37. John CP, Sangster LT, Jones OH: *Crotolaria spectablis* poisoning in swine. *J Am Vet Med Assoc* 1974, 165:633–638.

38. Knight AP, Kimberling CV, Stermitz FR, Roby MR: *Cynoglossum officinale* (hound's tongue)—a cause of pyrrolizidine alkaloid poisoning in horses. *J Am Vet Med Assoc* 1984, 185:647–650.

39. Baker DC, Smart RA, Ralphs M, Molyneux RJ: Hound's-tongue (*Cynoglossum officinale*) poisoning in a calf. *J Am Vet Med Assoc* 1989, 194:929–930.

40. Baker DC, Pfister JA, Molyneux RJ, Kechele P: *Cynoglossum officinale* toxicity in calves *J Comp Pathol* 1991, 104:403–410.

41. Stegelmeier BL, Gardner DR, Molyneux RJ, *et al.*: The clinicopathologic changes of *Cynoglossum officinale* (houndstongue) intoxication in horses. *In* Plant-Associated Toxins. Colegate SM, Dorling PR, Eds. Wallingford, Oxon, UK: Cab International; 1994:297–302.

42. Ketterer PJ, Glover PE, Smith LW: Blue heliotrope (*Heliotropium amplexicaule*) poisoning in cattle. *J S Afr Vet Assoc* 1975, 46:121–122.

43. Birecka H, Frolich MW, Hull L, Chaskes MJ: Pyrrolizidine alkaloids of Heliotropium from Mexico and adjacent USA. *Phytochemistry* 1980, 19:421–436.

44. Seaman JT, Turvey WS, Ottaway SJ, *et al.*: Investigations into the toxicity of *Echium plantagineum* in sheep. 1. Field grazing experiments. *Aust Vet J* 1989, 66:279–285.

45. Seaman JT, Dixon RJ: Investigations into the toxicity of *Echium plantagineum* in sheep. 2. Pen feeding experiments. *Aust Vet J* 1989, 66:286–292.

46. Culvenor CCJ, Edgar JA, Smith LW: Pyrrolizidine alkaloids in honey from *Echium plantagineum* L. *J Agric Food Chem* 1981, 29:958–960.

47. Culvenor CCJ *et al.*: Toxicity of *Echium plantagineum* (Paterson's curse). Marginal toxic effects in Merino wethers from long-term feeding. *Aust J Agric Res* 1984, 35:293–304.

48. Stillman AE, Huxtable R, Consroe P, *et al.*: Hepatic veno-occlusive diseases due to pyrrolizidine (*Senecio*) poisoning in Arizona. *Gastroenterology* 1977, 73:349–352.

49. Huxtable RJ. Herbal teas and toxins: Novel aspects of pyrrolizidine poisoning in the United States. *Perspect Biol Med* 1980, 24:1–14.

50. Huxtable RJ: Herbal teas and pyrrolizidine alkaloids. *In* Symposium on Pyrrolizidine Alkaloids: Toxicity, Metabolism and Poisonous Plant Control Measures. Cheeke PR, Ed. Corvalis, OR: Nutrition Research Institute, Oregon State University; 1979.

51. Mattocks AR: Toxic pyrrolizidine alkaloids in comfrey. *Lancet* 1980, November:1136–1137.

52. Hirono I, Mori H, Haga M: Carcinogenic activity of *Symphytum officinale*. *J Nat Cancer Inst* 1978, 61:865–869.

53. Geoger DE, Cheeke PR, Schmitz JA, Buhler DR: Toxicity of tansy ragwort (*Senecio jacobaea*) to goats. *Am J Vet Res* 1982, 43:252–254.

54. Mattocks AR: Recent studies on mechanisms of cytotoxic action of pyrrolizidine alkaloids. *In* Effects of Poisonous Plants on Livestock. Keeler RF, VanKampen KR, James LF, Eds. New York: Academic Press; 1978:177–187.

55. Swick RA: Hepatic metabolism and bioactivation of mycotoxins and plant toxins. *J Anim Sci* 1984, 58:1017–1028.

56. Hoper PT: Pyrrolizidine alkaloid poisoning: pathology with particular reference to differences in animal and plant species. *In* Effects of Poisonous Plants on Livestock. Keeler KR, VanKampen KR, James L, Eds. New York: Academic Press; 1978:161–176.

57. Segal HG, Wilson DW, Dallas JL, Haddon WF: Trans-4-hydroxy-2-hexenal: a reactive metabolite from the macrocyclic pyrrolizidine alkaloid senecioine. *Science* 1985, 229:472–475.

58. Everist SL, Ed: *Poisonous Plants of Australia*. Sydney: Angus and Robertson; 1981:405.

59. Mattocks AR: Toxicity of pyrrolizidine alkaloids. *Nature* 1968, 217:723–728.

60. Jago MV: The development of the hepatic megalocytosis of chronic pyrrolizidine alkaloid poisoning. *Am J Pathol* 1969, 56:405–422.

61. Huxtable RJ, Paplanus S, Laugharn J: The prevention of monocrotaline-induced right ventricular hypertrophy. *Chest* 1977, 71S: 308–310.

62. Roth LA, Dotzlaf LA, Kuo CH, *et al.*: Effect of monocrotaline ingestion on liver, kidney, and lung of rats. *Pharm Appl Pharmacol* 1981, 60:193–200.

63. Mattocks AR: Pyrrolizidine alkaloids: what metabolites are responsible for extrahepatic tissue damage in animals. *In* Poisonous Plants: Proceedings of the 3rd International Symposium. James LF, Keeler RF, Bailey EM, *et al..*, Eds. Ames, IA: Iowa State University Press; 1992:192–197.

64. Nobre D, Dagii MLZ, Haraguchi M: *Crotolaria juncea* intoxication in horses. *Vet Hum Toxicol* 1994, 36:445–448.

65. Schoental R: Toxicology and carcinogenic action of pyrrolizidine alkaloids. *Cancer Res* 1968, 28:2237–2246.

66. Johnson AE, Smart RA: Effects on cattle and their calves of tansy ragwort (*Senecio jacobaea*) fed in early gestation. *Am J Vet Res* 1983, 44:1215–1219.

67. Dickinson JO, Cooke MP, King RR, Mohamed PA: Milk transfer of pyrrolizidine alkaloids in cattle. *J Am Vet Med Assoc* 1976, 169:1192–1196.

66. Johnson AE: Toxicologic aspects of photosensitization in livestock. *J Natl Cancer Inst* 1982, 69:253–258.

67. Johnson AE: Changes in calves and rats consuming milk from cows fed chronic lethal doses of *Senecio jacobaea* (tansy ragwort). *Am J Vet Res* 1976, 37:107–110.

68. Goeger DE, Cheeke PR, Schmitz JA, Buhler DR: Effect of feeding milk from goats fed tansy ragwort (*Senecio jacobaea*) to rats and calves. *Am J Vet Res* 1982, 43:1631–1633.

69. Johnson AE, Molyneux RJ, Merrill GB: Chemistry of toxic range plants. Variation in pyrrolizidine alkaloid content of *Senecio*, *Amsinckia*, and *Crotolaria* species. *J Agric Food Chem* 1985, 33:50–55.

70. Molyneux RJ: Extraordinary levels of production of pyrrolizidine alkaloids in *Senecio ridelii*. *J Natural Products* 1984, 47:1030–1032.

71. Candrian U, Luthy J, Schlatter C, Gallasz E: Stability of pyrrolizidine alkaloids in hay and silage. *J Agric Food Chem* 1984, 32:935–937.

72. Goeger DE, Cheeke PR, Schmitz JA, Buhler DR: Toxicity of tansy ragwort (*Senecio jacobaea*) to goats. *Am J Vet Res* 1982, 43:252–254.

73. Lanigan GW: *Peptococcus heliotrinreducans*: a cytochrome producing anaerobe which metabolizes pyrrolizidine alkaloids. *J Gen Microbiol* 1976, 94:1–10.

74. Craig AM, Latham CJ, Blythe LL, *et al.*: Metabolism of toxic pyrrolizidine alkaloids from tansy ragwort (*Senecio jacobaea*) in ovine ruminal fluid under anaerobic conditions. *Appl Environ Microbiol* 1992, 58: 2730–2736.

75. Craig MA, Blythe LL, Lassen ED: Anaerobic metabolism of toxic compounds by ruminant bacteria. *In* Poisonous Plants: Proceedings of the 3rd International Symposium. James LF, Keeler RF, Bailey EM, *et al*, Eds.: Ames, IA: Iowa State University Press; 1992:208–214.

76. Cheeke PR: Comparative toxicity and metabolism of pyrrolizidine alkaloids (tansy ragwort [*Senecio jacobaea*]) in ruminants and nonruminant herbivores. *Can J Anim Sci* 1984, 84(suppl):201, 202.

77. Goeger DE, Dheeke PR, Buhler DR, Schmitz JA: Effect of tansy ragwort consumption. *In* Symposium on Pyrrolizidine (*Senecio*) Alkaloids: Toxicity, Metabolism and Poisonous Plant Control Measures. Cheeke PR, Eds. Corvalis, OR: Oregon State University; 1979:77–83.

78. Johnson AE: Toxicity of tansy ragwort to cattle. *In* Symposium on Pyrrolizidine (*Senecio*) Alkaloids: Toxicity, Metabolism and Poisonous Plant Control Measures. Cheeke PR, Eds. Corvalis, OR: Oregon State University; 1979:129–134.

79. Johnson AE: Tolerance of cattle to tansy ragwort (*Senecio jacobaea*). *Am J Vet Res* 1978, 39:1542–1544.

80. King PR, Dickinson JD: Comparative aspects of pyrrolizidine alkaloid toxicity in cattle and goats. *In* Symposium on Pyrrolizidine (*Senecio*) Alkaloids: Toxicity, Metabolism and Poisonous Plant Control Measures. Cheeke PR, Eds. Corvalis, OR: Oregon State University;1979:69–76.

81. Radostits OM, Blood DC, Gay CC, Eds: *Veterinary Medicine*, edn. 8. London: Bailliere Tindall; 1994:1600–1602.

82. Glasonbury JRW, Doughty FR, *et al*: A syndrome of hepatogenous photosensitization

resembling geeldikkop in sheep grazing *Tribulus terrestris. Aust Vet J* 1984, 61:314–316.

83. Coetzer JAW, Kellerman TS, Sadler W, Bath GF: Photosensitivity in South Africa. V. A comparative study of the pathology of the ovine hepatogenous photosensitivity diseases, facial eczema and geeldikkop (*Tribulosis ovis*) with special reference to their pathogenesis. *Ondersteport J Vet Res* 1983, 50:59–71.

84. Kellerman TS, Miles CO, Erasmus GL, *et al.*: The possible role of steroidal saponins in the pathogenisis of geeldikkop, a major hepatogenous photosensitization of small stock in South Africa. *In* Plant-Associated Toxins. Colegate SM, Dorling PR, Eds. Wallingford, Oxon, UK: Cab International; 1994:287–292.

85. Dollahite JW, Younger RL: Photosensitization in lambs grazing kleingrass. *J Am Vet Med Assoc* 1977, 171:1264–1267.

86. Muchiri DJ, Bridges CH, Ueckert DN, Bailey EM. Photosensitization of sheep on Kleingrass pasture. *J Am Vet Med Assoc* 1980, 177:353–354.

87. Mathews FP: Lechuguilla (*Agave lecheguilla*) poisoning in sheep, goats and laboratory animals. *Texas Agricultural Experimental Station Bulletin* 1937, 554:1–36.

88. Burrows G: Apparent Agave lecheguilla poisoning in Angora goats. *Vet Hum Toxicol* 1990, 32:259–260.

89. Abdelkader SV, Ceh L, Dishington IW, Hauge JG: Alveld-producing saponins. II Toxicological studies. *Acta Vet Scand* 1984, 25:76–85.

90. Pass MA, Heath T: Gallbladder paralysis in sheep during lantana poisoning. *J Comp Pathol* 1977, 87:301–306.

91. Pass MA: The mechanism and treatment of lantana poisoning. *Vet Clin Toxicol Proc* 1987, 103:19–22.

92. Pass MA, Seawright AA, Heath TJ, Gemmell RT: Lantana poisoning: a cholestatic disease of cattle and sheep. *In* Poisonous Plants: Effects on Livestock. Edited by Keeler RF, Van Kampen K, James LF, Eds. New York: Academic Press; 1978: 229–237.

93. Sharma OP, Makkar HPS, Dawra RK: A review of the noxious plant *Lantana camara*. *Toxicon* 1988; 26:975–987.

94. Nation PN: Alsike clover poisoning: a review. *Can Vet J* 1989, 30:410–415.

95. Nation PN: Hepatic disease in Alberta horses: a retrospective study of "alsike clover poisoning" (1973–1988). *Can Vet J* 1991, 32:602–607.

96. Traub JL, Potter KA, Bayley WM, *et al.*: Alsike clover poisoning. *Mod Vet Prac*, 1982; April:307–309.

97. Colon JL, Jackson CA, Del Piero F: Hepatic dysfunction and photodermatitis secondary to alsike clover poisoning. *Compendium of Continuing Education* 1996, 18:1022–1029.

98. Fincher MG, Fuller HK: Photosensitization-trifoliosis-light sensitization. *Cornell Vet* 1942, 32:95–98.

99. Morrill CC: Clover sickness or trifoliosis. *North Am Vet* 1943, 24:731–732.

100. Ames T, Angelos J, Gould S, *et al.*: Secondary photosensitization in horses eating *Cymodothea trifolli*-infested clover. *Am Assoc Vet Lab Diagn* 1994, 37:45.

101. Jubb KVF, Kennedy PC, Palmer N: *Trifolium hybridum* (alsike clover), *In* Pathology of Domestic Animals, vol 2, edn. 4. Jubb KVF, Kennedy PC, Palmer N, Eds. New York: Academic Press; 1993:398.

102. Richard JL: Mycotoxin photosensitivity. *J Am Vet Med Assoc* 1973, 163:1298–1299

103. Monlux AW, Glen BL, Panciera RJ, *et al.*: Bovine hepatogenous photosensitivity associated with feeding of alfalfa hay. *J Am Vet Med Assoc* 1963, 143:989–994.

104. Bagley CV, McKinnon JB, Asay CS: Photosensitization associated with exposure

of cattle to moldy straw. *J Am Vet Med Assoc* 1983, 183:802–803.

105. Putnam MR, Qualls CW, Rice LE, *et al.:* Hepatic enzyme changes in bovine hepatogenous photosensitivity caused by water-damaged alfalfa hay. *J Am Vet Med Assoc* 1986, 189:77–82.

106. Scruggs DW, Blue GK: Toxic hepatopathy and photosensitization in cattle fed moldy alfalfa hay. *J Am Vet Med Assoc* 1994, 204:264–266.

107. House JK, George LW, Oslund KL, *et al.:* Primary photosensitization related to ingestion of alfalfa silage by cattle. *J Am Vet Med Assoc* 1996, 209:1604–1607.

108. Casteel SW, Rottinghaus GE, Johnson GE, *et al.:* Hepatotoxicosis in cattle induced by consumption of alfalfa-grass hay. *In* Plant-Associated Toxins. Colegate SM, Dorling PR, Eds. Wallingford, Oxon: Cab International; 1994:307–312.

109. Beasley VR, Cook WO, Dahlem AM, *et al.:* Algae intoxication in livestock and waterfowl. *Vet Clin North Am* 1989, 5:345–361.

110. Beasley VR, Dahlem AM, Cook WO, *et al.*: Diagnostic and clinically important aspects of cyanobacterial (blue-green algae) toxicoses. *J Vet Diagn Invest* 1989, 1:359–365.

111. Osweiler G, Carson TL, Buck WB, Van Gelder GA, Eds: Toxic blue-green algae. *In* Clinical and Diagnostic Veterinary Toxicology, edn. 3. Ames, IA: Iowa State University Press; 1989:451–452.

112. Jackson ARB, McInnes A, Falconer IR, Runnnegar MTC: Clinical and pathological changes in sheep experimentally poisoned by the blue-green alga *Microcystis aeruginosa*. *Vet Pathol* 1984, 21:102–113.

113. Galey FD, Beasley VR, Carmichael WW, *et al.:* Blue-green algae (*Microcystis aeruginosa*) hepatotoxicosis in dairy cows. *Am J Vet Res* 1987, 48:1415–1420.

114. Hooser SB, Beasley VR, Basgall EJ, *et al.:* Microcystin-LR-induced ultrastructural changes in rats. *Vet Pathol* 1990, 27:9–15.

115. Hoover JP, Smith TA: Investigating a case of suspected cyanobacteria (blue-green algae) intoxication in a dog. *Vet Med* 1995, 90:1028–1032.

116. Ford EJH: The clinical aspects of ragwort poisoning in horses. *Vet Annu* 1973, 14:86–88.

117. Mendez MC, Riet-Correa F: Intoxication by *Senecio tweediei* in cattle in southern Brazil. *Vet Hum Toxicol* 1993, 35:55.

118. Smith BP: Pyrrolizidine alkaloid-induced hepatic disease in a group of calves. *Compendium of Continuing Education* 1982, 4:531–533.

119. McGinnes JP: *Senecio jacobaea* as a cause of hepatic encephalopathy. *Calif Vet* 1980, 34:20–22.

120. Seaman JT: Hepatogenous chronic copper poisoning in sheep associated with grazing *Echium plantagineum*. *Aust Vet J* 1985, 63:247–248.

121. Pearson EG: Liver failure attributable to pyrrolizidine alkaloid toxicosis and associated with inspiratory dyspnea in ponies: three cases (1982–1988). *J Am Vet Med Assoc* 1991, 198:1651–1654.

122. White RD, Swick RA, Cheeke PR: Effects of dietary copper and molybdenum on tansy ragwort (*Senecio jacobaea*) toxicity in sheep. *Am J Vet Res* 1984, 45:159–161.

123. Dewes HF, Lowe MD: Hemolytic crisis associated with ragwort poisoning and rail chewing in two thoroughbred fillies. *N Z Vet J* 1985, 33:159–160.

124. Milne CM, Podgson DM, Doxey DL: Secondary gastric impaction associated with ragwort poisoning in three ponies. *Vet Rec* 1990, 128:502–504.

125.   Bull LB: The histological evidence of liver damage from pyrrolizidine alkaloids. *Aust Vet J* 1955, 31:33–40.

126.   Craig AM, Pearson EG, Meyer C, Schmitz JA: Serum liver enzyme and histopathologic changes in calves with chronic and chronic-delayed *Senecio jacobaea* toxicosis. *Am J Vet Res* 1991, 52:1969–1978.

127.   Mendel VE, Witt MR, Gitchell BS, *et al.*: Pyrrolizidine alkaloid-induced liver disease in horses: an early diagnosis. *Am J Vet Res* 1988, 49:572–578.

128.   Ford EJH, Ritchie HE, Thorpe E: Serum changes following the feeding of ragwort (*Senecio jacobaea*) to calves. *J Comp Pathol* 1968, 78:207–218.

129.   Lessard P, Wilson WD, Olander HJ, *et al.*: Clinicopathologic study of horses surviving pyrrolizidine alkaloid (*Senecio vulgaris*) toxicosis. *Am J Vet Res* 1986, 47:1776–1780.

130.   Craig AM, Meter C, Koller LD, Schmitz JA: Serum enzyme tests for pyrrolizidine alkaloid toxicosis. *American Association of Veterinary Laboratory Diagnosticians* 21st Proceedings. 1978:161–178.

131.   Curran JM, Sutherland RJ, Peet RL: A screening test for subclinical liver disease in horses affected by pyrrolizidine alkaloid toxicosis. *Aust Vet J* 1996, 74:236–240.

132.   Giesecke PR: Serum biochemistry in horses with Echium poisoning. *Aust Vet J* 1986, 63:90–91.

133.   Rogers GR, Knight H, Gulick BA: Proposed method of diagnosis and treatment of pyrrolizidine alkaloid toxicity in horses. *In* Symposium of Pyrrolizidine (*Senecio*) Alkaloids: Toxicity, Metabolism and Poisonous Plant Control Measures Cheeke PR, Eds. Corvalis, OR: Oregon State University; 1979:129–134.

134.   Gulick BA, Liul KM, Qualis CW, *et al.*: Effect of pyrrolizidine alkaloid induced hepatic disease on plasma amino acid patterns in the horse. *Am J Vet Res* 1980, 41:1894–1898.

135.   Mattocks AR. Jukes R: Recovery of the pyrrolic nucleus of pyrrolizidine alkaloid metabolites from sulphur conjugates in tissues and body fluids. *Chemico-Biological Interactions* 1990, 75:225–239.

136.   Winter H, Seawright AA, Hrdlicka J, *et al.*: Pyrrolizidine alkaloid poisoning of yaks: diagnosis of pyrrolizidine alkaloid exposure by the demonstration of sulphur-conjugated pyrrolic metabolytes of the alkaloid in circulating hemoglobin. *Aust Vet J* 1993, 70:312-313.

137.   Stegelmeier BL, Gardner DR, James LF, Molyneux RJ: Pyrrole detection and pathologic progression of *Cynoglossum officinale* (houndstongue) poisoning in horses. *J Vet Diagn Invest* 1996, 8:81-90.

138.   Dickinson JO, Ball RA: Management of pyrrolizidine alkaloid toxicity. *Calif Vet* 1985, January:15-17.

139.   Garrett BJ, Holtan DW, Cheeke PR, *et al.*: Effects of dietary supplementation with butylated hydroxyanisole, cysteine, and vitamins B on tansy ragwort (*Senecio jacobaea*) toxicosis in ponies. *Am J Vet Res* 1984, 45:459–464.

140.   Cheeke PR, Schmitz JA, Lassen ED, Pearson EG. Effects of dietary supplementation with ethoxyquin, magnesium oxide, methionine hydroxy analog, and B vitamins on tansy ragwort (*Senecio jacobaea*) toxicosis in beef cattle. *Am J Vet Res* 1985, 46:2179–2183.

141.   Pemberton RW, Turner CE: Biological control of *Senecio jacobaea* in northern California, an enduring success. *Entomophga* 1990,35:71–77.

142.   Witzel DA, Dollahite JW, Jones LP: Photosensitization in sheep fed *Ammi majus* (bishop's weed) seed. *Am J Vet Res* 1978, 39:319–320.

143. Dollahite JW, Younger RL, Hoffman GO: Photosensitization in cattle and sheep caused by feeding *Ammi majus* (greater ammi, bishop's weed). *Am J Vet Re*s 1978, 39:193–197.

144. Eilat A, Malkinson M, Schlosberg A, Egyed MN: A field outbreak of photosensitization in goslings caused by the ingestion of *Ammi majus*. *Refuah Vet* 1974, 31:83–86.

145. Eyged MN, Schlosberg A, Eilat A, *et al.:* Photosensitization in dairy cattle with the ingestion of *Ammi majus*. *Refuah Vet* 1974, 31:128–131.

146. Kingsbury JM: *Poisonous Plants of the United States and Canada.* Englewood Cliffs, NJ: Prentice-Hall;1964:228–230.

147. Kako MDN, Al-Sultan II, Saleem AN: Studies of sheep experimentally poisoned with *Hypericum perforatum*. *Vet Hum Toxicol* 1993, 35:298–300.

148. Southwell IA, Campbell MH: Hypericin content variation in *Hypericum perforatum* in Australia. *Phytochemistry* 1991, 30:475–478.

149. Lenard J, Rabson A, Vanderoef R: Photodynamic inactivation of infectivity of human immunodeficiency virus and other enveloped viruses using hypericin and rose bengal: Inhibition of fusion and syncytia formation. *Proc Natl Acad Sci U S A* 1993, 90:158–162.

150. Lopez-Bazzoli I, Hudson JB, Towers GHN: Antiviral activity of the plant pigment hypericin. Photochem. *Photobiol* 1991, 54:95–98.

151. Jungherr E: Lechuguilla fever of sheep and goats, a form of swell-head in west Texas. *Cornell Vet* 1931, 21:227–242.

152. Mathews FP: Poisoning in sheep and goats by sacahuiste (*Nolina texana*) buds and blooms. *Texas Agricultural Experimental Station Bulletin* 1940, 585.

153. Johnson AE: Experimental photosensitization and toxicity in sheep produced by *Tetradymia glabrata*. *Can J Comp Med* 1974, 38:405–410.

154. Johnson AE: Tetradymia toxicity—a new look at an old problem. *In* Effects of Poisonous Plants on Livestock. Keeler RF, Van Kampen KR, James LF, Eds. New York: Academic Press; 1978:209—216.

155. McDonough SP, Woodbury AM, Galey FD, *et al:* Hepatogenous photosensitization of sheep in California associated with ingestion of *Tribulus terrestris* (puncture vine). *J Vet Diagn Invest* 1994, 6:392-395.

156. Bourke CA: Staggers in sheep associated with the ingestion of *Tribulis terrestris*. *Aust Vet J* 1984, 61:360—364.

157. Mathews FP: The toxicity of *Kallstroemia hirsutissima* (carpetweed) for cattle sheep and goats. *J Am Vet Med Assoc* 1944, 10:152—155.

158. Bourke CA, Stevens GR, Carrigan MJ: Locomotor effects in sheep of alkaloids identified in Australian *Tribulis terrestris*. *Aust Vet J* 1992, 69:163-165.

159. Patamalai B, Hejtmancik D, Bridges CH, *et al:* The isolation and identification of steroidal sapogenins in kleingrass. *Vet Hum Toxicol* 1990, 32:314-318.

160. Pfister JA, Baker DC, Lacey JR, Brownson R: Photosensitization of cattle in Montana: is Descurainia pinnata the culprit? *Vet Hum Toxicol* 1989, 31:225—227.

161. Pfister JA, Baker DC, Lacey JR, Brownson R: Is tansy mustard causing photosensitization in Montana? *Rangelands* 190, 12:170—172.

162. Galitzer SJ, Oehme FW: Kochia scoparia (L) Schrad toxicity in cattle: a literature review. *Vet Hum Toxicol* 1978, 20:421—423

163. Sprowls RW: Problems observed in horses, cattle and sheep grazing kochia.

Proceedings of the 24th Annual Meeting of the American Association Veterinary Laboratory Diagnosticians 1981: 397–406.

164. Dickie CW, Gerlach ML Hamar DW: Kochia scoparia oxalate content. *Vet Hum Toxicol* 1989, 31:240–241.

165. Sherrod LB: Nutritive value of Kochia scoparia. Yeild and chemical composition at three stages of maturity. *Agron J* 1971, 63:343–344.

166. Kirkpatrick JG, Helman RG, Burrows GE, *et al.*: Evaluation of hepatic changes and weight gains in sheep grazing Kochia scoparia. *Vet Hum Toxicol* 1999, 41:67–70.

167. Thilsted J, Hibbs C, Kiesling H, *et al.*: Kochia (*Kochia scoparia*) toxicosis in cattle: results of four experimental grazing trials. *Vet Hum Toxicol* 1989, 31:34–41.

168. Rankins DL, Smith GS: Nutritional and toxicological evaluations of kochia hay (*Kochia scoparia*) fed to lambs. *J Anim Sci* 1991, 69:2925–1931.

169. Dickie CW, Berryman JR: Polioencephalomalacia and photosensitization associate with *Kochia scoparia* consumption in range cattle. *J Am Vet Med Assoc* 1979, 175:463–465.

170. Dickie CW, James LF: *Kochia scoparia* poisoning in cattle. *J Am Vet Med Assoc* 1983, 183:765–768.

171. Loneragan GH, Gould DH, Callan RJ, *et al.*: Association of excess sulfur intake and an increase in hydrogen sulfide concentrations in the ruminal gas cap of recently weaned beef calves with polioencephalomalacia. *J Am Vet Med Assoc* 1998, 213:1599–604.

172. Craig JC, Mole ML, Billets S, El-Feraly F: Isolation and identification of the hypoglycemic agent, carboxyatractylate, from *Xanthium strumarium*. *Phytochemistry* 1976, 15:1178.

173. Cole RJ, Stuart BP, Lansden JA, Cox RH: Isolation and redefinition of the toxic agent from cocklebur (*Xanthium strumarium*). *J Agric Food Chem* 1980, 28:1330–1332.

174. Martin T, Stair EL, Dawson L: Cocklebur poisoning in cattle. *J Am Vet Med Assoc* 1986, 189:562–563.

175. Witte ST, Osweiler GD, Stahr HM, Mobley G: Cocklebur toxicosis in cattle associated with the consumption of mature *Xanthium strumarium*. *J Vet Diagn Invest* 1990, 2:263–267.

176. Stuart BP, Cole RJ, Gosser HS: Cocklebur (*Xanthium strumarium. L.* var. *strumarium*) intoxication in swine: review and definition of the toxic principle. *Vet Pathol* 1981, 18:368–383.

177. Hatch RC, Jain AV, Weiss R, Clark JD: Toxicologic study of carboxyactractyloside (active principle in cocklebur, *Xanthium strumarium*) in rats treated with enzyme inducers and inhibitors and glutathione precursor and depletor. *Am J Vet Res* 1982, 43:111–116.

178. Panciera RJ, Johnson L, Osburn BI: A disease of cattle grazing hairy vetch pasture. *J Am Vet Med Assoc* 1966, 148:804-808.

179. Anderson CA, Divers TJ: Systemic granulomatous inflammation in a horse grazing hairy vetch. *J Am Vet Med Assoc* 1983, 183:569-570.

180. Kerr LA, Edwards WC. *Vet Med Sm Anim Clinician* 1982,77:257-258.

181. Harper PAW, Cook RW, Gill PA *et al*: Vetch toxicosis in cattle grazing *Vicia villosa* ssp. *dassycarpa* and *V. benghalensis*. *Australian Vet J* 1993, 70: 140-144.

182. Odriozola E, Paloma E, Lopez T, Campero C: An outbreak of *Vicia villosa* (hairy vetch) poisoning in grazing Aberdeen Angus bulls in Argentina. *Vet Hum Toxicol* 1991, 33:278-280.

183. Panciera RJ, Mosier DA, Ritchey JW: Hairy vetch (*Vicia villosa* Roth) poisoning in cattle: update and experimental induction of disease. *J Vet Diagn Invest* 1992,4:318-325.

184. Johnson B, Moore J, Woods LW, Galey FD: Systemic granulomatous disease in cattle in California associated with grazing hairy vetch (*Vicia villosa*). *J Vet Diagn Invest* 1992, 4:360-362.

185. Tollefsen SE, Kornfeld R: Isolation and characterization of lectins from *Vicia villosa*. *J Biol Chem* 1983, 258:5165-5171.

186. Claughton WP, Claughton HD: Vetch seed poisoning. *Auburn Vet* 1954, 10:125-126.

187. Hegarty MP, Court RD, Christie GS, Lee CP: Mimosine in *Leucaena leucocephala* is metabolized to a goitrogen in ruminants. *Aust Vet J* 1976, 52:490-492.

188. Hegarty MP: Toxic amino acids of plant origin. *In* Effects of poisonous plants on livestock. KeelerRF, Van Kampen KR, James LR, Eds. 1978, 575-585.

189. Jones RJ, Blunt CG, Nurnberg BI: Toxicity of *Leucaena leucocephala*: the effect of iodine and mineral supplements on penned steers fed a sole diet of *Leucaena*. *Aust Vet J* 1978, 54:387-392.

190. Allison MJ, Hammond AC, Jones RJ: Detection of ruminal bacteria that degrade toxic dihydroxypyridine compounds produce from mimosine. *Appl Environ Microbiol* 1990, 56:590-594.

191. Allison MJ, Mayberry WR, McSweeny CS, Stahl DA: *Synergistes jonesii* gen. Nov, sp.nov.: a rumen bacterium that degrades toxic pyridinediols. *Syst Appl Microbiol.* 1992, 15:522-529.

192. McSweeny CS, Allison MJ, Mackie RI: Amino acid utilization by the ruminal bacterium *Synergistes jonesii* strain 78-1. *Arch Microbiol* 1993, 159:131-135.

193. Hammond AC: Leucaena toxicosis and its control in ruminants. *J Anim Sci* 1995, 73:1487-1492.

194. Holmes JH, Humphrey JD, Walton EA, O'Shea JD: Cataracts, goiter, and infertility in cattle grazed on an exclusive diet of *Leucaena leucocephala*. *Aust Vet J* 1981, 57: 257-261.

195. Wayman O, Iwanaga , Hugh II: Fetal resorbtion in swine caused by *Leucaena leucocephala* (Lam) in the diet. *J Anim Sci* 1970, 30:583-588.

196. Reynolds GW, Rodriguez E: Dermatotoxic phenolis from glandular trichomes of *Phacelia campanularia*, and *P. pedicellata*. *Phytochemistry* 1986, 25:1617–1619.

197. Baer H: Allergic contact dermatitis from plants. *In* Plant and Fungal Toxins, vol. 1. Keeler RF, Tu AT, Eds. New York: Marcel Dekker; 1983:421–442.

CHAPTER 5

Plants Affecting the Blood

A variety of plants contain compounds that damage red blood cell (RBCs) metabolism and cell membrane integrity causing them to be removed from the circulation by the spleen. As a result, the excess hemoglobin from the damaged RBCs is filtered by the kidneys causing hemoglobinuria. If present in large quantity, the hemoglobin causes the urine to become red or dark brown in color. Plants that contain toxic compounds capable of affecting the RBCs include onions (*Allium* spp.); common crop plants such as kale, rape, turnips (*Brassica* spp.); and red maple (*Acer rubrum*). Red colored urine may also be encountered with hairy vetch (*Vicia villosa*) poisoning. In cattle, plants such as bracken fern (*Pteridium aquilinum*), and sweet clover (*Melilotus officinalis*) inhibit normal blood clotting with resulting hemorrhage into the urinary system (hematuria). Bracken fern is also capable of inducing the formation of urinary bladder tumors in cattle which may cause red-colored urine.

Onion Poisoning

Allium **spp.**
A. cepa Domesticated onion
A. canadensis Wild onion
A. validum Wild onion
Liliaceae (Lily family)

Habitat

Wild onions are usually found in moist meadows, open hillsides, and in sandy bottomlands throughout North America.

Description

Onions are herbaceous plants with bulbs and narrowly linear leaves that smell of "onion." The leaves are sheathing, usually basal, and hollow. The stem is simple and erect with a terminal umbel subtended by two or three membranous bracts. The six-parted flowers may be white, purple, pink, or green and are borne on slender pedicels. (Figure 5-1). All parts of the plant smell of onion if crushed.

Principal Toxin

An alkaloid, *N*-propyl disulphide, present in both cultivated and wild onions, chives, and garlic, affects the enzyme, glucose-6-phosphate dehydrogenase in RBCs thereby interfering with the hexose monophosphate pathway.[1,2] Oxidation of hemoglobin results because there is insufficient phosphate dehydrogenase or glutathione to protect the RBCs from oxidative injury.[1] The oxidized hemoglobin precipitates in the RBCs to form Heinz bodies. The cells containing Heinz bodies are removed by the spleen with the resulting anemia being proportional to the number of Heinz bodies formed and the rate at which the spleen removes the damaged cells. Cattle are the most susceptible to onion poisoning; horses and dogs are intermediate; and sheep and goats are the most resistant.[3-8] Pregnant ewes are able to eat a diet consisting of 90 to 100 percent cull onions without developing a severe anemia as would cattle on a ration with more than 25 percent dry matter of onions. The sheep's adaptation to onions appears to be related to the ability of the animal's rumen microflora to rapidly change to a population of organisms capable of reducing the sulfide in the onions.[9] Sheep are therefore able to metabolize the sulfide in the rumen more effectively than can cattle, thus preventing a progressive Heinz body anemia from developing.

Dog breeds such as Akitas and Shibas with high erythrocyte levels of reduced glutathione and potassium are especially susceptible to the hemolytic effects of oxidants such as *N*-propyl disulfide.[10] Dogs may also be poisoned after eating quantities of onion even after the plants are cooked.[11]

Figure 5-1 Wild onion (*Allium* spp.).

The severity of Heinz body anemia that develops will vary with the quantity and the rate at which onions are consumed and the species of animal. Diets containing more than 25 percent dry matter of onion have the potential to cause clinical signs of anemia. Calves 6 to 12 months old consuming 8 to 15 kg/day of onions for 5 days develop characteristic Heinz body anemia.[12] The formation of Heinz bodies in sufficient numbers to cause anemia and hemoglobinuria may occur within 1 to 3 weeks of eating onions. Small numbers of Heinz bodies will be formed in cattle even though the amount of onion consumed may be less than that necessary to induce anemia.

Clinical Signs

Most onion poisoning is associated with the feeding of cull domestic onions to animals. It is rare to encounter poisoning from wild onions. The first noticeable sign of onion poisoning is often the presence of dark red-brown urine (hemoglobinuria). Affected animals have pale mucous membranes and a fast, weak pulse; they may stagger and collapse as a result of anemia. There is frequently a distinct odor of onion on the breath, feces, urine, and milk of poisoned animals. In severely anemic animals, additional stress and heavy parasite infestations may be sufficient to cause death of the animal. Lactating animals eating onions may have an onion flavor to their milk making it undesirable for human consumption. The onion flavor disappears from the milk after the lactating animal has been off of onions for 24 hours.

Treatment

Animals that are anemic from onion poisoning should not be stressed, and onion feeding should be discontinued. Whole blood transfusions may be necessary in severely anemic animals. Sheep being fed onions rarely require any treatment even though they are anemic because they adapt to an onion diet and do not develop a progressive Heinz body anemia that is fatal.[9]

Cull onions can be a valuable food source for livestock in onion-growing areas. Up to 25 percent dry matter of onions and can be successfully fed to cattle in a balanced ration.[12] The onions should be chopped and well mixed in cattle rations to avoid choking the animals. Sheep are able to eat far higher quantities of onion than cattle without detriment, although weight gains in feeder lambs on diets containing more than 50 percent dry matter of onions tend to be reduced.[9] Because cull onions consist of about 90 percent water, the total caloric intake necessary for growth will be limited due to the limited capacity of the rumen. Growth rates and milk production in cows may therefore be reduced if onions comprise the bulk of an animal's diet.

Brassica Poisoning

Members of the Brassicaceae (*Cruciferae*), especially those grown as crop plants (turnips, kale, rape, cabbage, cauliflower, broccoli, and brussels sprouts) contain a variety of toxic compounds that have different effects on animals that eat them.[13-18] Glucosinolates and sulfur-containing amino acids in these plants cause signs of poisoning ranging from goiter, hypothyroidism, blindness, diarrhea, red-colored urine, and pulmonary emphysema. These compounds are found in greatest quantity in the young green plant and in the seeds. However, signs of poisoning produced depends on the glucosinolate unique to the particular brasicca or mustard. Hemolytic anemia occurs in cattle due to the presence of the compound S-methy-L-cystein sulfoxide (SMCO), a sulfur-containing amino acid unique to the family Brassicaceae.[19] is converted in the rumen to dimethyl disulphide that oxidizes hemoglobin in a similar manner to N-propyl disulphide, a closely allied compound found in onions. The denatured hemoglobin is detectable as Heinz bodies in the RBCs that are removed from circulation by the spleen's reticuloendothelial system. Hemolysis of the cells also occurs as a result of oxidative damage to the cell membranes that result in the hemoglobinuria seen in affected animals. The severity of poisoning is greatest in cattle that are fed kale (*B. oleracea*), rape (*B. napus*), and turnips (*B. campestris*) grown as forage crops; sheep appear to be less susceptible.[19]

Cattle grazing various brassicas including turnips, rape, and canola may experience up to a 10 percent death loss as a result of bloat and acute pulmonary emphysema.[18] Green turnip tops are a rich source of tryptophan that is converted in the rumen to 3-methylindole (a 3-substituted furan). The bioactivated furans bind with the protein of type I cells of the lung airways, causing acute pulmonary emphysema and edema. If the animal survives the acute stage, rapid proliferation of cells in the lungs (type II alveolar cells) decrease normal air exchange and cause death from an interstitial or proliferative pneumonia. Other plants that contain similar 3-substituted furans capable of causing acute pulmonary emphysema and edema in cattle include perilla mint (*Perilla frutescen*) (Figure 5-2), and sweet potato (*Ipomoea batata*).

The toxicity of the turnip tops is markedly reduced after they have been frozen. The turnips themselves are relatively low in tryptopohan and are more likely the cause of thiamine deficiency induced polioencepahalomalacia a disease characterized by blindness, depression, and death.[18] Cattle may concurrently develop hemolytic anemia that can be overlooked in the presence of the other accompanying diseases. Nitrate poisoning may also result from feeding turnips and other brassicas.

Glucosinolate Poisoning

Brassica spp. contain various glucosinolates

Figure 5-2 Perilla, or purple mint (*Perilla frutescens*).

that can cause goiter and hypothyroidism, poor growth rates, and reproductive failure.[20] Glucosinolates are the precursors of active metabolites such as isothiocyanates and nitriles, irritants that can cause colic and diarrhea.[21] Glucosinolate levels are highest in seeds of brassicas, especially of the mustards, and it is for this reason that caution should be exercised when feeding rape seed meal to animals.[20] Decreased feed consumption, poor growth rates, and decreased milk production have been credited to the feeding of high levels of rape seed.[22] Acute fatal poisoning of cattle characterized by severe edema of the forestomachs and abomasum has been reported following consumption of large quantities of discarded mustard seed.[23,24] The irritant effects of the isothiocyanates presumably causes vascular damage and edema of the stomachs.[23,24]

Clinical Signs

If cattle are given sudden access to turnip fields after being on relatively dry, high roughage diets, they may develop nonfrothy bloat and acute respiratory distress as a result of pulmonary emphysema and edema.[18] Panting, open-mouthed breathing, coughing, and frothing at the mouth are typical of acute pulmonary emphysema. Mortality is usually high in animals with the acute respiratory involvement. Blindness and marked depression due to polioencephalomalacia may develop in some cattle after about a week of eating turnips. As with onion poisoning, cattle consuming brassicas develop hemoglobinuria, hematuria, and jaundice depending on the quantity of brassica consumed.[1] Cattle become progressively weaker and may eventually die from severe anemia unless they are removed from the source of the plants. Death of anemic animals can be accelerated by stress and by concurrent parasitic diseases that may worsen the anemia. Lactating cows fed *Brassica* spp. may also have off-flavored milk due to the presence of glucosinolates.

Tansy mustard (*Descurainia pinnata*), a member of the Brassica family and a common weed throughout North America, is sporadically associated with a syndrome in cattle characterized by photosensitization involving the nonpigmented skin around the face, neck, and udder (see chapter 4).[26] Feeding trials with tansy mustard, however, have been unsuccessful in creating the disease, suggesting other factors are involved.[26] Symptoms of poisoning appear to coincide with years in which timely rainfall results in lush growth of tansy mustard. Blindness, protrusion of the tongue, and difficulty in eating suggestive of polioencepalomalacia have also been associated with cattle forced to eat nothing but tansy mustard.

Treatment

When possible the affected cattle should be removed from the source of the brassica and provided a good-quality roughage diet. Severely anemic animals should be carefully handled to avoid stress. Blood transfusions may be lifesaving in animals that are severely anemic. For similar reasons, animals with pulmonary emphysema should be handled very carefully. Large doses of corticosteroids and diuretics, and oxygen therapy, if available and practical, may help reduce the acute lung edema and emphysema in the early stages of poisoning.

Cattle showing signs of blindness should be injected with large doses of thiamin (10 mg/kg body weight intravenously every 3 to 4 hours for 2 to 3 days). Corticosteroids may be beneficial in reducing brain edema. Rehydrating severely affected animals with large volumes of oral fluids and providing a good-quality grass/alfalfa hay is beneficial in treating animals with brassica poisoning.

Preventive Measures

In situations where brassicas and mustards are intended for cattle forage, problems with bloat and pulmonary emphysema can be reduced by gradually introducing the animals to the new food source. Waiting until after a hard freeze will markedly reduce the risk of acute pulmonary emphysema. To reduce the rapid consumption of the turnip tops, cattle should be fed a good-quality roughage beforehand so as to limit turnip intake. Feeding a few pounds of grain per head per day, along with roughages such as corn stalks prior to the turnips works well in some circumstances. The use of antibiotics such as monensin and lasalocid are also effective in reducing the conversion of tryptophan to 3-methylindole, the instigator of pulmonary emphysema. Additionally the antibiotics reduce rumen lactic acidosis and subsequent thiamin deficiency. Monensin can be administered as part of the grain or as a component of a liquid mineral/protein supplement.[18]

Red Maple Poisoning

Acer rubrum
Aceraceae (Maple family)

Habitat

Red maple trees are common throughout most of eastern North America and south to Florida and Texas. They adapt to moist or dry areas and are often planted as ornamental trees for their striking fall colors.

Description

These large trees attain heights of 100 feet (30 meters) at maturity (Figure 5-3A). Leaves have three to five lobes and are simple, opposite with red petiole, shiny green topside and white/gray underside (Figure 5-3B). Leaves turn bright red in the fall. Dense clusters of red flowers appear before leaves, the male and female flowers being on separate trees. Fruits are red in color and have two wings 0.75 to 1 inch (2 to 2.5 cm) long.

Principal Toxin

An unidentified toxin with oxidant properties is present in the wilted or dried leaves of red maples.[27-30] Only the red maple (*A. rubrum*) and possibly closely related hybrids are known to be toxic. Horses, ponies, and zebras appear to be the only animals affected by the toxin in red maples.[27,28,31] The toxin causes oxidant damage to hemoglobin resulting in the precipitation of the oxidized hemoglobin as Heinz bodies in the RBCs. Damage apparently also occurs to the RBC membranes, which results in hemolytic anemia.[28] Poisoning is especially likely in the fall or after a storm when leaves of fallen branches become accessible to horses.[32]

Figure 5-3A Red maple tree (*Acer rubrum*). The fresh green leaves apparently are not toxic, but once dried they may remain toxic for up to 30 days. The bark from red maple trees is also toxic. Fatal poisoning of ponies fed 3.0 kg of dried red maple leaves occurred in 1 to 5 days.[32] As little as 1.5 kg of dried red maple leaves will induce formation of Heinz bodies and anemia.[32]

Figure 5-3B Red maple leaves (*A. rubrum*).

Clinical Signs

After they eat relatively small amounts of dried red maple leaves, horses exhibit clinical signs within 1 to 2 days. Poisoning is characterized by an acute hemolytic anemia that causes weakness, increased respiratory and heart rates, cyanosis, icterus, and a red-brown coloration of the urine.[27-30] Pregnant mares may abort without showing signs of hemolytic anemia.[32] Blood changes include a marked reduction in the hematocrit, methemoglobinemia, Heinz bodies in the erythrocytes, and depletion of erythrocyte glutathione, the product essential to maintain hemoglobin in its reduced state.[28,30] Serum aspartate aminotransferase, sorbitol dehydrogenase, protein, and bilirubin blood levels are usually elevated.[28-30]

A diagnosis of red maple poisoning can generally made when horses develop an acute hemolytic anemia, with Heinz body formation, and evidence indicates that they have had access to and have eaten wilted or dried red maple leaves. The prognosis is always guarded to poor for horses with red maple poisoning because of the rapid development of intravascular hemolysis, coagulopathy, precipitation of hemoglobin in the kidneys, and vascular thrombosis.[32,34] Postmortem examination of horses fatally poisoned by red maple leaves reveals pale organ color due to the anemia, hemorrhages on serosal surfaces, and splenic enlargement.[32] In acutely poisoned horses, there may be dark brown blood and brownish discoloration of tissues due to the severe methemoglobinemia.[32] In the acute case, there is often evidence of liver lipidosis and necrosis.[31,32]

Treatment

Affected horses should be denied further access to red maple leaves and blood transfusions given as necessary. Administration of intravenous fluids is of benefit in preventing dehydration and maintaining kidney function.[30] Concurrent use of large doses of vitamin C (ascorbic acid) are also of benefit.[35] Methylene blue advocated for the treatment of erythrocyte oxidant damage should be used with caution in horses. Methylene blue is contraindicated if Heinz bodies are already formed because it induces Heinz body formation. The dosage of methylene blue should not exceed 8 mg/kg body weight and should be administered slowly intravenously as a 1 percent solution.

Prevention of red maple poisoning is best accomplished by maintaining a good feeding program for horses and removing red maple leaves and fallen branches from horse pens. It is inadvisable to plant red maple trees in or closely surrounding horse enclosures.

Yellow Sweet Clover

Melilotus officinalis
M. alba. (White sweet clover)
Fabaceae (Legume family)

Habitat

Yellow sweet clover is commonly grown in the northwestern United States and western Canada as a forage for livestock. It has, however, become established as a drought-tolerant weed and grows wild over much of the continent especially along roadsides and waste areas.

Description

As biennials, both species grow to 5 feet (1.5 meters) in height and have compound leaves with three leaflets that have serrated edges, with the terminal leaflet on a stalk. The pea-like yellow flowers are produced in axillary racemes up to 5 inches (12 to 13 cm) in length (Figure 5-4A). White sweet clover is very similar except for its white flowers (Figure 5-4B). Smooth yellow seeds are produced in pods 0.2 to 0.3 cm (2 to 3 mm) long.

Principal Toxin

Yellow sweet clover in itself is not poisonous. If, however, it becomes moldy, a variety of different fungi including *Penicillium* and *Aspergillus*, and *Mucor* spp. growing on sweet clover are capable of converting coumarin in the plant to dicoumarol (dicoumarin or dihydroxycoumarin), a potent anticoagulant.[36,37] Other plants containing coumarin and melilotin include sweet vernal grass (*Anthoxanthum odoratum*), *Lespedeza stipulacea*

Figure 5-4A Yellow sweet clover (*Melilotus officinalis*).

and other species of melilotus (melilot). It is the coumarin in grass that is responsible for the smell of freshly cut hay.[37,38] Sweet clover poisoning is clinically identical to warfarin poisoning, a common rodenticide. Moldy sweet clover poisoning is most commonly encountered in cattle, but occasionally horses and other livestock are susceptible to the effects of dicoumarol.[39-44] Sheep appear quite resistant to the toxic effects of dicoumarol.[41,42,44,45] Signs of poisoning may not appear for up to 3 weeks after feeding moldy sweet clover hay and depend on the quantity of dicoumarol consumed. Poisoning is likely to occur when dicoumarol levels exceed 10 mg/kg of hay.[45] Hay containing dicoumarol levels of 10 to 20 mg/kg of feed can be fed for 100 days before poisoning develops. Feeds containing 60 to 70 mg/kg of feed can cause poisoning in as little as 21 days.[45]

Dicoumarol has strong anticoagulant properties that interferes with the production of vitamin K and therefore affects vitamin K-dependent coagulation factors VII, IX, and

X and prothrombin. The rate of depletion of these factors is directly related to the duration and amount of dicoumarol ingested, usually over a period of several weeks. Affected animals, unable to synthesize these factors, fail to stabilize fibrin necessary for normal clotting of blood with resulting internal and external hemorrhaging. Calves are usually more severely affected than adult cattle, and dicoumarol can cross the placenta to affect the newborn calf.[46]

Sweet clover, being a legume, can also cause acute rumen bloat in cattle, especially if it is lush and leafy, and cattle are not accustomed to it. Like alfalfa, sweet clover produces a frothy bloat that can cause high mortality unless treated early.

Clinical Signs

Early signs of sweet clover poisoning are not easily recognized because affected animals may only appear weak and depressed. The sudden appearance of subcutaneous swellings, bleeding from the nose, and melena are common in sweet clover poisoning. Subcutaneous hematomas, especially ventrally and over areas that are easily traumatized, frequently develop. These fluctuant swellings are painless and are not hot to the touch, thus differentiating them from abscesses. Hematomas in the mesentery

Figure 5-4B White sweet clover (*Melilotus alba*).

with resulting colic may also develop.[38] Hemarthrosis is often seen in the carpal and hock joints. Lameness due to massive intramuscular hemorrhage can be the primary presenting sign, and a swollen leg can resemble blackleg, a severe myositis caused by *Clostridium chauvoei*.[37] Hemorrhaging may be observed in the anterior chamber of the eye, and hemorrhages may be seen in the mucous membranes. Vaginal hemorrhaging may be a presenting sign in dairy cattle with sweet clover poisoning.[43] Fatal hemorrhaging in cows at calving is a common occurrence.[41] Abdominocentesis often reveals intra-abdominal hemorrhage. The hemorrhagic syndrome of sweet clover poisoning has been likened to hemorrhagic septicemia due to Pasteurella hemolytica infections without the signs of inflammation.[47] Signs of anemia including pale mucous membranes and a rapid weak pulse are a consistent finding in sweet clover poisoning. Affected animals are afebrile and maintain a good appetite. Mortality is usually high.

Sweet clover poisoning should be suspected when hemorrhaging is excessive after surgical procedures such as castration and dehorning. Before any surgical procedures are performed, it is important to ensure that animals have not been fed sweet clover in the last month. Prothrombin times should be determined if animals have been eating sweet clover and surgery is contemplated. Prothrombin times above 40 seconds suggest decreased clotting ability. Blood prothrombin, activated partial thromboplastin times, and clotting times are markedly increased from their normal values of 9 to 12 seconds, 30 to 45 seconds, and 3 to 15 minutes, respectively.

Improperly cured or spoiled sweet clover hay and haylage is not always toxic, but should only be used for animal feed after it has been tested for the presence of dicoumarol. In one survey of 272 cured sweet clover hay samples, over one-third contained more than 10 mg/kg of dicoumarol (range, 0 to 164.7 mg/kg).[48] The concentration of dicoumarol tends to be higher in large round bales where the hay is more likely to have a higher moisture content.[38] Dicoumarol-containing sweet clover may be fed to livestock as long as it does not constitute more than 25 percent of the animals' total diet, although feeding any moldy feed is not recommended. Sweet clover hay containing less than 20 µg/g is safe to feed to cattle. Concentrations of dicoumarol exceeding 50 µg/g are toxic and will severely affect an animal's blood clotting ability after 3 weeks of feeding the affected hay.[52] To prevent the risk of moldy sweet clover poisoning, sweet clover hay or haylage should not be fed for at least 3 weeks before parturition or elective surgery such as castration, dehorning or tail docking. There are select varieties of sweet clover that contain very low levels of coumarin and, therefore, hay from these varieties is safe, even if moldy.[49] Properly cured sweet clover silage is low in dicoumarol because dicoumarol-producing fungi require oxygen.

Treatment

Affected animals should be treated with whole blood transfusions as necessary at the rate of 10 mL/kg body weight. Ideally, 1 mg/lb body wt of vitamin K_1 should be injected to restore prothrombin time to normal within 24 hours.[50,51] However, the cost may be prohibitive and vitamin K_3 (menadione sodium bisulfite) although less effective is often administered.[5] Beneficial results can be obtained if treatment with large doses of vitamin K_3 is continued for 4 to 6 days.[40,50] Supplementing the ration with vitamin K_3 where dicoumarol is present is not effective in preventing poisoning.[52] If vitamin K_3 is given parenterally, the dosage should be 1 mg/kg body weight and should not exceed 2 mg/kg body weight. Greater than 2 mg/kg of vitamin K_3 administered parenterally greatly increases the risk of vitamin K renal toxicosis, especially in horses.[50]

Bracken Fern, Brake Fern, Eagle Fern
Pteridium aquilinum
Polypodiacae (Fern family)

Habitat

Bracken fern is found throughout the United States and has been associated with poisoning in cattle, sheep, pigs, horses, and humans.[53,54] In North America bracken fern is commonly found in the eastern, intermountain and western states, from Canada to Mexico. Its growth in the midwestern states is sparse. Bracken fern prefers to grow in moist open woodlands with sandy soils, often forming dense stands following clear-cutting or burning of forests. It will grow in relatively dry soils, and because of its prolific root system spreads rapidly to form dense monocultures to the exclusion of any other plants.

Bracken fern is also found in many parts of the world with the majority of animal poisoning reported in England and Europe.[53,54]

Description

Bracken fern is a perennial fern with a black, horizontal branching root system often extending for several meters. Leaves arise directly from the rhizome, are broadly triangular, up to 6.5 feet (2 meters) in height, bipinnately compound, and heavily haired on the underside (Figure 5-5A). The characteristic brown reproductive spores are produced under the rolled edge of the leaflets in late summer (Figure 5-5B).

Bracken Fern Poisoning

Bracken fern has been associated with a variety of syndromes in animals, the best recognized of which include:

- Thiamin deficiency
- Retinal degeneration and blindness
- Hemorrhaging and bone marrow destruction (thrombocytopenia)
- Urinary bladder cancer (enzootic hematuria)
- Digestive tract cancers

In the interest of continuity, all the syndromes of bracken fern poisoning will be discussed in this chapter and only mentioned in subsequent chapters on plants causing digestive and urinary system diseases.

Principal Toxins

Figure 5-5A Bracken fern (*Pteridium aquilinum*).

Bracken fern contains an enzyme thiaminase, which splits the essential vitamin thiamin (B₁) into its two inactive components pyrimidine and thiazole.[55] Thiamin is essential in energy metabolism, especially in the conversion of pyruvate to acetyl-coenzyme A, and the oxidation of α-ketoglutarate to succinyl-coenzyme A in the citric acid cycle.[55] Horses and pigs are most susceptible to the effects of thiaminase. Ruminants are rarely affected because they produce ample thiamin in the rumen.[56-59] Horses have to consume a diet containing 3 to 5 percent bracken fern for at least 30 days before clinical signs appear.[60] Sheep can be experimentally poisoned if they are fed large quantities of bracken fern for prolonged periods.[61] Affected animals develop a thiamin deficiency that is characterized by central nervous system depression and polioencephalomalacia. A similar thiaminase enzyme is also found in other plants including horsetail (*Equisetum arvense*), the Australian nardoo fern (*Marsilea drummondii*), and the rock fern (*Cheilanthes sieberi*).[62,63]

Bracken fern and rock ferns (*Cheilanthes* spp.) also contain ptaquiloside, a norsesquiterpene glycoside that has carcinogenic and bone marrow depressant activity.[64-69] All species of bracken fern should probably be considered toxic because at least *P. aquilinum*, *P. esculentum*, and *P. revolutum* are known to be toxic.[70-72] The root rhizome is the most toxic part of the plant; the leaves are poisonous whether green or dried. The concentration of ptaquiloside varies in bracken fern depending on its geographic distribution. In one survey of 77 samples from plants collected worldwide, 43 percent had ptaquiloside concentrations of over 1000 μg/g bracken.[73] Some bracken fern samples from Australia have had as high as 12,000 μg ptaquiloside/g of plant.[73] Ptaquiloside is transferred through the milk of animals eating bracken fern and

Figure 5-5B Spores lining the underside of bracken fern leaflets.

consequently has the potential of affecting the suckling young.[73] In addition to ptaquiloside, bracken fern also contains various pterosides, the toxicity of which have not been determined.[55,75]

Cattle appear to be the most susceptible to the effects of ptaquiloside; sheep are minimally affected, and horses are resistant.[75] The newly emerging fiddleheads and fronds of bracken fern are five times as toxic as the mature fronds and are quite palatable to cattle if other forage is scarce. In general, cattle have to eat their weight in bracken fern over several months to develop disease. If large quantities of fern are eaten in a short period, cattle develop an acute, usually fatal, hemorrhagic disease due to severe bone marrow destruction probably induced by ptaquiloside.[76] Long-term consumption of bracken fern leads to the development of tumors in the urinary bladder (enzootic hematuria or red water disease)[72] and possibly other parts of the digestive tract.[77-79] The susceptibility of cattle to the carcinogenic effects of ptaquiloside is possibly due to the fact they generally have alkaline urine. Under alkaline conditions, ptaquiloside is converted to the active carcinogen dienone that alkylates DNA leading to tumor formation.[75]

In cattle evidence suggests that bracken fern may predispose papilloma (wart) type 4 virus to produce malignant tumors in the mouth, esophagus, and rumen. The compound quercetin found in bracken fern acts as a cocarcinogen with the papilloma virus to produce the tumor.[79-83] A variety of different tumors in people have been associated with carcinogens in bracken fern.[84-86]

Clinical Signs of Bracken Fern Poisoning in Cattle and Sheep

Bracken fern poisoning in cattle and sheep may present in various forms depending on the quantity and duration of consumption of the plant.[53,83,87,88] In acute poisoning, cattle develop severe bone marrow depression that decreases blood platelets (thrombocytopenia) and produces anemia and leukopenia. The clinical signs that appear after animals have eaten the plant 1 to 2 months include depression and hemorrhages on the mucous membranes of the nose and mouth. Hemorrhaging may occur from the nose, mouth, and vagina. The anterior chamber of the eyes may fill with blood (hyphema). Hemorrhagic diarrhea, melena, and red urine (hematuria) are indicative of hemorrhaging into the digestive and urinary tracts.[87] Anemia results both from blood loss and bone marrow depression. Affected animals may have a high temperature due to secondary bacterial infection resulting from the severely depleted bone marrow and decreased circulating neutrophils. Mortality is very high in animals whose leukocyte count is below 2000/μL and platelet count less than 50,000/μL.

Enzootic Hematuria of Cattle

Enzootic hematuria (red water disease) occurs worldwide in cattle wherever bracken fern is grazed.[72,87] Ptaquiloside is the primary carcinogen in bracken fern, and has been shown experimentally to produce cancer of the urinary bladder if cattle consume bracken fern for extended periods.[64,71,74,80,88-91] Small polyp-like tumors develop in the bladder and form bleeding tumor masses (hemangiomas) that result in the formation of red urine (hematuria).

Cattle with enzootic hematuria are often first noticed when voiding red-colored urine, which over time leads to severe blood loss and anemia. There is no effective treatment for the bladder tumors, and usually the animal will die from severe anemia or local tumor invasion of the tissue around the bladder. A variety of tumor types including hemangiomas, hemangiosarcomas, papillomas, fibromas, adenomas, and transitional cell carcinomas have been associated with bracken fern carcinogenicity.[72,92,93]

Treatment

Early treatment with blood transfusions and bone marrow stimulants may be beneficial.[87] Batyl alcohol has been used as a bone marrow stimulant in cattle, but it is not consistently effective when thrombocytes and leukocytes are severely depleted.[88] Cattle have also been treated effectively before severe bone marrow depletion has occurred by administering protamine sulfate and blood intravenously.[89] Protamine sulfate is a heparin antagonist and therefore will counteract the increase in heparin that may occur in bracken fern poisoning. Broad-spectrum antibiotics are indicated to help protect the animal against secondary bacterial infections that may develop as a result of bone marrow depletion.

Diagnosis of bracken fern poisoning should be based on the history of the fern being eaten for an extended period of time, a hemorrhagic syndrome caused by bone marrow depletion resulting in thrombocytopenia, and leukopenia. At postmortem examination, there is usually diffuse hemorrhaging involving multiple organs.

Bracken Fern Poisoning in Sheep

"Bright blindness" of sheep is a syndrome of retinal degeneration and blindness associated with grazing of bracken fern in England.[94,95] The disease has been produced experimentally in sheep by feeding them a diet of 50 percent bracken fern for a period of 63 weeks.[75] The blind sheep have a dilated pupil that reflects light from the depigmented retina giving the syndrome its name. The exact cause of the retinal depigmentation is not known, and blindness is permanent.

Bracken Fern Poisoning in Horses

Bracken fern poisoning in horses is uncommon. When encountered it is characterized by a nervous system disease resulting from depletion of thiamin, a vitamin essential for normal energy metabolism.[53,94] Affected horses refuse to eat and consequently lose weight. Depression, muscle tremors, uncoordinated gait, especially of the hind legs and paralysis are typical of bracken fern poisoning. Horses may show colic, constipation, hemoglobinuria, severe anemia, elevated temperature, and rapid heart rate.[96]

Diagnosis of bracken fern poisoning should be based on evidence that horses have eaten the fern, the clinical signs, and the animal's response to thiamin therapy. Elevated serum pyruvic acid levels (normal 2 to 3 µg/dL) and decreased thiamin levels (normal 8 to 10 µg/dL) are helpful in confirming the diagnosis. Bracken fern poisoning in horses should be differentiated from viral encephalitis and hepatic encephalopathy, which have similar clinical signs.

Treatment

Horses with thiamin deficiency should be treated with intravenous thiamin, 5 mg/kg body weight. This dose should be repeated intramuscularly for several days. Horses should be provided with a balanced diet that is free of bracken fern.

Bracken Fern Poisoning in People

The young bracken fern shoots (fiddleheads) have for years been considered a delicacy in many Asian countries. A strong correlation appears to exist between the chronic consumption of bracken fern shoots and the high incidence of stomach cancer, especially in Japan.[84,85] Cooking and treating bracken fern by boiling in an alkaline solution reduces but does not eliminate the carcinogenic properties of bracken fern.[86] Because good evidence also indicates that bracken fern causes tumors in the esophagus, stomach, and intestine of animals, it is recommended that people do not eat bracken fern under any circumstances.

REFERENCES

Onion

1. Smith RH: Kale poisoning: the *Brassica anaemia* factor. *Vet Rec* 1980, 107:12–15.

2. Verhoeff J, Hajer R, Van Den Ingh TSGAM: Onion poisoning in young cattle. *Vet Rec* 1985, 117:497–498

3. Koger LM: Onion poisoning in cattle. *J Am Vet Med Assoc* 1956, 129:75.

4. James LF, Binns W: Effects of feeding wild onions (*Allium validum*) to bred ewes. *J Am Vet Med Assoc* 1966, 149:512–514.

5. Hutchison TWS: Onions as a cause of Heinz body anemia and death in cattle. *Can Vet J* 1977, 18:358–360.

6. Hutchison TWS: Onion toxicosis. *J Am Vet Med Assoc* 1978, 172:1440.

7. Kirk JH, Bulgin MS: Effects of feeding cull domestic onions (*Allium cepa*) to sheep. *Am J Vet Res* 1979, 40:397–399.

8. Pierce KR, Joyce JR, Jones LP: Acute hemolytic anemia caused by wild onion poisoning in horses. *J Am Vet Med Assoc* 1977, 160:323–327.

9. Knight AP, Lassen D, McBride T, *et al.*: Adaptation of pregnant ewes to an exclusive onion diet. *J Vet Hum Toxicol* Feb 2000, 42:1-4.

10. Yamoto O, Maede Y: Susceptibility to onion induced hemolysis in dogs with hereditary high erythrocyte reduced glutathione and potassium concentrations. *Am J Vet Res* 1992, 53:134–137.

11. Harvey JW, Rackear D: Experimental onion-induced hemolytic anemia in dogs. *Vet Pathol* 1985, 22:387–392.

12. Lincoln SD, Howell ME, Combs JJ, Hinman DD: Hematologic effects and feeding performance in cattle fed cull domestic onions (*Allium cepa*). *J Am Vet Med Assoc* 1992, 200:1090–1094.

Brassica

13. Perrett DR: Suspected rape poisoning in cattle. *Vet Rec* 1947, 59:674.

14. Clegg FG, Evans RK: Hemoglobinemia of cattle associated with the feeding of Brassicae species. *Vet Rec* 1962, 74:1169–1176.

15. Breeze RG, Gay CC, Turk JR: Turnip toxicity in cattle. *Bov Clin* 1982, 2:7.

16. Evans ETR: Kale and rape poisoning in cattle. *Vet Rec* 1951, 63:348–349.

17. Grant CA, Holtenius P, Jonsson G, *et al.*: Kale anemia in ruminants. 1. Survey of the literature and experimental induction of kale anemia in lactating cows. *Acta Vet Scand* 1968, 9:126–140.

18. Wikse SE, Leathers CW, Parish SM: Diseases of cattle that graze turnips. *Compendium Food Animal* 1987, 9:112–121.

19. Carlson J, Breeze RG: Ruminal metabolism of plant toxins with emphasis on indolic compounds. *J Anim Sci* 1984, 58:1040–1049.

20. Bell JM: Nutrients and toxicants in rapeseed meal: a review. *J Anim Sci* 1984, 58:996–1010.

21. Fenwick GR, Heaney RK, Mawson R: Glucosinolates. *In* Toxicants of Plant Origin 2. Cheeke PR, Eds. Boca Raton, FL: CRC Press; 1989:2–31.

22. Sharma HR, Ingalls JR, McKirdy JA: Effects of feeding a high level of tower rapeseed meal in dairy rations on feed intake and milk production. *Can J Anim Sci* 1977, 653–662.

23. Kernaleguen A, Smith RA: Acute mustard seed toxicosis in beef cattle. *Can J Vet J* 1989, 30:524.

24. Mason RW, Lucas P: Acute poisoning in cattle after eating old non-viable seed of chou moellier (*Brassica oleracea* convar. *acephala*). *Aust Vet J* 1983, 60:272–273.

25. Schofield FW: Acute pulmonary emphysema of cattle. *J Am Vet Med Assoc* 1948, 112:254–258.

26. Pfister JA, Baker DC, Lacey JR, Browson R: Photosensitization of cattle in Montana: is *Descurania pinnata* the culprit? *Vet Hum Toxicol* 1989, 31:225–227.

Red Maple

27. Tennant B, Dill SG, Glickman LT, *et al.*: Acute hemolytic anemia, methemoglobinemia, and Heinz body formation associated with the ingestion of red maple leaves by horses. *J Am Vet Med Assoc* 1981, 179:143–150.

28. Divers TJ, George LW, George JW: Hemolytic anemia in horses after the ingestion of red maple leaves. *J Am Vet Med Assoc* 1982, 180:300–302.

29. Plumlee KH: Red maple toxicity in a horse. *Vet Hum Toxicol* 1991, 33:66–67.

30. Semrad SD: Acute hemolytic anemia from the ingestion of red maple leaves. *Compendium of Continuing Education* 1993, 15:261–264.

31. Weber M, Miller RE: Presumptive red maple (*Acer rubrum*) toxicosis in Grevy's zebras (*Equus grevyi*). *J Zoo Widlife Med* 1997, 28:105–108.

32. George LW, Divers TJ, Mahaffey EA, Suarez MJH: Heinz body anemia and methemoglobinemia in ponies given red maple (*Acer rubrum*) leaves. *Vet Pathol* 1982, 19:521–533.

33. Stair EL, Edwards WC, Burrows GE, Torbeck K: Suspected red maple (*Acer rubrum*) toxicosis with abortion in two Percheron mares. *Vet Hum Toxicol* 1993, 35:229–230.

34. Long PH, Payne JW: Red maple-associated pulmonary thrombosis in a horse. *J Am Vet Med Assoc* 1984,184:977–978.

35. McConnico RS, Brownie CF: The use of ascorbic acid in the treatment of 2 cases of red maple (*Acer rubrum*) poisoned horses. *Cornell Vet* 1992, 82:293–300.

Sweet Clover

36. Scheel LD: Mycotoxicosis in cattle. *In* Mycotoxic Fungi, Mycotoxins, Mycotoxicoses—An Encyclopedic Handbook, Vol. 2. Wyllie TD, Morehouse LG, Eds. New York: Marcell Dekker, 1978:121—142.

37. Radostits OM, Blood DC, Gay CC, Eds. *Veterinary Medicine*, edn 8. Philadelphia: Bailliere Tindall; 1994:1546–1548.

38. Bartol JM, Thompson LJ, Minnier S, Divers JD: Hemorrhagic diathesis, mesenteric hematoma, and colic associated with ingestion of sweet vernal grass in a cow. *J Am Vet Med Assoc* 2000, 216:1605-1608.

39. Radostits OM, Searcy GP, Mitchall KG: Moldy sweet clover poisoning in cattle. *Can Vet J* 1980, 21:155–158.

40. McDonald GK: Moldy sweet clover poisoning in a horse. *Can Vet J* 1980, 21:250–251.

41. Kingsbury JM, Ed. *Poisonous Plants of the United States and Canada*. Englewood Cliffs, NJ: Prentice-Hall; 1964:342–346.

42. Linton JH, Goplen BP, Bell JM, Jaques LB: Dicoumarol studies I. Oral administration of synthetic dicoumarol to various classes of sheep and cattle. *Can J Anim Sci* 1963, 43: 344–352.

43. Puschner B, Galey FD, Holstege DM, Palazoglu M: Sweet clover poisoning in dairy cattle in California. *J Am Vet Med Assoc* 1998, 212:857–859.

44. Linton JH, Goplen BP, Bell JM, Jaques LB: Dicoumarol studies II. The prothrombin time response of sheep to various levels of contamination in low coumarin sweet clover varieties. *Can J Anim Sci* 1963, 43:353–360.

45. Casper HH, Alstad AD, Monson SB, Johnson LJ: Dicoumarol levels in sweet clover toxic to cattle. *Proceedings of the 25th Annual Veterinary Laboratory Diagnosticians* 1982, 25:41–48.

46. Frazer CM, Nelson J: Sweet clover poisoning in newborn calves. *J Am Vet Med Assoc* 1959, 135:283–286.

47. Schoefield FW: A brief account of a disease of cattle simulating hemorrhagic septicemia due to feeding sweet clover. *Can Vet J* 1984, 25:453–455.

48. Benson ME, Casper HH, Johnson LJ: Occurrence and range of dicoumarol concentrations in sweet clover. *Am J Vet Res* 1981, 42:2014–2015.

49. Goplen BP: Sweet clover production and agronomy. *Can Vet J* 1980, 21:149–151.

50. Rebhun WC, Tennant BC, DillSG, *et al.:* Vitamin K3 induced renal toxicosis in a horse. *J Am Vet Med Assoc* 1984, 184:1237–1239.

51. Alstad AD, Casper HH, Johnson LJ: Vitamin K treatment of sweet clover poisoning in calves. *J Am Vet Med Assoc* 1985, 187:729–731.

52. Casper HH, Alstad AD, Tacke DB, *et al.:* Evaluation of vitamin K3-- feed additive for prevention of sweet clover disease. *J Vet Diagn Invest* 1989, 1:116–119.

Bracken Fern

53. Evans WC: Bracken poisoning of farm animals. *Vet Rec* 1964, 76:365–369.

54. Kingsbury JM, Ed: *Pteridium aquilinum. Poisonous Plants of the United States and Canada.* Englewood Cliffs, NJ: Prentice-Hall; 1964:105–113.

55. Cheeke PR, Ed: Carcinogens and metabolic inhibitors. *In* Natural Toxicants in Feeds, Forages, and Poisonous Plants. Danville, IL: Interstate Publishers; 1998:423–444.

56. Evans WC, Widdop B, Harding JD: Experimental poisoning by bracken rhizomes in pigs. *Vet Res* 1972, 90:471–475.

57. Evans WC: Bracken thiaminase-mediated neurotoxic syndromes. *Bot J Linn Soc* 1976, 73:113–131.

58. Roberts HE, Evans ET, Evans WC: The production of "bracken staggers" in the horse, and its treatment with vitamin B1 therapy. *Vet Rec* 1949, 61:549–550.

59. Konishi K, Ichijo S: Experimentally induced equine bracken poisoning by thermostable anti-thiamine factor (SF factor) extracted from dried bracken. *J Japan Vet Med Assoc* 1984, 37:730–734.

60. Carpenter KJ, Phillipson AT, Thomson W: Experiments with dried bracken (*Pteris aquilina*). *Br Vet J* 1950, 106:292–308.

61. Bakker HJ, Dickinson J, Steele P, Nottle MC: Experimental induction of ovine polioencephalomalacia. *Vet Rec* 1980, 107:364–366.

62. Henderson JA, Evans EV, McIntosh RA: The antithiamine action of Equisetum. *J Am Vet Med Assoc* 1952, 120:375–378.

63. Meyer P: Thiaminase activities and thiamine content of *Pteridium aquilinum*, *Equisetum ramosissimum*, *Malva parviflora*, *Pennisetum clandestinum*, and *Medicago sativa*. *Onderstepoort J Vet Res* 1989, 56:145–146.

64. Saito K, Nagao T, Matoba M, *et al.:* Chemical assay of ptaquiloside, the carcinogen of *Pteridium aquilinum*, and the distribution of related compounds in the pteridaceae. *Phytochemisty* 1989, 28:1605–1611.

65. Hirono I: Carcinogenic principles isolated from bracken fern. *Crit Rev Toxicol* 1986, 17:1–22.

66. Hirono I, Ogino H, Fujimoto M, *et al.:* Induction of tumors in ACI rats given a diet containing ptaquiloside, a bracken carcinogen. *J Nat Cancer Inst* 1987, 79:1143–1149.

67. Prakash AS, Pereira TN, Smith BL, *et al.:* Mechanism of bracken fern carcinogenisis: evidence for H-ras activation via initial adenine alkylation by ptaquiloside. *Natural Toxins* 1996, 4:221–227.

68. Oyamada T, Yoshikawa T: Histopathogenisis of intestinal tumors induced by oral administration of bracken fern, *Pteridium aquilinum* in rats. *Jap J Vet Sci* 1987, 49:687–696.

69. Santos RC, Brasileiro-Filho G, Hojo ES: Induction of tumors in rats by bracken fern (*Pteridium aquilinum*) from Ouro Preto (Minas Gerais, Brazil). *Brazilian J Med Biol Res* 1987, 20:73.

70. Smith BL, Embling PP, Lauren DR, *et al.:* Carcinogen in rock fern (*Cheilanthes sieberi*) from New Zealand and Australia. *Aust Vet J* 1989, 66:154–155.

71. Xu LR: Bracken poisoning and enzootic haematuria in cattle in China. *Res Vet Sci* 1992, 53:116–121.

72. McKenzie RA: Bovine enzootic haematuria in Queensland. *Aust Vet J* 1978, 54:61–64.

73. Smith BL, Seawright AA, Ng JC, *et al.:* Concentration of ptaquiloside, a major carcinogen in bracken fern (*Pteridium* spp.), from eastern Australia and from a cultivated worldwide collection held in Sydney, Australia. *Natural Toxins* 1994, 2:347–353.

74. Pamukcu AM, Erturk E, Yalciner S, *et al.:* Carcinogenic and mutagenic activities of milk from cows fed bracken fern (*Pteridium aquilinum*). *Cancer Res* 1978, 38:1556–1560.

75. Fenwick GR: Bracken (*Pteridium aquilinum*)—toxic effects and toxic constituents. *J Sci Food Agric* 1988, 46:14 7–173.

76. Hirono I, Kono K, Takahashi K, *et al.:* Reproduction of acute bracken poisoning in a calf with ptaquiloside, a bracken constituent. *Vet Rec* 1984a, 115:239–246.

77. Jarrett WFH, McNeil PE, Grimshaw WTR, *et al.:* High incidence area of cattle cancer with a possible interaction between environmental carcinogen and a papilloma virus. *Nature* 1978, 274:215–217.

78. Moura JW, Stocco dos Santos RC, Dagli MLZ, *et al.:* Chromosome aberrations in cattle raised on bracken fern pasture. *Experientia* 1988, 44:785–788.

79. Hopkins DM: Bracken (*Pteridium aquilinum*): its distribution and animal health implications. *Br Vet J* 1990, 146:316–326.

80. Pamucku AM, Yalciner S, Hatcher JF, Bryan GT: Quercetin, a rat intestinal and bladder carcinogen present in bracken fern (*Pteridium aquilinum*). *Cancer Res* 1980, 40:3466–3472.

81. Pamucku AM, Wang CY, Hatcher JF, Bryan GT: Oncogenicity of tannin from bracken fern. *J Nat Cancer Inst* 1980, 65:131–136.

82. Pennie WD, Saveria Campo M: Synergism between bovine papilloma virus type 4 and the flavenoid quercetin in cell transformation in vitro. *Virology* 1992, 190:861–865.

83. Hopkins NCG: Aetiology of enzootic haematuria, *Vet Rec* 1986, 118:715–717.

84. Alonso-Amelot ME, Perez-Mena M, Calcagno MP, *et al.:* Ontogenic variation of biologically active metabolites of *Pteridium aquilinum* (L. Kuhn), pterosins A and B, and ptaquiloside in a bracken population of the tropical Andes. *J Chem Ecol* 1992, 18:1405–1420.

85. Smith BL, Seawright AA: Bracken fern (*Pteridium* spp.) carcinogenicity and human health-a brief review. *Natural Toxins* 1995, 3:1–5.

86. Hirono I: Carcinogenic bracken glycosides. *In* Toxicants of Plant Origin. Cheeke PR, Ed. Boca Raton, FL: CRC Press; 1989:239–251.

87. Radostits OM, Blood DC, Gay CC: *Veterinary Medicine* edn 8. Philadelphia: Bailliere Tindall; 1994:1559–1563.

88. Osweiler GD, Ruhr LP: Plants affecting blood coagulation. *In* Current Veterinary Therapy   Food Animal Practice 2. Howard JL, Ed. Philadelphia: WB Saunders; 1986:404–406.

89. Tustin RC, Adelaar TF, Medal-Johnsen CM: Bracken poisoning in cattle in the Natal medlands. *J South Afr Vet Med Assoc* 1969, 39:91.

90. Price JM, Pamucku AM: The induction of neoplasms of the urinary bladder of the cow and the small intestine of the rat by feeding bracken fern. *Cancer Res* 1968, 28:2247–2251.

91. Pamukcu AM, Price JM, Bryan GT: Naturally occurring and bracken fern induced bovine urinary bladder tumors. *Vet Pathol* 1976, 13:110–122.

92. Schacham P, Philp RB, Gowdey CW: Antihematopoietic and carcinogenic effects of bracken fern (*Pteridium aquilinum*) in rats. *Am J Vet Res* 1970, 31:191–197.

93. Smith BL, Embling PP, Agnew MP, *et al.:* Carcinogenicity of bracken fern (*Pteridium esculentum*) in New Zealand. *N Z Vet J* 1988, 36:56–58.

94. Barnett KC, Watson WA: Bright blindness in sheep. A primary retinopathy due to feeding bracken fern (*Pteris aquilina*). *Res Vet Sci* 1970, 11:289–290.

95. Watson WA, Terlecki S, Patterson SP, *et al.:* Experimentally produced progressive retinal degeneration (bright blindness) in sheep. *Br Vet J* 1972, 128:457–468.

96. Kelleway RA, Geovjian L: Acute bracken fern poisoning in a 14-month-old horse. *Vet Med/Small Anim Clin* 1978, 73:295–296.

CHAPTER 6

Plants Affecting the Nervous System

A variety of indigenous and exotic plants found in North American affect the nervous system of animals, some being associated with considerable economic losses to the livestock industry. Arguably the most important group of plants affecting the nervous system are those belonging to the genera *Astragalus* and *Oxytropis*, collectively known as locoweeds. These plants, found in vast areas of western North America and in many parts of the world, have long been recognized as a problem to livestock.[1,2]

There are numerous other plants affecting the nervous system such as yellow star thistle (*Centaurea solstitialis*) and Russian knapweed (*Acroptilon repens*) that will be discussed in the second section of this chapter. These and a variety of other plants are significant not only because they are poisonous and cause neurological disease in animals, but they also aggressively displace indigenous plants and reduce the value of natural ranges for grazing. Others like white snakeroot (*Eupatorium rugosum*) have historical and contemporary significance because of their effects on the nervous system. For example, deaths of early settlers in eastern North America who drank the milk from cows with"milk sickness," a fatal disease characterized by severe muscle tremors, is attributed to the toxin in white snakeroot. Once the neurotoxic effects of the white snakeroot were recognized, preventative management measures have largely eliminated this form of plant poisoning in people. The plants listed in Table 6-1 cause poisoning when normal forages are scarce, or the plants are accidentally incorporated in hay and grain fed to animals.

Table 6-1 Neurotoxic Plants	
SCIENTIFIC NAME	COMMON NAME
Aesculus spp.	Buckeye, horse chestnut
Artemisia filifolia	Sand sage
Astragalus spp.	Locoweeds
Centaurea solstitialis	Yellow star thistle
Acroptilon repens	Russian knapweed
Corydalis spp.	Fitweed
Equisetum arvense	Horsetail
Eupatorium rugosum	Snake root
Haplopappus heterophyllus	Rayless goldenrod
Karwinskia humboldtiana	Coyotillo
Kochia scoparia	Kochia weed
Oxytropis spp.	Locoweed
Pteridium aquilinum	Bracken fern
Sophora secundiflora	Mescal bean

Locoweeds

Locoweed Poisoning

More than 2000 species of the genera *Astragalus* and *Oxytropis* exist throughout the world, of which at least 370 occur in North America.[3] Collectively these plants are commonly called locoweeds, vetches, or milk vetches, and are similar in appearance. Only the experienced botanical taxonomist can readily differentiate the species. Livestock poisoning attributable to these genera has been recognized since the early part of this century, and the locoweeds continue to cause more economic losses to the livestock industry than all other plant-induced toxicities combined.[1,4-7] Although all species of *Astragalus* and *Oxytropis* in North America should be considered poisonous unless proven otherwise, some species of milk vetch such as *A. cicer* and *A. tenellus* are nontoxic, and serve as useful forages for range animals.[8] In one extensive survey of 1690 species of *Astragalus*, 221 (13%) tested positive for toxic nitro compounds.[9] This points out the necessity of testing exotic species of *Astragalus* to ensure they are not toxic before they are introduced as forages for livestock in arid areas, or as plants for erosion control

The locoweeds or vetches have been associated with three general syndromes of livestock poisoning. The best known is that of locoism due to the effects of the alkaloid swainsonine[1,10-12] In addition, the locoweeds/vetches cause respiratory problems and peripheral nerve degeneration due to nitroglycoside compounds in the plants.[9,10,13] A third syndrome is caused by chronic selenium toxicity due to the ingestion of locoweeds or vetches that accumulate selenium.[10] Selenium toxicity is discussed in Chapter 9. Some species of *Astragalus* may contain one or more of the toxins and consequently may cause a combination of clinical signs in animals that have eaten them.[10,14]

Principal Toxin

Early researchers suspected a toxic substance in locoweeds they called locoine,[1,15] an alkaloid that has now been named swainsonine because it was first isolated and characterized from the gray swainson pea (*Swainsona canescens*).[16,17] The *Swainsona* spp. are leguminous herbs mostly confined to Australia and are closely related to plants in the genus *Astragalus* found in other parts of the world.[18] Swainsonine is one of a group of indolizidine alkaloids found in locoweeds that appears to play a major role in the pathogenesis of locoweed poisoning.[16,17] In North America, swainsonine has been found in variety of *Astragalus* and *Oxytropis* spp. listed in Table 6-2.[19-21] Horses, cattle, sheep, goats, elk, and domestic cats are both naturally and experimentally susceptible to the toxic effects of swainsonine.[22-25] All animals, however, should be considered susceptible to the toxic effects of the alkaloid.

Signs of poisoning do not become evident until animals have consumed significant quantities of locoweeds over many weeks and the toxic threshold is reached. Although horses, cattle, and sheep were thought to develop an addiction for locoweeds, it is more correctly termed habituation because there is no dependence on the plants as there would be in the case of addiction.[26,27] Locoweeds are palatable and of similar nutrient value to alfalfa, which may explain why animals eat locoweed even when normal forages are present.[28]

Table 6-2 *Astragalus* and *Oxytropis* Species Containing Swainsonine[2,14,21]	
A. allochrous	A. oxyphysus
A. argillophilus	A. playanus
A. assymetricus	A. praelongus
A. bisulcatus	A. pubentissumus
A. didymocarpus	A. pycnostachysus
A. dyphysus	A. tephrodes
A. earlei	A. thurberi
A. emoryanus	A. wootonii
A. flavus	O. besseyi
A. humistratus	O. campestris
A. lentiginosus	O. condensata
A. lonchocarpus	O. lambertii
A. mollissimus	O. saximontana
A. missouriensis	O. sericea
A. nothoxys	

The quantity of swainsonine in locoweeds varies according to the species, stage of growth, and the growing conditions. The succulent preseed-stage plants appear to be the most palatable, although cattle appear to relish the flowers and immature seed pods.[28] The palatability of locoweed does not have any relationship to the quantity of swainsonine in the plant.[28]

Swainsonine inhibits the action of two lysosomal enzymes (α-D-mannosidase and Golgi mannosidase II) that aid in the metabolism of saccharides.[29,30] The inhibition of α-mannosidase therefore causes cells to accumulate complex sugars or oligosaccharides.[31] When Golgi mannosidase II is inhibited, the normal structure of oligosaccharide components of glycoproteins is affected, thus furthering their accumulation.[30] As a result, oligosaccharides accumulate in the cells of the brain and many other organs and interfere with normal cellular function.[11,12,32] In effect swainsonine causes a generalized lysosomal storage disease similar to the genetically transmitted disease mannosidosis.[33-35] The indolizidine alkaloid castanospermine isolated from the seeds of the Australian Morton Bay chestnut tree (*Castanospermum australe*) is similar to swainsone in its inhibitory effects on glucosidases.[31] In Brazil and Mozambique, lysosomal storage disease in goats, similar to that caused by swainsonine, has also been attributed to (*Sida carpinifolia*), and a morning glory species (*Ipomoea carnea*).[36,37] Young animals are most severely affected by swainsonine because maturing neurons are more vulnerable to the effects of the toxin. In addition, swainsonine is passed through the milk, thereby increasing the dose of alkaloid a suckling animal may acquire to that which it ingests through the plants it eats.[24,38]

The concentration of swainsonine in locoweed (0.09-0.23 percent dry weight) varies with the stage of growth. The highest concentrations of swainsonine are found in the flowers and seeds (0.28 and 0.36 percent, respectively) and consequently the quantity of plant that an animal has to eat to receive a toxic dose will vary.[31,39] However, the potency of swainsonine is such that it is not necessary for high doses to be ingested to induce poisoning. Experimentally, locoism can be produced by feeding 0.75 to 1.0 lb/day of dried locoweed to horses and cattle for 75 to 85 days.[39] In general, the greater the quantity of locoweed consumed, the more rapid the onset of poisoning.

Clinical Syndromes Attributable to Locoweed Poisoning

Locoism

Locoism, the term originally given to locoweed poisoning, is derived from the Spanish word "loco" meaning crazy.[1,2] This aptly describes the abnormal neurologic behavior shown by horses and other animals intoxicated by locoweeds. Horses show the nervous signs of locoweed poisoning more commonly than do cattle or sheep. Affected animals may exhibit a variety of signs including depression, circling, incoordination, staggering gait, and unpredictable behavior especially if the animal is stressed or excited.[1,2,7,22,24,39] Some horses become totally unpredictable in their response to handling and may fall down when being haltered or ridden. Poor vision, incoordination, and sudden changes in behavior such as rearing and falling over backward, make these horses dangerous and unsafe to ride. Cattle may become aggressive when handled, and are more difficult to manage and work with when intoxicated. Weight loss, despite ample forage and grain being available, is typical.[40] In young animals, weight loss and poor growth rates may be a result of nervous system depression and apparent inability to eat normally.[27] If removed from the source of the locoweeds and fed a nutritious diet, animals will improve and appear relatively normal after several months. However, regeneration of affected neurons in the brain and spinal cord may not occur completely making horses a potential liability to human safety if ridden.[39] The prognosis for locoed horses should therefore always be guarded.

Locoweed-Induced Infertility and Reproduction Failure

In cattle, locoweed poisoning is more commonly associated with abortions, infertility, subcutaneous edema in the fetus, and fetal deformities.[41-45] Cows consuming locoweed may exhibit abnormal and lengthened estrus cycles and fail to conceive.[45] The reproductive problems and congenital defects commonly associated with locoweeds most probably result from the effects of swainsonine acting directly on the uterus, placenta and fetus.[11,43] Cytoplasmic vacuolation identical to that seen in many other organs is also evident in the placental tissues.[11] The placenta appears to be most susceptible to the effects of locoweed during the first 90 days of pregnancy, but it is vulnerable at any stage.[42] Normal placentation may be interrupted causing fetal resorption, abortion, or hydrops allantois.[45] The latter condition has been observed in sheep and cattle and is commonly referred to as water belly because the uterus becomes greatly distended with fluid.[4,46,47] Cows with hydrops may carry the pregnancy to term, but more frequently abort or become recumbent and unable to rise owing to the massive weight of the fluid in the uterus. The severely distended uterus also compresses the abdominal organs and interferes with normal digestion. Retention of fetal membranes and subsequent infertility are common sequels to hydrops.

Lambs and calves born to locoweed-poisoned dams may be born alive but weak and often die after a few days. Others may be smaller than normal or have deformities of the limbs or head.[39,48] Pregnant mares that consume quantities of woolly loco (*A. mollisimus*) in early gestation may produce foals with crooked legs especially involving the bones of the fetlocks.[49] The formation of congenital abnormalities such as crooked legs is probably due to fetal immobilization induced by the locoweed.[50] These teratogenic effects of locoweeds and other plants are discussed further in Chapter 8.

Rams having consumed locoweed (*A. lentiginosus*) for prolonged periods have been observed to undergo testicular atrophy with decreases in spermatogenesis.[51] Similar changes in bulls grazing locoweeds can be anticipated. The formation of abnormal sperm and decreased motility of sperm are attributable to the cytoplasmic vacuolation of the cells of the seminiferous tubules, epididymis, and vas deferens.[52] The vacuoles result from the accumulation of oligosaccharides that impair normal sperm maturation and function.[53] The vacuolation, however, is transitory and disappears after 70 days of a locoweed-free diet.[54]

Locoweed-Induced Heart Failure

There is correlation between the incidence of "high mountain disease" or congestive right heart failure in cattle and the consumption of locoweed (*O. sericea*).[55-58] Cytoplasmic vacuolation characteristic of swainsonine toxicity, observed in the lungs of cattle with "high mountain disease", possibly compounds the effects of high-altitude hypoxia on the pulmonary vasculature, causing pulmonary hypertension and eventual right heart failure. Experimentally, the fetuses of ewes fed locoweed for 2 to 3 months developed heart irregularities and right ventricular hypertrophy, suggesting there may also be a direct effect of swainsonine on the heart.[47,48]

Weight Loss and Ill-Thrift

The continuous consumption of locoweeds by young animals over the course of the growing season, and even when the plants are dry or dormant during the winter, can result in stunted, poorly growing animals. This affect on the growth of animals is probably due to the effects of swainsonine on the pituitary and thyroid glands affecting production of growth hormone and thyroxine, and decreased nutrient absorption from the intestinal cells affected by the alkaloid. Many calves with locoism exhibit symptoms identical to those produced by bovine virus diarrhea; poor body condition, rough hair coat, lower than normal weaning weights. Locoweed poisoned calves have significantly lower weaning weights and gain poorly when placed on a concentrate ration under feedlot conditions. However, once the effects of swainsonine have diminished after calves have been off all sources of locoweed for about 60 days, compensatory weight gains can be expected.[59]

Animals poisoned by locoweed often appear to have a greater incidence of common diseases such as foot rot, pneumonia, warts, and other infections. This suggests that affected animals may have a compromised immune system.[60] Evidence for this is detectable in sheep, which when repeatedly fed locoweed had lymphocytes with decreased responsiveness to mitogens.[60] Peripheral lymphocytes in the affected sheep contained cytoplasmic vacuoles characteristic of swainsonine toxicity. Similarly calves chronically eating locoweed and concurrently vaccinated with a modified live virus vaccine develop a poor titer to the vaccine. It is therefore important to remove calves from locoweed for at least a month to allow recovery of their immune system with normal lymphocytes capable of response to vaccines.[59]

Diagnosis of Locoism

Locoweed poisoning may be confirmed in animals by demonstrating the presence of swainsonine in the serum, coupled with decreased serum "α-mannosidase activity.[61] The half-life of swainsonine in the serum is 16 to 20 hours.[62] This means that an animal suspected of locoweed poisoning must have a blood sample taken within 2 days of eating locoweed for swainsonine to be detected. Similarly serum α-mannosidase activity will return to normal within 6 days of locoweed being eliminated from the diet.[62] Elevated levels of the serum enzymes alkaline phosphatase, aspartate aminotransferase, and lactate dehydrogenase are likely in locoweed poisoned animals.[60,61,63] Serum protein and thyroid levels (T4 and T3) are usually decreased in animals poisoned by locoweed.[62] The presence of vacuoles in the cytoplasm of peripheral lymphocytes is a characteristic finding in locoweed poisoning in some animals.[39,60] Vacuolated lymphocytes may persist in the circulation for several weeks after an animal has ceased eating locoweed.

At postmortem examination there no specific gross lesions characteristic of locoweed poisoning. Emaciation, occasional stomach ulcers, thyroid hypertrophy, and pale coloration to the liver and kidneys may be seen.[11,64] Animals exhibiting neurologic signs, often have cytoplasmic vacuolation of neurons in the brain. Similar vacuolation is often present in the pituitary gland, thyroid, pancreas, kidneys, liver, lymph nodes, retina and other structures of the eye.[11,39,65]

In animals chronically affected with locoism, and which have not eaten locoweed for more than a month, the vacuolation is restricted to hepatocytes and neurons of the brain.[39] Purkinje cells of the cerebellum retain vacuoles for over a year. The noticeable loss of these cells over time helps to explain the residual neurologic abnormalities typical of locoweed poisoning.[39] Experimentally, vacuoles can be first demonstrated in the kidney tubules 4 days after initiating locoweed feeding. Vacuoles in the neurons are detectable by day 8, and placental lesions develop 8 to 16 days after the start of locoweed consumption.[65,66] The vacuoles present in cells that are characteristic of swainsonine toxicity are similar to those found with other lysosomal storage diseases in animals such as "α- and β-D-mannosidosis reported in cattle and goats.[34,67,68]

Treatment of Locoism

An effective treatment for locoweed poisoning has yet to be developed. Fowler's solution and reserpine, once recommended for treating locoism, are unwarranted in light of current knowledge regarding the cause of locoweed poisoning.[69] Animals will recover from locoism provided they are removed from the locoweed before extensive cellular degeneration has occurred in the brain. Animals that have aborted generally will conceive in subsequent seasons provided they do not develop severe secondary infections of the reproductive system. Compensatory weight gains can also be anticipated in animals once they are removed from locoweed and are fed a balanced nutritious diet. Further consumption of locoweed should be prevented immediately, and every year thereafter because animals may retain a preference for the plants from year to year.

Prevention and Management of Locoweed Poisoning

To manage livestock on rangeland or pasture containing locoweeds, it is important to recognize the growth characteristics of locoweeds and the conditions under which livestock eat the plants. Cattle and probably other herbivores eat locoweed particularly when grazing pressure is high and other forages are depleted.[70-72] However, cattle will voluntarily eat locoweed when it is in the flower and early seed pod stage.[73] Although this corresponds to when swainsonine levels are highest in the plant, it is not the swainsonine level in the plant that appears to attract the animals to eat the plant at this stage of growth. The early seed pod stage coincides with the highest levels of protein and carbohydrate in the plant, which may increase palatability.[74] As other forages green up in the spring and summer, cattle will usually stop eating locoweed until grasses are depleted as a result of the season or overgrazing.[73] If temperature and moisture conditions are ideal as occurs in mild winters, locoweeds may retain green leaves in winter that make the plant attractive to livestock. It is important to pay attention to the rainfall pattern because a wet summer can mean more growth of locoweed, which in turn can mean that more plants are available both in the summer and winter.[75] Locoweeds retain some of their toxicity even when dried.

It has long been observed that once a cow starts eating locoweed, she will teach others in the herd to eat it. Experimentally it has been shown that cattle naive to locoweed will greatly increase their consumption of locoweed when placed with locoweed eaters.[76] This phenomenon is referred to as social facilitation, and it is an important influence in a herd of cattle. Social facilitation is probably the reason that over a period of seasons the percentage of cattle grazing locoweed in the herd grows to the point that all animals will eat it. Cattle and sheep do not become addicted to locoweed as was once thought because they will voluntarily stop eating the plant when other, more palatable forages are available.[77] Animals develop a preference for locoweed, apparently because the plants taste good to them!

Because locoweed poisoning is dependent upon the dose of locoweed an animal eats, and it usually takes several weeks of grazing the plants to become intoxicated, grazing management strategies can be developed to take advantage of this. Through close observation of the cattle, it is possible to determine when the consumption of locoweed is becoming excessive and poisoning is likely. At this point the cattle should be moved to a "safe area" for several weeks, an area with little or no locoweed. By rotating the cattle from pastures with locoweed to those without, effective use of the rangeland can be accomplished without severely poisoning the cattle or sheep.

Locoweeds can be controlled by the use of herbicides, but it is important to recognize that the seeds of many locoweeds can persist for at least 50 years.[78] Repeated application of appropriate herbicides will generally be needed to control locoweed reemergence. A variety of herbicides are effective in controlling *Astragalus* and *Oxytropis* spp., including the approved chemicals clopyralid, dicamba, picloram, and triclopyr.[79] The phenoxy herbicides 2,4-dichlorophenoxyacetic acid (2,4-D) and 2,4,5-trichlorophenoxyacetic acid (2,4,5-T) do control locoweeds as long as they are applied when the plants are actively growing. In all cases, it is most important to ensure the herbicide being used is approved for application in the area and that the manufacturer's recommendations are followed.

Using the principle of conditioned food aversion, it is possible to train cattle to avoid eating locoweed.[80] Locoweed eaters can be averted from eating locoweed by associating the eating of locoweed with a distasteful experience. This is accomplished by feeding the cows locoweed and at the same time dosing them with a bolus of lithium chloride, a chemically irritating compound. The cattle develop "colic" from the lithium chloride and associate this unpleasant experience with eating locoweed. This aversion to eating the locoweed persists even into subsequent growing seasons provided the averted animals are not mingled with locoweed eaters. Through social facilitation averted cows relearn to eat locoweed from the locoweed eaters. Food aversion methods have the potential for preventing locoweed poisoning under some controlled management systems, and provide an alternative means of grazing cattle on rangeland heavily infested with locoweed.

Although various mineral and protein supplements have been recommended for the prevention of locoweed poisoning in livestock, there is no proven preventative value of feeding them.[81,82] Animals, however, that are deficient in protein and minerals will potentially benefit nutritionally from such supplements.

Astragalus Species Containing Nitro Compounds

As early as 1932, milk vetches were recognized as plants poisonous to cattle and sheep in the Rocky Mountain states.[83] Some 263 species of *Astragalus* (milk vetches) located principally in western North America from Canada to northern Mexico contain toxic nitroglycoside compounds capable of causing livestock poisoning.[3] Most *Astragalus* spp. containing nitro compounds do not contain the alkaloid swainsonine responsible for locoism nor do they accumulate toxic levels of selenium. The only known species to accumulate selenium and nitro compounds is *A. toanus* (Toano milk vetch), a relatively uncommon plant of eastern Utah, Nevada, and Idaho.[84] The more common milk vetches that contain nitro compounds associated with livestock poisoning are listed in Table 6-3.[2,14,85,86]

Principal Toxins

The toxicity of the milk vetches is primarily attributed to the nitro compound called miserotoxin, so called because it was first recognized in timber milk vetch (*A. miser*).[87-90] Miserotoxin, a glycoside 3-nitro-1-propanol, is hydrolyzed in the rumen to the toxic 3-nitro-1-propanol (3-NPOH).[91] Other species of milk vetch contain 3-

Table 6-3
Common *Astragalus* Species Containing Nitroglycosides
A. atropubescens
A. campestris
A. canadensis
A. convallarius
A. cibarius
A. falcatus
A. flexuosus
A. emoryanus
A. miser var. oblongifolius
A. miser var. serotinus
A. miser var. hylophilus
A. praelongus
A. pterocarpus
A. tetrapterus
A. toanus
A. whitnii

nitropropionic acid (3-NPA) and not 3-NPOH. Misertoxin (3-NPOH) is rapidly absorbed into the blood of cattle and sheep where it is converted to 3-NPA.[92] Although the mechanism is not fully understood, poisoning by these nitro compounds appears to occur in two ways. Nitrite (NO_2) from 3-NPA oxidizes hemoglobin to methemoglobin (up to 33 percent), which causes severe respiratory distress.[93-96] Secondly, 3-NPA or other unidentified metabolite also affects the brain and spinal cord causing muscular weakness and collapse.[94-96] Cattle and sheep are most frequently poisoned but horses are also susceptible. Other legumes, *Indigofera spicata* (creeping indigo)[97] and *Coronilla varia* (crown vetch),[98,99] contain 3-NPA and have been associated with poisoning in horses.[100,101] Dogs that have eaten the meat from affected horses may also become intoxicated.[102] Crown vetch is toxic to nonruminants, and if diets contain more than 5 percent of the plant, growth and development retardation in the young can be expected.[103]

The nitro compound content of milk vetches varies with the species and the stage of growth, being highest during the flower and seed pod stage.[86] Years with high rainfall not only produce a flush of milk vetch growth, but the level of misertoxin is also increased in the plants.[87] The nitro compound content of Emory milk vetch (measured as nitrite) is in the range of 6 to 9 mg NO_2/g plant during the bud to flowering stage.[96] Nitro compounds are stable in the dried green plant but are lost rapidly from the dried, bleached-out plant.[96] The nature and severity of poisoning depends on the quantity and rate of absorption of 3-NPOH from the rumen. Cattle fed 2.2 g dry weight of Emory milk vetch per kilogram body weight daily develop signs of poisoning in 3 to 4 days.[93] Sheep appear to be much more tolerant of nitro compounds in milk vetch than are cattle.[94]

Clinical Signs

The nitro-containing *Astragalus* spp. produce both acute and chronic syndromes of poisoning in cattle, sheep, and occasionally horses. The quantity and rate at which the plants are eaten determines the clinical course of the poisoning. Signs of acute poisoning include general weakness, depression, knuckling of the fetlock joints, ataxia, respiratory difficulty, cyanosis, and sudden collapse before death.[93-96] The cyanosis and respiratory distress result from the rapid oxidation of hemoglobin to form methemaglobin by nitrite derived from 3-NPA. Severe respiratory distress occurs when 20 percent or more of the animal's total hemoglobin is converted to methemoglobin. Any form of stress or forced movement often causes the animal to collapse. Death of the animal is, however, not solely due to the formation of methemoglobin, because serum levels of 30 percent methemoglobin are not fatal in animals with acute nitrate poisoning. Fatalities occurred when methemoglobin levels approached 80 percent.[104] That death is not solely due to methemoglobin formation is born out by the fact that treatment with methylene blue, the treatment for nitrate poisoning, did not prevent death of animals fed nitroglycoside containing *Astragalus* spp.[94,105]

Chronic milk vetch poisoning in cattle is referred to as "cracker heels" because affected animals click their hooves together while walking as a result of weakness in the hind legs and knuckling-over at the fetlocks. As the condition progresses, the animal may show posterior paralysis and be unable to stand. Milk vetch poisoning has also been referred to as "roaring disease" because of the noisy respiratory sounds associated with difficulty in breathing.[93] In chronic poisoning the first noticeable signs are those of rapid and labored breathing.[95,96] Sheep with chronic poisoning develop emphysema and pulmonary edema with marked respiratory difficulty with less muscular weakness and uncoordination.[95,96] All animals with chronic milk vetch poisoning show marked weight loss, and eventually become recumbent before dying. Recovery rarely occurs in the severely affected animal.

A diagnosis of milk vetch poisoning is generally based on the clinical signs of respiratory difficulty coupled with the characteristic hind leg weakness and evidence that the animals had been eating one of the nitroglycoside-containing milk vetches.

Treatment

There is no known specific treatment for milk vetch poisoning. Affected animals should be removed from the source of the toxic plants. New methylene blue may be given intravenously to help convert methemaglobin back to hemoglobin and reduce the respiratory problems.[96] This will not, however, alter the effects of the nitro compounds on the nervous system. If the animals have not consumed lethal quantities of the milk vetch, they may slowly recover but usually do not grow well, and many times these animals are an overall economic loss to the producer.

Postmortem Findings

At autopsy there are no definitive lesions of milk vetch poisoning. Pulmonary edema and emphysema may be present in animals showing respiratory signs. Congestion and swelling of the liver are common. Most affected animals have excessive amounts of cerebrospinal fluid.[96] Microscopically, the principal lesions include degeneration of the spinal cord and peripheral nerves, brain hemorrhages, edema, and emphysema of the lungs.[96]

Prevention and Control

As with the locoweeds, the milk vetches are difficult to control and eradicate because they cover vast geographic areas and have seeds that exist in the soil for many years before germinating when optimum conditions of moisture and temperature exist. Careful monitoring of rangeland conditions to determine when milk vetches are abundant and moving animals to areas where there is minimal risk from the plants can be helpful in reducing losses yet enable the range to be used. In the fall and winter, when the plant has dried and lost its green color, levels of nitroglycosides in the milk vetches are negligible.[96]

Appropriate use of herbicides such as 2,4-D, triclopyr, and picloram eventually kill milk vetches. Herbicides significantly lower the quantity of the nitroglycosides in the plants once dried and bleached out, thereby reducing the risk of poisoning to livestock.[106]

Blue or Spotted Locoweed
Astragalus lentiginosus
Fabaceae (Legume family)

Habitat

Blue locoweed is found typically in the plains, dunes, and slopes at lower elevations from western Colorado, central New Mexico, westward to Arizona, California, Nevada, northern Idaho, Montana, and Utah.

Description

These are leafy-stemmed perennial plants with pinnately compound, alternate leaves. The leaflets number 13 to 21. The pealike purple flowers are in clusters on stalks that arise in the leaf axils (Figure 6-1A). The pods are inflated two-celled pods with variable red-brown spotting. (Figure 6-1B).

Figure 6-1A Blue or spotted locoweed (*Astragalus lentiginosus*).

Figure 6-1B Blue or spotted locoweed showing flowers and spotted seed pods (*A. lentiginosus*).

Principal Toxin

The indolizidine alkaloid swainsonine is present in all parts of the plant, and the plant remains toxic when dried.

Woolly or Purple Locoweed
Astragalus mollissimus
Fabacea (Legume family)

Habitat

Woolly locoweed is found in the dry plains and foothills at lower elevations 2000-4000 feet (610 to 1,219 meters) from southwestern South Dakota, western Nebraska, and southeastern Wyoming, western Kansas, southward to northwest Texas and eastern New Mexico.

Description

Woolly locoweed is a leafy perennial plant, with relatively short stems, usually less than 10 cm tall (Figure 6-2A). The leaves are pinnately compound with 21 to 31 leaflets and are densely haired. The flowers are rose-purple, in dense spike-like racemes (Figure 6-2B). The fruit is a pod, oblong in shape and not hairy.

Figure 6-2A Wooly locoweed (*Astragalus mollissimus*).

Figure 6-2B Wooly locoweed showing hairy leaves and purple flowers (*A. mollissimus*).

Principal Toxin

The indolizidine alkaloid swainsonine is present in all parts of the plant.

White Locoweed, White Point Vetch, Crazy Weed
Oxytropis sericea
Fabaceae (Legume family)

Habitat

White locoweed is typically found in open gravelly or well-drained slopes and hills at lower to middle elevations from western Canada, Montana, North Dakota, and western Minnesota, south to Arizona, New Mexico, and Texas.

Description

White locoweed is a perennial, herbaceous plant growing in a tuft with a stout tap root. The leaves are usually alternate and basal, odd-pinnately compound, covered with silvery hairs. The flowers are borne on leafless stems above the leaves in a raceme (Figure 6-3). The flowers are white with a distinctive purple-tipped point to the fused front petals or keel. The pealike pods are stalkless and have a short curved beak.

Figure 6-3 White locoweed (*Oxytropis sericea*).

Principal Toxin

The indolizidine alkaloid swainsonine is present all parts of the plant, even when it is dried. White locoweed seems to be most palatable when it is in bloom, although it is readily consumed at all times if other forage is absent. Overgrazing will therefore increase the probability that livestock will eat locoweed.

Purple Locoweed, Lambert's Red Loco
Oxytropis lambertii
Fabaceae (Legume family)

Habitat

Purple locoweed is commonly found on the prairies and mountains usually in drier situations, at lower to middle elevations from Minnesota to Saskatchewan and southward to Arizona, New Mexico, Texas, and Oklahoma.

Description

This is a perennial herbaceous plant that grows in clumps with a tap root. The leaves are produced basally, odd-pinnately compound, and sparsely haired. The flowers are produced on a leafless stalk in a raceme above the leaves (Figure 6-4). The corolla is purple to rose in color and has a pointed keel. The pealike seed pods are 1.5 to 2.5 cm (15 to 25 mm) long, tapering to a point or divergent beak. Purple locoweed often grows in the same area as white locoweed (*O. sericea*), but tends to bloom just as the white locoweed goes out of bloom and starts producing seed pods.

Principal Toxin

Indolizidine alkaloids are believed to be the primary cause of poisoning. Swainsonine, an indolizidine alkaloid, is not consistently present in *O. lambertii*. The plants are consumed readily by livestock especially if there is little else for them to eat. Overgrazing will increase the probability of locoweed consumption.

Note: Swainson pea (*Sphaerophysa salsula*), introduced from Asia, has become established in waste ground, along roadways, and in cultivated fields of the western United States, where its seeds have contaminated alfalfa seed and other grains.

It is an erect perennial plant reproducing from lateral roots and seeds, and forming dense stands. Leaves are pinnately compound, with numerous alternate, hairy leaflets. Orange-red showy pealike flowers are produced terminally on the branches. Many green-colored seeds are produced in translucent bladder-like pods. It should not be confused with the swainson pea (*Swainsona* spp.) found in Australia and that has the same toxin (swainsonine) as found in the locoweeds. A toxin in *S. salsula* has not been determined. It does not, however, contain the alkaloid swainsonine.

Figure 6-4 Purple or Lambert's locoweed (*Oxytropis lambertii*).

Timber Milk Vetch
Astragalus miser
Fabaceae (Legume family)

Habitat

Timber milk vetch is found growing in a variety of soil types, often under aspen trees, or in areas where sage brush has been cleared. Varieties of timber milk vetch range from southern Canada, Washington, Idaho, Wyoming, Utah, Colorado, and northern Arizona.

Description

These perennial, multistemmed plants arise from a slender taproot, the crown of which is at or just below the soil surface. The alternate, odd-pinnate leaves are composed of 7 to 21 leaflets. The flowers are produced on leafless stems from leaf axils and range in color from creamy white with purple veins to purple. The pealike pods 0.75 to 1 inch (2 to 3 cm long) lack stalks, hang downward, and open from the apex along the lines joining the halves (Figure 6-5).

The three varieties of timber milk vetch are difficult to differentiate and have been named according to their geographic location. *Astragalus miser* var. *oblongifolius* (Wasatch milk vetch), *A. miser* var. *hylophilus* (Yellowstone milk vetch), *A. miser* var. *serotinus* (Columbia milk vetch).[85]

Figure 6-5 Timber milk vetch leaves and pods (*Astragalus miser*). (Courtesy Dr. Michael H. Ralphs, USDA Poisonous Plants Research Laboratory, Logan, Utah.)

Principal Toxin

Miserotoxin (3-NPOH), a nitroglycoside, is found in timber milk vetch and some 263 other members of the genus.

Crown Vetch
Coronilla varia
Fabaceae (Legume family)

Habitat
Crown vetch is frequently used as a drought-tolerant ornamental ground cover to help control erosion in the United States especially in the Midwest and Northeast.

Description
Crown vetch is a perennial herb with trailing or ascending, branching stems arising from a taproot. Leaves are alternate and pinnate with a terminal leaflet. The pealike flowers are produced from the leaf axils on a stalk 3 to 5 inches (7.5 to 12.5 cm) in length (Figure 6-6A). Flower color varies from white to dark pink. Many brown, cylindrical seeds are produced in pods 1 to 2 inches (2.5 to 5 cm) in length.

Figure 6-6A Crown vetch (*Coronilla varia*).

Figure 6-6B Creeping Indigo (Courtesy of Gerald D. Carr, University of Hawaii).

Principal Toxin
Crown vetch contains nitroglycosides, the most toxic of which, coronarian is poisonous to horses and other nonruminants.[99] Ruminants are not affected because they are able to convert the coronarian to 3-NPA that is readily detoxified in the rumen.[98] Toxicity may be due to the development of methemoglobinemia and the inhibition of succinate dehydrogenase, an important enzyme necessary for energy metabolism.[101] Crown vetch does not cause bloat in ruminants because it contains tannins that precipitate soluble proteins, which contribute to the rapid formation of a frothy foam in the rumen. Creeping indigo (*Indigofera spicata*), (Figure 6-6B) a legume introduced into southern Florida and Hawaii, also contains nitroglycosides that have caused poisoning in horses.[97-100]

Clinical Signs
Nonruminants may show weight loss, poor growth rates, depression, ataxia, and posterior paralysis leading eventually to death of the animal.[103]

Other Plants Affecting the Nervous System

In this section a wide variety of indigenous and introduced plants with different effects on the brain and nervous system will be discussed. All are important when considering the cause of nervous system disorders in animals that have access to the plants. Many of these neurotoxic plants also have effects on other organ systems that can cause a variety of clinical signs in addition to those involving the nervous system.

Sagebrush Poisoning

More than 200 species of sagebrush (*Artemisia* spp.) grow in North America, many of which are commonly eaten by both wild and domestic ruminants. Some of the more common sages have the potential to cause poisoning in horses especially if they eat large amounts of sagebrush when other forages are unavailable. From a clinical perspective, horses with sage poisoning exhibit very similar neurologic signs to those encountered with locoweed poisoning. Consequently it is important to consider sage poisoning along with locoweed poisoning when determining the cause of the neurologic disease in horses that are grazed on pasture or range containing both locoweeds and sage. In western North America both locoweeds and sages commonly grow in the same habitat.

Sand sage (*A. filifolia*), common in the sandy soils along the eastern side of the Rocky Mountains and south into Mexico, has been associated with a syndrome in horses called "sage-sickness."[107] Budsage (*A. spinescens*) has caused similar problems in animals in California and Nevada.[108,109] Recently the author encountered a neurologic disease of horses that were wintered on overgrazed range in Colorado where fringed sage (*A. frigida*) was the predominant forage being eaten by the horses.

Although the actual toxin that causes sage sickness has not been defined, some monoterpenes present in sagebrush are known to be neurotoxins. Thujone, a terpene present in wormwood (*A. absinthium*), has been associated with a neurologic syndrome in humans that chronically consume absinthe, the alcoholic beverage produced from wormwood.[111] A similar toxicity is presumed to develop in horses that consume a sufficient quantity of sage. Wild and domestic ruminants are not affected by eating sagebrush.

Horses appear to develop neurologic signs after they are forced to eat sagebrush when other forage is depleted or unavailable either as a result of deep snow cover or pasture overgrazing. After eating sage for several days, horses suddenly exhibit abnormal behavior characterized by ataxia and a tendency to fall down or act abnormally to stimuli that would not normally elicit such responses.[112] Tying a horse to a fence, for example, will cause the animal to pull back violently, eventually throwing itself to the ground in panic. If left undisturbed, the animal will recover and will act relatively normal. Ataxia is particularly noticeable in the front forequarters with the hindquarters seemly normal. Some animals may circle incessantly, and others may become excitable and unpredictable. The characteristic smell of sage is often noticeable on the breath and in the feces. Horses poisoned by sagebrush maintain an appetite and have a normal temperature, pulse, and respiration. It is the author's observation that the clinical signs resemble those of a horse that has been poisoned by locoweeds.[110]

However, unlike "locoed" horses that will not recover fully, "saged" horses tend to recover 1 to 2 months after they stop eating sage and are fed a nutritious diet. Some horses, after adapting to eating sage for a period of weeks, seem to be able to tolerate it without problem.[112] Supportive therapy, including protection from extreme climatic conditions, will aid in the recovery. A poisoned horse should not be ridden until it is fully recovered and evaluated for normal behavior and neurologic function.

The only lesion observable in severely "saged" horses is a nonspecific degenerative toxic encephalopathy, with intraneuronal pigment accumulation and degeneration, especially in the medulla, brain stem, and cerebellum.[110]

Sand Sage
Artemisia filifolia
Asteraceae (Sunflower family)

Habitat

Sand sage prefers sandy soils especially of the central plains from the Dakotas, south to north Texas, and west to Arizona and Nevada.

Description

A perennial woody plant growing to a height of 4 feet (1 meter) (Figure 6-7). The multiple stems and leaves are covered with gray hairs, giving the plant a silvery green appearance. The leaves are up to 4 inches (10 cm) in length and are divided linearly into fine segments. The inconspicuous flower clusters are produced at the ends of the stems in the leaf axils.

Principal Toxin

Sagebrush species contain both sesquiterpene lactones and essential oils or monoterpenes that are potentially toxic to horses, but not to cattle, sheep, goats, and wild ruminants.[112] The principal toxins in sagebrush vary considerably in quantity depending on growing conditions and season, being greatest in the fall and winter months.[113-116]

Figure 6-7 Sand sage (*Artemisia filifolia*).

Fringed Sage, Sagewort, Estafiata
Artemisia frigida
Asteraceae (Sunflower family)

Habitat

Fringed sagebrush is found in a vast area extending from Alaska, south through the western United States, into Mexico. Its range also extends into Siberia, Asia, and Europe. Fringed sage is common in the Rocky Mountain states, and ranges from low semidesert areas to altitudes of 11,000 feet (3,352 meters). Fringed sage can become invasive to the point of becoming a monoculture and as such its presence over large areas is indicative of overgrazing. In Colorado it is considered a noxious weed in may counties. In the southwestern States, fringed sage forms a palatable forage for cattle and sheep when other forages are depleted.

Description

Fringed sage is a low-growing, perennial, with a woody base and deep taproot. Roots also form where the spreading stems touch the ground. Leaves are small up to 0.5 inch (1 cm), divided, and silvery haired. Numerous erect, densely haired, flowering stems up to 2 feet (0.2 meters) in height are produced annually. The stems contain many leaves and numerous drooping, globe-shaped structures with many small yellow flowers (Figure 6-8). Crushing the leaves and stems yields a distinctive sage smell.

Figure 6-8 Fringed sage or sagewort (*Artemisia frigida*).

Principal Toxin

Sagebrushes contain both sesquiterpene lactones and essential oils or monoterpenes that are potentially toxic to horses, but not ruminants.

Big Sagebrush
Artemisia tridentata
Asteraceae (Sunflower family)

Habitat

Big sagebrush is found in large areas throughout the intermountain region of the western United States, from California to the Dakotas and south to Mexico. It prefers dry plains and hillsides and does not tolerate irrigation. It is commonly used as a food source for both domestic and wild ruminants, but it has occasionally been associated with poisoning in sheep.[108]

Description

Big sage is an erect, branching, woody, perennial shrub growing 4 to 12 feet (1.2 to 3.5 meters) in height. The leaves are narrow at the base and have three distinctive lobes at the apex (Figure 6-9). The leaves are covered with fine gray hairs, and when crushed, have a strong sage smell. The flowers are numerous, in a loose terminal panicles, each flower head having up to eight inconspicuous flowers surrounded by numerous bracts.

Figure 6-9 Big sage showing 3-toothed leaves (*Artemisia tridentata*).

Principal Toxin

Sagebrushes contain both sesquiterpene lactones and essential oils or monoterpenes that are potentially toxic to animals. Big sagebrush has been suspected of being toxic to sheep. An experimental dosage of 3/4 lb of *A. tridentata* for 1 to 3 days was lethal to sheep.[108] It has also been associated with abortions in sheep.[116] However, most of the many species of *Artemisia* are useful forage plants for range livestock.

Coyotillo, Tullidora
Karwinskia humboldtiana
Rhamnaceae (Buckthorn family)

Habitat

Coyotillo grows in the dry, gravelly soils of hillsides and canyons of southwestern North America and is found in the dry plains of Texas, New Mexico, and Mexico.

Description

Coyotillo is a large shrub or small tree of the buckthorn family, growing up to 24 feet (7 meters) in height. The leaves are opposite, ovate or elliptic, and are dark green above and pale green beneath. The leaf veins and young twigs have black spots. The flowers are in axillary cymes, are perfect and actinomorphic. The petals, sepals and stamens are all five in number and are inserted on a disk. The flowers are sparsely pubescent. The fruits are ovoid and berry-like, turning brown-black when ripe with one seed.

Principal Toxin

The principal toxin responsible for causing the neurologic signs is unknown. The plant leaves and especially the brownish black fruits are toxic to cattle, sheep, goats, horses, pigs, chickens, and humans.[117-120] Poisoning can develop following a single feeding on the fruits (0.05-0.3 percent body weight of the dried fruit), with cattle being the most susceptible and chickens the least.[121] The leaves are also toxic. Clinical signs of poisoning may take from several days to several weeks to develop. The toxins cause nervous symptoms characterized by progressive impaired function of the cerebellum and peripheral nerves.

Clinical Signs

Initially goats poisoned with coyotillo show increased alertness, hypersensitivity to stimuli, muscle tremors, abnormal gait, and a hunched back.[119,120] Rapid breathing and a high stepping gait with exaggerated flexing of the legs develops in some goats. A progressive polyneuropathy causes a progressive decrease in neuromuscular reflexes, with paralysis developing especially in the rear quarters. Recumbency precedes death in poisoned animals. Appetite, defecation, and urination are unaffected until shortly before death. Pulmonary edema, which is usually fatal, develops in some animals. Segmental demyelination, degeneration of axons in peripheral nerves, and myodegeneration are characteristic of *Karwinskia* toxicity.[119-121,122]

There is no specific treatment for coyotillo poisoning other than to prevent the animal from eating any more of the plant and providing a nutritious diet. Mildly affected animals will recover with time, but those showing severe muscle weakness and recumbency seldom survive.

Bracken Fern, Brake Fern, Eagle Fern
Pterdium aquilinum
Polypodiacae (Fern family)

Habitat

Bracken fern is found throughout the United States and has been associated with poisoning in cattle, sheep, pigs, horses, and humans.[122,123] In North America bracken fern is commonly found in the eastern, intermountain, and western states, from Canada to Mexico. Its growth in the midwestern states is sparse. Bracken fern prefers to grow in moist open woodlands with sandy soils, often forming dense stands following clear-cutting or burning of forests. It will grow in relatively dry soils and because of its prolific root system spreads rapidly to form dense monocultures to the exclusion of any other plants.

Bracken fern is also found in most parts of the world with most cases of animal poisoning occurring in England and Europe.[123,124]

Description

Bracken fern is a perennial fern with a black horizontal branching root system often extending for several meters. Leaves arise directly from the rhizome, are broadly triangular, up to 6.5 feet (2 meters) in height, bipinnately compound, and heavily haired on the underside (see Figure 5-5A). Characteristic brown reproductive spores are produced under the rolled edge of the leaflets in late summer (see Figure 5-5B).

Bracken Fern Poisoning

Bracken fern has been associated with a variety of different syndromes in animals, the best recognized of which include:

- Thiamin deficiency
- Retinal degeneration and blindness
- Hemorrhaging and bone marrow destruction (thrombocytopenia)
- Urinary bladder cancer (enzootic hematuria)
- Digestive tract cancers

In the interest of continuity, all the syndromes of bracken fern poisoning will be discussed in Chapter 8, and only the neurologic signs of bracken fern poisoning are briefly mentioned here.

Principal Toxins

Bracken fern contains an enzyme thiaminase, which splits the essential vitamin thiamin (B_1) into its two inactive components pyrimidine and thiazole.[125] Thiamin is essential in energy metabolism, especially in the conversion of pyruvate to acetyl-coenzyme A (CoA), and the oxidation of α-ketoglutarate to succinyl-CoA in the citric acid cycle.[125] Horses and pigs are most susceptible to the effects of thiaminase. Ruminants are rarely affected because they produce ample thiamin in the rumen.[126-129] Horses have to consume a diet containing 3 to 5 percent bracken fern for at least 30 days before clinical signs appear.[130] Sheep can be experimentally poisoned if they are fed large quantities of bracken fern for prolonged periods.[131] Affected animals develop a thiamin deficiency that is characterized by central nervous system depression and polioencephalomalacia. A similar thiaminase enzyme is also found in other plants including horsetail (*Equisetum arvense*), the Australian nardoo fern (*Marsilea drummondii*), and the rock fern (*Cheilanthes sieberi*).[132,133]

Bracken Fern Poisoning in Horses

Bracken fern poisoning in horses is uncommon. When encountered it is characterized by a nervous system disease resulting from depletion of thiamin.[123] Affected horses refuse to eat and consequently lose weight. Depression, muscle tremors, uncoordinated gait, especially of the hindlegs and paralysis are typical of bracken fern poisoning. Horses may show colic, constipation, hemoglobinuria, severe anemia, elevated temperature, and rapid heart rate.[123]

Diagnosis of bracken fern poisoning should be based on evidence that horses have eaten the fern, the clinical signs, and the animal's response to thiamin therapy. Elevated serum pyruvic acid levels (normal 2-3 :g/dL) and decreased thiamin levels (normal 8-10 :g/dL) are helpful in confirming the diagnosis. Bracken fern poisoning in horses should be differentiated from viral encephalitis and hepatic encephalopathy, which have similar clinical signs.

Treatment

Horses with thiamine deficiency should be treated with intravenous thiamine in a dose of 5 mg/kg body weight. This dose should be repeated intramuscularly for several days. Horses should be provided with a balanced diet that is free of bracken fern.

Common Horsetail, Scouring Rush
Equisetum spp.
Equisetaceae (Adder's tongue family)

Habitat

The common horsetail is a weed of moist, sandy soils in fields, roadsides, and along banks of rivers and lakes throughout most of North America. Six species (or varieties) of Equisetum occur throughout the United States.

Figure 6-10A Horsetail or scouring rush fertile stems with terminal spore caps (*Equisetum arvense*).

Description

Horsetails are perennial rushlike plants with characteristic jointed, hollow stems that readily separate at the nodes (Figure 6-10A). The leaves have been reduced over time to black-tipped papery scales surrounding the stems at each node. The stems are cylindrical, ridged, and coarse to touch owing to their high silicate content. Multiple whorled branching occurs at the nodes in some species (Figure 6-10B). The plant reproduces from a deeply buried rhizome and from terminal spore-bearing cones.

Principal Toxin

Although a variety of substances have been identified, including silica, various alkaloids, and organic acids, the primary toxicity of bracken fern appears to be related to its antithiamine effect.[134-136] Successful treatment of the affected horses with thiamine hydrochloride (vitamin B_1) indicated that the primary toxin in horsetail is a thiaminase enzyme.[135,136] Thiamin is essential in energy metabolism, particularly in the conversion of pyruvate to acetyl-CoA, and the oxidation of α-ketoglutarate to succinyl-CoA in the citric acid cycle.[137] Horses are most frequently poisoned by horsetail, but cattle and sheep are occasionally affected especially by *E. palustre*.[134] The plants are toxic at all stages and remain toxic in hay. Hay containing 20 percent horsetail fed to horses for over 2 weeks has caused clinical signs.[135]

Clinical Signs

Young horses are most susceptible to *Equisetum* poisoning; clinical signs develop after the horse has eaten the plants for several days. Initially weight loss is most evident followed several weeks later by incoordination of the hind legs. Diarrhea may precede the onset of weight loss and incoordination. Affected animals become progressively weaker and eventually recumbent, but continue to eat relatively well throughout. Serum pyruvate levels increase, while thiamin levels are depleted. Once horses are down and cannot get up, they usually die in 1 to 2 weeks. Cattle poisoned with horsetail show weight loss, decreased milk production, diarrhea, and hyperexcitability.[134]

Treatment

Horses suspected of being poisoned by *Equisetum* should be taken off of hay or pasture containing the plant and fed a nutritious diet that includes cereal grains that are rich in thiamine.

Figure 6-10B *E. arvense* with sterile branching stems.

Treatment with a large dose of thiamine hydrochloride (3-5 mg/kg body weight) intravenously followed by several days of thiamine intramuscularly (1-2 mg/kg body weight) provides rapid recovery and restores thiamine levels to normal.[137,138]

Yellow Star Thistle and Russian Knapweed Poisoning

Both yellow star thistle and Russian knapweed produce a unique poisoning of horses that is generally fatal. The plants occur in areas of the western United States, Australia, and Argentina.[139,140] Both plants were introduced to the United States when their seeds contaminated imported grains, and since have become well established as noxious weeds.[139,140] Unfortunately the range of these noxious weeds continues to spread.

Yellow Star Thistle
Centaurea solstitialis
Asteraceae (Sunflower family)

Habitat

Yellow star thistle is a weed that was introduced from the Mediterranean area and has become well established in California, Oregon, Idaho, Washington, and in some areas throughout the southwestern, and southeastern states. It is an invasive noxious weed of cultivated areas and spreads along roadsides and waste areas. Malta star-this-

Figure 6-11A Yellow star thistle immature plant (*Centaurea solstitialis*).

tle or tocalote (*C. melitensis*) is a similarly introduced plant that has become a noxious weed in some areas of the southwest. It has not been reported as toxic.[141]

Description

The plant is an annual herbaceous weed, branching from the base up to 12 inches (30 cm) tall (Figure 6-11A). The branches are winged and ascending. Leaves are covered with cottony hair, the basal leaves being deeply lobed; the stem leaves are linear and entire (Figure 6-11B). The disc flowers are yellow and fertile. The bracts are tipped with characteristic stiff yellow spines 0.5 to 1 inch (1 to 2 cm) long (Figure 6-11C).

Malta star-thistle or tocalote (*C. melitensis*) is a similar plant but is smaller and the spines on the bracts are small and dark tipped (Figure 6-11D). It is not reported to be poisonous.

Figure 6-11B Yellow star thistle mature blooming plant (*C. solstitialis*).

Principal Toxin

Several neurotoxic components of yellow star thistle have been identified including aspartic and glutamic acids and two sesquiterpene lactones, solstitialin A 13-acetate and cynaropicrin.[142-144] A dopaminergic neurotoxin, 2,3 dihydro-3, 5 dihydroxy-6-methyl-4 (H) pyran-4-1, very similar to the compound that induces Parkinson's disease in people, has been isolated from yellow star thistle.[145] This compound specifically destroys the dopaminergic nigrostriatal pathway that has coordinating and inhibiting effects on the cerebral cortex pathways that control prehension and chewing of food (cranial nerves V, VII, IX).[146-148] The plant is poisonous only to horses and is toxic in both its green and dried states. Horses will eat yellow star thistle in all its stages of growth and will acquire a preferential liking for the plant. Cattle and sheep graze the plants without problem and have been used as a means of controlling the weed.

Figure 6-11C Yellow star thistle with yellow ray flowers and long spiny bracts (*C. solstitialis*).

Figure 6-11D Malta star thistle (*C. meletensis*).

Signs of toxicity in horses do not occur until the plant has been eaten in large quantities for 30 to 60 days. Experimentally it takes approximately the animal's body weight in green plant over 33 days to produce clinical signs.[143] It has been calculated that horses have to consume an amount of green yellow star thistle equal to 86 to 200 percent of their body weight before clinical signs develop.[147,148] Although this is a relatively large amount of plant, horses apparently develop a preference for the plants and will eat them even in the mature spiny state. In California there are two periods of the year, June/July and October/November, when yellow star thistle poisoning is most prevalent, suggesting some seasonal variation in palatability or toxin content of the plants.

Russian Knapweed
Acroptilon repens, (Centaurea repens)
Asteraceae (Sunflower family)

Habitat

Introduced from Russia in alfalfa seed, the plant has established itself in cultivated fields, pastures, and roadsides.[138,140] It is considered a noxious weed in many areas of Colorado and other Rocky Mountain states where it has become a highly invasive weed. Russian knapweed grows in all soil types and spreads by an extensive root system. It is also allelopathic, meaning it is capable of producing an inhibitory substance from its roots that retards other plant growth in its vicinity.

Description

Russian knapweed is a creeping perennial plant with black horizontal roots. It is erect, rather stiff, and branched, usually ranging from 1 to 3 feet (1 meter) high. The young stems are covered with soft gray hair or nap. The lower leaves are alternate with toothed margins, which become entire margined at the top of the plant and are covered with short stiff hairs. The flowers are in heads like some thistles, one-third to one-half inch in diameter and are lavender to whitish in color (Figure 6-12). The bracts are

Figure 6-12 Russian knapweed in bloom (*Acroptilon repens*).

papery and have no spines. The seeds are chalky white or grayish in color and about one-eighth inch long with bristles at one end. The seeds are rarely shed from the seed head itself that may aid in their dispersal. Rodents readily carry off the seed heads and hide them in their burrows, and by so doing inadvertently plant the seeds and spread the weed.

Principal Toxin

Russian knapweed contains a sesquiterpene lactone, repin, that has shown neurotoxic properties.[149,151] However, a dopaminergic neurotoxin isolated from yellow star thistle that interferes with the dopamine pathway may also be identified in Russian knapweed in due course.[145] Russian knapweed is toxic only to horses and causes lesions and clinical signs identical to those produced by yellow star thistle (*C. solstitialis*).[147,150] Russian knapweed appears to be more toxic than yellow star thistle requiring less plant mass (1.8-2.6 kg/100 kg body weight) and a shorter feeding period (28 to 35 days) to produce disease in horses.[147] It has been calculated that horses have to consume an amount of green Russian knapweed or yellow star thistle equal to 59 to 71 percent and 86 to 200 percent of their body weight, respectively, before clinical signs develop.[147,148]

Clinical Signs

Prolonged consumption of yellow star thistle and Russian knapweed results in a disease of horses called "chewing disease," characterized by increased tonicity and incoordination of the muscles that enable prehension and chewing of food.[145-148,152-154] The hypertonicity of the facial muscles produces a "wooden" expression to the face. Food is often held in the mouth because it cannot be chewed normally. The continual chewing movements cause frothing of the saliva, which can resemble that seen in rabies. Some horses may wander about with their lips brushing through the grass, which to the unobservant, could be mistaken for normal grazing.

Even though prehension and mastication are severely affected, swallowing is unaffected. Some horses may learn to submerge their heads far enough into a deep trough of water to allow water to reach the pharyngeal area where it can be swallowed. The tongue has increased tone and the horse will often curl the tongue from side to side. Some horses may show more involvement of one side so that the lips, tongue, and head movements are to one side. Circling to the same side may also occur. Other abnormal behavior may include violent head tossing and excessive yawning. Clinical signs are most severe initially and after a few days the horse may show some improvement. Weight loss and depression are common. Pneumonia resulting from inhalation of feed is a serious sequel to the disease.

Although the clinical signs of yellow star thistle poisoning are unlike any other, they should be differentiated from viral encephalitis (sleeping sickness), rabies, glossopharyngeal and hypoglossal nerve injuries, pyrrolizidine alkaloid poisoning, and foreign bodies lodged in the mouth or esophagus.[154]

There is no effective treatment for either yellow star thistle or Russian knapweed poisoning because the affected areas in the brain undergo liquefactive necrosis and do not regenerate. Affected horses may be kept alive by administering water, electrolytes, and a high-energy liquid diet through a nasogastric tube or an esophagotomy. Euthanasia of affected horses is eventually necessary because of the debilitating and irreversible effects of the brain lesions and complicating inhalation pneumonia.

Postmortem Findings

The lesions of yellow star thistle and Russian knapweed poisoning are unlike any other plant-induced poisoning and consist of bilaterally symmetrical foci of liquefactive necrosis in the globus pallidus and the substantia nigra.[145,147,148,153] Occasionally the lesions are unilateral. The lesion of nigropallidal encephalomalacia are microscopically discernable within 2 days of the appearance of clinical signs and are grossly visible after 2 days as yellowish necrotic lesions.

White Snakeroot Poisoning

White snakeroot is the plant that was responsible for the syndromes of "milk sickness" in humans and "trembles" in animals that early settlers encountered in the eastern and central United States.[155-157] As early as 1777, there were reports of an unknown poisoning of humans and cattle in the Carolinas and Tennessee, now known to be due to white snakeroot or richweed (*Eupatorium rugosum*).[158] Abraham Lincoln's mother reportedly died of milk sickness in 1818 after drinking milk from a cow that had access to white snakeroot growing at the Lincoln cabin site.[159] Many human and animal deaths were attributed to white snakeroot poisoning. Human poisoning virtually ceased once the plant was identified as the source of the problem and farmers learned to keep their animals away from the plant. Animals continue to be poisoned by snakeroot sporadically if they gain access to the plant.[160-162] The incidence of white snakeroot poisoning has the potential to increase in the future as many new small acreage farms are being established in areas of white snakeroot. These new farmers may well be unaware of the lethal potential of the plant.

White Snakeroot
Eupatorium rugosum
Asteraceae (Sunflower family)

Habitat

White snakeroot is found in low, moist areas or bordering streams, often on rich or basic soils of open woodlands, from eastern Canada, west to the Dakotas, south to Georgia, and west to Texas.[163-165] Synonyms for white snakeroot include Eupatorium urticaefolium and Ageratina altissima.[166]

Description

This showy, herbaceous perennial forms erect, stiff stems as much as 39 to 55 inches (100 to 140 cm) tall, developing from a shallow mat of fibrous roots. Leaves are opposite, long petioled, ovate to cordate, and range from 3 to 6 inches (7.5 to 15 cm) in length. The leaf margins are coarsely and sharply serrate; the tip sharply pointed, and the leaves with three distinct veins. The tubular flowers are showy, snow white, and small in composite heads of 10 to 30 flowers (Figure 6-13). The heads are grouped in open corymbs.

Principal Toxin

The toxic component of white snakeroot has been named tremetone (tremetol) and requires microsomal activation before it becomes toxic to mammalian cells.[163,164] All animals including man are susceptible to poisoning.[166-169] The toxin has its highest content in the green plant and remains toxic when dried in hay. The tremetone is cumulative in animals and is secreted in milk from cows that have eaten snakeroot. Humans drinking the milk develop a severe nervous syndrome known as "milk sickness."[160] In animals the toxicity is known as "trembles" because of the muscle tremors induced by the toxin. Poisoning develops after an animal has eaten from 0.5 to 1.5 percent of an its body weight in green plant.[163] Feeding horses 1 to 2 percent of their body weight of the plant induced poisoning in 1 to 2 weeks.[170] Approximately 20 lb of green plant consumed over several days induces poisoning in horses.[170] The toxin of white snakeroot is readily secreted through the milk and is therefore a hazard to humans or animals that drink it.[171] Foals may develop white snakeroot poisoning when they drink the milk from their dam that has been eating the plant.[160] E. wrightii of Texas, Arizona, and Mexico has been reported to cause sudden death in cattle. Other species of *Eupatorium* including *E. adenophorum* (crofton weed) and *E. riparium* from Australia are toxic to horses.[172-175] In addition to white snakeroot, tremorgenic toxins are also found in rayless goldenrod (*Haplopappus heterophylus*), Jimmy fern (*Notholaena sinuata*), western mountain laurel (*Sophora secundiflora*), and silky sophora (*Sophora sericea*).[176]

Various fungi growing on plants produce tremorgenic toxins that cause muscle tremors in livestock that can resemble those encountered in white snakeroot poisoning. Fungal tremorgens include those produced by the fungus Claviceps paspalli in Dallis grass (*Paspalum dilitatum*) and Bahia grass (*Paspalum notatum*).[176,177] Perennial ryegrass (*Lolium perene*) staggers is caused by the endophytic fungus *Acremonium loliae*, whereas annual ryegrass (*Lolium rigidum*) staggers is caused by a toxin produced by a bacterium *Corynebacterium rathayi* growing on the grass. Bermuda grass (*Cynodon dactylon*) and Phalaris or canary grass (*Phalaris* spp.) staggers are similar to other tremorgenic syndromes.[179]

Clinical Signs

Horses, cattle, sheep, and goats poisoned with white snakeroot are initially listless,

Figure 6-13 White snakeroot (*Eupatorium rugosum*).

depressed, lethargic, and disinclined to move much.[157,158,168,178,179] Cattle, in particular, develop muscle tremors, especially after exercise, and may show signs of colic, constipation, blood in the feces, and a peculiar acetone-like odor to the breath.[166,176,177] Once muscle tremors begin, animals are reluctant to move, showing marked stiffness and eventual recumbency.

Horses may show signs of choking due to the paralysis of the pharyngeal muscles. Nursing animals may have milk run out the nostrils.[162,167,168] Patchy sweating may be evident. The urine may become dark brown due to myoglobinurea.[170] A rapid and irregular heart rate and signs of congestive right heart failure may precede death.[167] Electrocardiograms taken from horses with white snakeroot poisoning may reveal complete atrioventricular block, ventricular premature beats, and marked ST-segment depression indicative of myocardial ischemia.[170]

Poisoning in humans who have consumed milk containing tremetol is characterized by weakness, muscle tremors, depression, garlicky odor to the breath, abdominal pain and vomiting, constipation, delirium, and death. People who survive experience prolonged signs of weakness.[158,159]

Postmortem Findings

Mortality is usually high in livestock showing "trembles." Death appears due to severe skeletal and myocardial degeneration.[168-170] Fatty degeneration of the liver and kidney is a prominent necropsy finding. In Australia, horses poisoned by crofton weed (*E. adenophorum*) develop a respiratory syndrome characterized by severe pulmonary fibrosis and infarction.[172-174]

Diagnosis

There is no specific diagnostic test for tremetone that does not give false-positive results due to normally occurring miscellaneous ketones secreted in milk.[168] Diagnosis therefore is based on the clinical signs and the presence of white snakeroot in the hay or pasture where the animal has been eating.

Treatment

There is no specific antidote for white snakeroot poisoning. Further consumption of the plant or affected milk should be stopped immediately. Excitement and exercise should be prevented. Where possible lactating animals should be repeatedly milked to help remove the toxin from the animal's system. The milk should be carefully discarded to ensure no animal drinks it. Laxatives and activated charcoal improve chances of recovery if administered early. Horses that exhibit difficulty in swallowing should be given water, electrolytes, and appropriate nutrition via nasogastric tube. Recumbent animals should be placed in well bedded stalls to prevent the development of pressure sores.

Poisoning in humans today is infrequent due to the practice of pooling milk from many cows, which dilutes any tremetone that may have been present. The individual family in rural areas that drinks raw milk from their cow have a greater potential for poisoning if there is white snakeroot in the animal's pasture. Pasteurization does not detoxify tremetone in milk.

Jimmy Weed,
Rayless Goldenrod, Burrow Weed
Isocoma pluraflora (Isocoma wrightii),
(Haplopappus heterophyllus)
Asteraceae (Sunflower family)

Habitat

Jimmy weed is commonly found in the alkaline soils of drier rangeland, river valleys, drainage areas, and irrigation canals in Texas, Arizona, and New Mexico.

Description

Jimmy weed is an erect, sparsely branched, woody perennial growing to 2 to 4 feet (0.5 to 1 meter). Leaves are sticky, linear, and alternate. Flowers are yellow, borne in numerous, small, terminal flat topped heads composed of 7 to 15 flowers each (Figure 6-14).

Principal Toxin

A compound tremetone (tremetol), similar to that found in white snakeroot (*E. rugosum*), appears to be responsible for causing poisoning in livestock.[174] Tremetone is secreted in milk and is therefore potentially hazardous to people or animals drinking affected milk. Poisoning of livestock most often occurs in the fall and winter when other forages are scarce and livestock are forced to eat the jimmy weed. The toxic dose of the plant that needs to be eaten to induce poisoning is approximately 1.5 percent of the animal's body weight of green plant over a period of a week.[180] The dried plant appears to be less toxic.

Figure 6-14 Burrow weed (*Isocoma pluraflora*) (Courtesy of Dr. Robert Glock, Veterinary Diagnostic Laboratory, Tucson, Arizona.)

Clinical Signs

The symptoms of *Haplopappus* and *Eupatorium* spp. poisoning are similar. Initially animals exhibit marked depression and are reluctant to move. Forced exercise often precipitates muscle tremors and collapse.[180-182] As the disease progresses the animals become weak and unable to walk. Death frequently occurs at this stage. Urinary incontinence and an acetone-like smell to the breath may be noted in severe cases.

Treatment

Further consumption of the plants or milk from affected animals should be prevented. Treatment with laxatives may help in preventing further absorption of the toxin. Food, water, and other supportive treatment should be given as necessary.

Miscellaneous Neurotoxic Plants

Mescal Bean, Frijolito, Mountain or Texas Laurel
Sophora secundiflora
Fabaceae (Legume family)

Habitat

Mountain laurel or mescal bean is commonly found growing in the limestone range-lands, dry hills, and canyons of western Texas, Arizona, and Mexico.

Description

Mescal bean is a woody perennial evergreen shrub or small tree up to 33 feet (10 meters) in height. Leathery leaves are alternate, pinnate, opposite, once compound with terminal leaflet. Leguminous blue to purple, showy, fragrant, flowers appear in a one-sided terminal raceme. The fruit is a multiple seeded woody pod up to 6 inches (15 cm) long (Figure 6-15A). Seeds are bright red to orange in color and have a very hard seed coat (Figure 6-15B).

Figure 6-15B Mescal beans (*S. secundiflora*).

Figure 6-15A Mescal bean or Texas laurel (*Sophora secundiflora*).

Principal Toxins

The toxicity of *S. secundiflora* has been recognized for both humans and animals for some time.[183] The red seeds have been used as hallucinogens by Indian tribes, and the leaves and seeds have been recognized as a cause of poisoning in cattle, sheep, and goats.[183] A suspected case of mescal bean poisoning has been reported in a dog that had been chewing the beans.[184] The quinolizidine alkaloid cytisine (sophorine), along with other alkaloids and amino acids in combination, are responsible for the toxicity.[185-87] The mature leaves and fruit are toxic. The seeds are poisonous if crushed but otherwise pass through the animal's digestive system intact and without effect. Livestock only eat the plant if other forage is unavailable in the late summer and winter months. Silky sophora (*S. sericea*) is similarly toxic. Other plants containing similar alkaloids include wild indigo (*Baptisia* spp.), scotch broom (*Cytisus scoparia*), golden banner (*Thermopsis* spp.), and white loco (*Sophora nuttalliana*). This last plant is not related to white locoweed (*Oxytropis sericea*).

Clinical Signs

Poisoned animals show minimal clinical signs until they are stressed or forced to exercise, in which case muscle tremors, incoordination, and a stiff gait are characteristic.[187] If left alone, affected animals appear to recover, but clinical signs can be induced when the animals are stressed. Severely poisoned animals become recumbent and comatose and eventually die. No known specific treatment for mescal bean poisoning is known, and therapy should be directed at providing supportive care and a nutritious diet.

Buckeye, Horse Chestnut
Aesculus spp.
Hippocastanaceae (Buckeye family)

Habitat

There are at least 25 species of horse chestnut, some of which have been introduced to North America from Europe. Seven species of *Aesculus* are indigenous to various areas of North America and have become widely cultivated as desirable shade trees. They generally prefer rich, moist soils of woodlands, although some species have adapted to drier conditions. The buckeye or horse chestnut is not related to the edible chestnut (*Castanea* spp.)

Description

These trees or shrubs have opposite, palmately compound leaves, with five to seven serrated leaflets (Figure 6-16A). The inflorescence is a panicle of large, erect flowers which are usually yellow, whitish yellow, or red in color depending on the species (Figure 6-16B). The fruit is a one- to three-seeded leathery capsule with or without sharp spines (Figure 6-16C). The seeds are large, 1 inch (2.5 cm) in diameter, glossy brown when newly exposed with a conspicuous lighter scar (Figure 6-16D).

Figure 6-16A Swamp buckeye flower and typical leaves (*Aesculus octandra*).

- *A. californica* (California buckeye)
- *A. glabra* (Ohio buckeye)
- *A. hippocastanum* (horse chestnut)
- *A. octandra* (yellow buckeye)
- *A. pavia* (red buckeye)

Principal Toxin

The glycosides aesculin and fraxin, and possibly a narcotic alkaloid, present in the young growing sprouts, leaves, and seeds are thought to be responsible for toxicity in animals.[188,189] Five bioactive triterpene oligoglycosides named escins have been isolated from the seeds of the horse chestnut, the toxic significance of which has yet to be determined.[190] Cattle, sheep, horses, swine, chickens, and humans have been poisoned naturally and experimentally by various species of buckeye.[188,189,191]

The nectar and sap of *A. californicum* is known to kill honey bees that feed on it, so much so that bee keepers recommend moving hives during the flowering period of the California buckeye.[192] Poisoning of livestock generally occurs when animals eat the leaves and sprouts of the buckeye because it generally leafs out before other plants in the spring. The green husk of the immature fruit is especially toxic.[193]

Clinical Signs

In simple-stomached animals, vomiting and gastroenteritis are the predominant effects of the toxic glycoside. In ruminants, the glycoside is converted in the rumen to the soluble aglycone, which following absorption, produces neurologic and not gastrointestinal signs.[194]   Animals develop signs of poisoning about 16 hours after consuming toxic quantities of the plant, and as little as 0.5 percent body weight of ground nuts fed to calves produced severe poisoning.[195] Affected animals initially show muscle twitching, weakness, and a peculiar "hopping" gait, especially involving the hind legs.[191,194,196,197]   In severe cases of buckeye poisoning, muscle tremors rapidly progress to muscle spasms and recumbency.   These spasms may occur every 30 seconds and can be induced by handling or stressing the animal. A dorsal-medial strabismus occurs in severe cases. Hyperglycemia, glucosurea, and proteinurea appear to be consistent features of severe toxicity. Vomiting and abdominal pain may also occur. Once animals become recumbent and progress into a coma they rarely recover.

Treatment

There is no specific therapy for buckeye poisoning.   Laxatives may be given to help remove the ingested plant as rapidly as possible from the intestines.   Supportive intravenous fluid therapy with calcium gluconate and dextrose may be beneficial.

Figure 6-16B California buckeye blooms (*A. california*).

Figure 6-16C Horse chestnut fruits (*A. hippocastanum*).

Figure 6-16D Horse chestnut seeds (*A. hippocastanum*).

Kentucky Coffee Tree
American Coffee Berry
Kentucky Mahogany
Gymnocladus dioica
Fabaceae (Legume family)

Habitat

The Kentucky coffee tree is indigenous to eastern North America, from Ontario south to Florida and west to Nebraska. It is grown as a large, semihardy shade tree in other areas.

Description

This is a deciduous, multibranched, large tree to 90 feet (27 meters) in height. Leaves are large, twice pinnate; each leaflet is oval (Figure 6-17A). Trees may have both male and female flowers that are small, greenish white, and fragrant. The characteristic fruits are hard, brown, leguminous pods 4 to 6 inches (10 to 15 cm) long, containing five to seven hard, dark brown seeds (Figure 6-17B).

Principal Toxin

Little is known about the toxin other than it is thought to be the quinolizidine alkaloid cytisine, which acts on the nervous system like nicotine.[198,199] New sprouts, leaves, and the fruits are toxic to animals and humans.[198] The dried seeds when roasted appear to be nontoxic and have been used as a coffee substitute.[198]

Figure 6-17A Kentucky coffee tree branch with leaves (*Gymnocladus dioica*).

Clinical Signs

Cases of poisoning have infrequently been reported in cattle, sheep, horses, and humans.[198] Signs of intense gastrointestinal irritation (vomiting, colic, and diarrhea) begin within an hour of eating the shoots, leaves, or chewed seeds.[193] Hypotension, decreased heart and respiratory rates, muscle paralysis, and convulsions may precede death depending on the quantity of plant consumed. No specific lesions are visible at necropsy other than those of gastroenteritis.

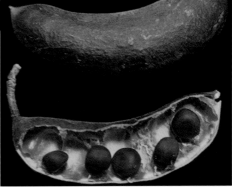

Figure 6-17B Kentucky coffee tree seed pods (*G. dioica*).

Treatment

No specific treatment is available, and therapy where possible should be directed at preventing further absorption of the toxin from the gastrointestinal tract. The use of oral activated charcoal and saline cathartics is helpful in this respect. Intravenous fluid therapy to counteract hypotension, atropine to reverse the bradycardia, and phenobarbital or diazepam to control convulsions should be administered as necessary.

Carolina Jessamine, False or Yellow Jessamine
Gelsemium sempervirens
Loganiaceae (Logania family)

Habitat

Carolina jessamine is a common indigenous plant of the southeastern states, where it is found growing in open woods, fence rows, and stream banks. It has been widely cultivated as an ornamental plant in other areas.

Description

The plant is a twining or trailing, evergreen perennial vine up to 20 feet (6 meters) in length, with opposite, simple, ovate, glossy leaves, and reddish brown stems. The flowers are fragrant, showy, and yellow and produced in clusters from the leaf axils (Figure 6-18). Many winged seeds are produced in two-celled, elliptical capsules.

Figure 6-18 Carolina jessamine (*Gelsemium sempervirens*).

Principal Toxin

Gelsemine, and several other indole alkaloids with properties similar to strychnine, are found in all parts of the plant including the flowers and nectar. All animals, birds, and humans are susceptible to the toxin. The honey made from the nectar of Carolina jessamine is reportedly toxic, and children have been poisoned after sucking the flowers.[200,201] Poultry have been poisoned in large numbers after eating the roots of *Gelsemium*.[201]

Clinical Signs

The usual presenting sign of *Gelsemium* poisoning in livestock is the finding of a prostrate animal. Other signs of poisoning, including hypothermia, dilated pupils, weakness or rigidity of the legs, incoordination, and respiratory failure, precede death. Most animals die within 2 days of developing signs of poisoning.

Treatment

No known specific treatment is known, and animals are best given supportive therapy including activated charcoal and saline cathartics orally, intravenous fluids, and anticonvulsants as needed.

Sweet Shrub, Carolina Allspice
Strawberry Bush
Calycanthus fertilis
C. floridus
Calycanthaceae

Habitat

Found mostly in the southern tier of States, species of Calycanthus grow either wild or are cultivated as ornamentals. Most cases of sweet shrub poisoning have been reported in Tennessee and surrounding area.

Description

This perennial, large shrub to small tree attains a height of 10 feet (3 meters). Leaves are alternate, 2 to 3 inches (5 to 7.5 cm) ovate with blunt or pointed tips, dark green in the summer, and yellow in the fall. Flowers have dark brown or maroon petals and sepals fused at their base into a cup, and sweet fragrance suggestive of strawberries (Figure 6-19). The seed pods are urn-shaped, turning brown when ripe, and containing several one-seeded achenes.

Principal Toxin

Little is known about the toxin in *Calycanthus* spp., but the indole alkaloid calycanthine, with strychnine-like properties is believed to be responsible for the clinical signs.[202,203] A dog experimentally poisoned with an aqueous extract of the seeds exhibited strychnine-like symptoms.[203]

Clinical Signs

There are very few confirmed reports of *Calycanthus* spp. poisoning.[202-204] Clinical signs that have been associated with *Calycanthus* poisoning include hyperexcitability, recumbency, tetanic muscle spasms resulting in rigid extension of the extremities, opisthotonus, coma, and death.[203] Death occurs shortly after onset of tetanic signs. Diagnosis is based on confirming that the animal had access to the plant

Figure 6-19 Sweet shrub (*Calycanthus floridus*).

and evidence that the plant has been eaten. The detection of the seeds in the rumen is helpful in diagnosis.

Treatment

There is no known specific antidote, and symptomatic treatment should be administered. Activated charcoal via stomach tube and sedation to manage the animal's convulsions are appropriate.

Golden Chain Tree, Bean Tree
Laburnum anagyroides
Fabaceae (Legume family)

Habitat

The golden chain tree grows in most areas except where winters are severe. It is frequently cultivated for its showy yellow flowers that hang in long chains.

Description

This large deciduous shrub or tree grows to 30 feet (9 meters) in height. The leaves are produced on long stalks, each with three leaflets, the undersides of which are covered with soft hairs. The pendulous racemes are up to 30 inches (76 cm) in length, with numerous bright yellow pealike flowers (Figure 6-20). Numerous legume pods have up to eight flat seeds. The root has a licorice taste.

Figure 6-20 Golden chain tree (*Laburnum anagyroides*).

Principal Toxin

The quinolizidine alkaloid cytisine with nicotine-like properties is thought to be the principal toxin. Horses, cattle, dogs, and humans have been poisoned by eating the seeds.[205-207] The oral toxic dose of seeds for a horse is 0.05 percent of the animal's body weight. The immature seed pods and seeds are highly toxic. Cytisine is secreted in milk.[205]

Clinical Signs

Affected animals may show loss of appetite, excitement, muscle tremors, irregular gait, convulsions, coma, and death. Vomiting, diarrhea, and dilated pupils may be observed before death. Degeneration of the muscles is observable microscopically.[205]

Treatment

There is no specific treatment, and therapy should be supportive. Activated charcoal orally may help to reduce further toxin absorption.

Bleeding Heart, Dutchman's Breeches
Squirrel Corn, Stagger Weed
Dicentra spp.
Fumariaceae (Fumatory family)

Habitat

About a dozen species of *Dicentra* are found in North America, occurring in the eastern, southeastern states, and northwestern states. Bleeding hearts usually grow in rich, well-drained soils of forests and woodlands. Cultivated varieties (*D. spectabilis*) are widely grown for their showy flowers and fernlike foliage.

- *D. cucullaria* (Dutchman's breeches)
- *D. canadensis* (squirrel corn)
- *D. formosa* (western bleeding heart)
- *D. eximia* (wild bleeding heart)

Dutchman's breeches is the name also given to plants of the western and southwestern United States belonging to the genus *Thamnosma* that are unrelated to the *Dicentra* spp., and cause photosenitization.[208]

Description

Arising from a perennial tuberous root, delicate fernlike leaves are produced in early spring. The leaves are hairless, broadly triangular, finely divided, giving a lacy appearance. The characteristic flowers are pendant, symmetrical, pink to white in color, and produced in racemes extending above the leaves (Figure 6-21).

Principal Toxin

A variety of isoquinolone alkaloids (protoberberines) have been demonstrated in *Dicentra* and *Corydalis* spp. that are poisonous to cattle and horses but apparently not to sheep. Most poisoning occurs in the early spring because *Dicentra* and *Corydalis* spp. emerge before most other plants. Fitweed (*C. aurea*) and the similar *Dicentra* spp. contain similar alkaloids and have produced poisoning in cattle and sheep.[209-212]

Figure 6-21 Bleeding heart plant in flower (*Dicentra* spp.).

Clinical Signs

Cattle may first exhibit muscle tremors, running back and forth, and incoordination that has been referred to as "spring staggers."[210,213] Depending on the amount of plant eaten, projectile vomiting, convulsions, and lateral recumbency with the head thrown back and legs rigidly extended may develop. Animals usually recover from the convulsive seizures.[210,213]

Treatment

There is no specific treatment, and animals tend to recover if removed from the area where the plant is present. Mineral oil or other saline cathartic may be given orally to prevent further absorption of the toxins.

Lobelia
Lobelia spp.
Campanulaceae (Bellflower family)

Habitat

Lobelias grow in moist, fertile soils, along water courses, and in open woods from North Dakota, south into Texas and Mexico, and in much of the southeast. Hybrids of lobelia are grown as ornamentals in most of North America. Species that have been associated with poisoning include:

- *L. berlandieri* Berlandier lobelia
- *L. siphilitica* Great blue lobelia
- *L. inflata* Indian tobacco
- *L. cardinalis* Cardinal flower

Description

Lobelia are perennial, erect plants that may reach 3 to 4 feet (1 meter) in height. Leaves are throughout the stem, being larger at the base of the stem, alternate, hairless, and irregularly serrate. The inflorescence is a raceme of tubular flowers, the lower lip having three lobes and the upper two lobes. The color ranges from scarlet (*L. cardinalis*) to pale blue (*L. siphilitica*) (Figure 6-22A), and various colored hybrids (Figure 6-22B). Fruits are ellipsoid capsules with many small brown seeds.

Figure 6-22A Cardinal flower (*Lobelia cardinalis*). Inset: Great blue lobelia (*L. siphilitica*).

Principal Toxin

Lobelias contain many nicotine-like alkaloids, lobeline being common to most species. Cattle, sheep, and goats are most frequently poisoned by *L. berlandieri*, especially in Mexico and Texas.[214-216] Humans are also susceptible to poisoning from the alkaloids.

Clinical Signs

Clinical signs include excessive salivation, vomiting, diarrhea, dilated pupils, coma and death.[214-216] Ulcers may be found on the cornea and in the mouth. A diagnosis of lobelia poisoning may be made on the basis of the clinical signs, evidence that the plants have been eaten, and the finding of identifiable plant parts in the animal's stomach.

Treatment

There is no specific treatment for lobelia poisoning, although atropine is helpful in relieving some of the signs. Administration of mineral oil and saline laxatives soon after an animal has eaten the plants may help in reducing the absorption of the toxic alkaloids.

Figure 6-22B Lobelia hybrid (*Lobelia* spp.).

Western Horse Nettle, Potato Weed
Solanum dimidiatum
Solanaceae (Nightshade family)

Habitat

Western horse nettle grows in prairies, fields, along roadsides and waste areas.

Description

These erect, perennial herbs grow to 1 to 2 feet (0.3 to 0.5 meters) tall, with a deep branching root system. Leaves are alternate, ovate to elliptical, 3 to 7 inches (7.5 to 17.5 cm) long and 2 to 4 inches (5 to 10 cm) wide, with stiff hairs along the stripes on both sides of the leaves.The stems have sparse, yellow spines. The flowers are produced terminally and consist of panicles of star-shaped, bluish purple to white flowers. The fruits are yellow round berries approximately 1 inch (2.5 cm) in diameter. *S. dimidiatum* closely resembles *S. carolinense*.

Principal Toxin

The toxin responsible for causing the neurologic signs associated with *S. dimidiatum* toxicity is not known. Calystegins, which are glycosidase inhibitors, have been identified in *S. dimidiatum* and *S. kwebense*, a plant from South Africa that produces similar cerebellar degeneration in cattle.[217-220] In Brazil, *S. fastigiatum* and *S. bonariensis* produce similar neurologic signs in cattle[220]; goats in Australia have been reported to develop cerebellar degeneration after eating *S. cinerum*.[221] Calystegins are similar to the indolizidine alkaloid swainsonine found in some of the locoweeds (*Astragalus* spp.), and it has been hypothesized that these various *Solanum* spp. induce a lysosomal storage disease similar to that of locoism.[219,221] The neurologic signs are primarily associated with cerebellar degeneration.

Clinical Signs

"Crazy cow syndrome," the name given to the disease in Texas, aptly describes the behavior of animals poisoned by western horse nettle. Cattle appear normal until they are stressed or excited, whereupon they fall over and struggle futilely to stand. If left alone the affected animal regains its normal neuromuscular control. Animals infrequently die directly from the disease, although many are prone to drowning after falling into waterholes or creeks.[223]

The most distinctive lesion found at postmortem examination of chronic cases is cerebellar atrophy.[220,221] Microscopically, there is severe depletion of the Purkinje neurons in the cerebellum, and those that are present contain vacuolated cytoplasm.[218,220,223]

Sudan and Johnson Grass Poisoning

In addition to the acute toxic effects of cyanide poisoning discussed in Chapter 1, it is well known that low levels of cyanogenic glycosides found in some plants are associated with a chronic neurologic disease in man and animals.[224-228] A chronic poisoning has also been recognized in people in Africa who consume cassava (*Manihot esculentum*) containing cyanogens. The result is a degenerative disease of the nervous system referred to as tropical ataxic neuropathy, and that is characterized by degeneration of the peripheral, optic, and auditory nerves.[229] Enlargement of the thyroid gland (goiter) is common and is due to increased levels of thiocyanate that inhibits the uptake of iodine by the thyroid gland. The precise cause of the degenerative process in the nervous system is not known and is assumed to be related to the chronic elevation of thiocyanate levels that result following the consumption of cyanogens.[230]

A similar syndrome of chronic cyanide poisoning occurs in horses, cattle, and sheep that are fed on pasture or hay containing sorghums such as Sudan and Johnson grass.[224-228] Animals consuming these grasses for prolonged periods develop a syndrome of posterior ataxia, urinary incontinence, cystitis, and weight loss. The disease results from lower spinal cord degeneration induced by low levels of cyanogenic glycosides in the plants that cause demyelinization of the peripheral nerves. It has been hypothesized that some cyanide may be converted to T-glutamyl-β-cyanoalanine, a known lathyrogen that interferes with the neurotransmitter activity of glutamate.[231] Lathyrogens affect normal development of nervous tissue causing signs of ataxia, urinary incontinence, and musculoskeletal deformities similar to those seen in foals and calves born to mares and cows grazing Sudan grass in early pregnancy.[226,227,231-233] In addition to limb deformities (arthrogryposis), calves also develop severe degeneration of the spinal cord and brain.[226]

Clinical Signs

Ataxia is most noticeable in affected horses when they are backed or turned, causing the horse to sit on its hindquarters or fall over. Paralysis of the urinary bladder results in the continual dribbling of urine, which causes scald and hair loss of the lower hind legs in the male and perineum area in mares. Paralysis of the perineum may cause the lips of the vulva to stay open resulting in vaginitis. Loss of tone to the rectum may occur causing fecal impaction and constipation. Flaccid paralysis of the tail develops in some cases. Affected animals generally retain a good appetite and physical parameters remain normal until cystitis and an ascending nephritis develop secondary to incontinence. As this occurs, the urine becomes thick and opaque and contains large amounts of amorphous sediment. On rectal palpation, the bladder is characteristically enlarged and flaccid and contains concretions adherent to the dependent portion of the bladder.[225,231]

The clinical signs of chronic cyanide poisoning should be differentiated from those of equine herpes virus-1 myeloencephalitis, equine protozoal encephalomyelitis, and equine viral encephalitis. The history of feeding sorghum grasses and the presence of cystitis, which is not usually a feature of the previously mentioned diseases, should help confirm the diagnosis of chronic Sudan grass poisoning.

Postmortem examination generally reveals a severe ulcerative necrotizing urethritis and cystitis extending to a pyelonephritis. Degeneration and demyelination of axons throughout the length of the spinal cord are evident histologically.[231] The predominant finding in sheep dying after grazing *Sorghum* pastures was the presence of focal axonal enlargement (spheroids) in the medulla, cerebellum, midbrain, and spinal cord.[228]

Animals slowly recover from the syndrome if they are removed from the toxic sorghum grasses before the cystitis and ataxia become complicated by serious secondary problems, such as the ascending infection of the kidneys and pylonephritis. Complete recovery seldom occurs once horses have developed severe signs of ataxia and cystitis. Some improvement in the animal's condition can be expected if it is removed from the source of the toxic sorghum grasses and the urinary system infections are treated with appropriate antibiotics.

Peas
Lathyrus sativus Chickling pea
L. odoratus Sweet pea
L. latifolius Perennial sweet pea
L. sylvestris Flat pea
L. hirsutus Rough pea
Fabaceae (Legume family)

Description

The perennial or everlasting pea (*L. latifolius*) has multiple climbing or trailing stems 3 to 6 feet (1 to 2 meters) in length developing from rhizomes. The stems are hairless, slightly glaucous, with two obvious wings. Leaves are alternate, pinnate, with the rachis ending in a branched tendril. The flowers, white, pink or purple, sometimes striped, are produced from leaf axils on a long peduncle that extends beyond the leaves (Figure 6-23). The pods are 2 to 3 inches (5 to 7.5 cm) long and contain up to seven seeds.

Lathyrism

Lathyrism is a disease of people in some eastern Asian and African countries where the seeds of certain peas (*Lathyrus* spp.) are eaten when other foods are scarce.[234-238] Lathyrism is an irreversible, nonprogressive, spastic, paraparesis associated with degeneration of the spinal cord caused by an as yet poorly defined neurotoxin. In humans there are two recognized forms of lathyrism: neurolathyrism and osteolathyrism. Neurolathyrism is the most common and is due to spinal cord degeneration caused by a neurotoxin in *L. sativus*. The disease affects mostly men who eat the uncooked or poorly fermented pea seeds for several months and are stressed through arduous work. Symptoms include muscular rigidity, weakness, and paralysis of the legs that results in people eventually only being able to crawl around on their knees.[238] Death eventually results unless the eating of the peas is stopped and effective nursing care is provided.[229]

Osteolathyrism is the most common form of lathyrism affecting animals. The perennial sweet pea (*L. latifolius*) and the annual sweet pea (*L. odoratus*) seeds contain lathyrogens capable of producing osteolathyrism in animals, especially horses.[229] This disease in horses is characterized by skeletal deformities and aortic rupture due to defective synthesis of cartilage and connective tissue.[229] A similar syndrome of musculoskeletal deformities in foals and calves has been associated with pregnant mares and cows having eaten Sudan grass (*Sorghum sudanense*) or sorghum hybrids over a period of months.[239,240] In addition to limb deformities (arthrogryposis), calves also develop severe degeneration of the spinal cord and brain.[240] Adult animals may develop posterior ataxia, urinary incontinence, and cystitis resulting from lower spinal cord degeneration.[245]

Principal Toxins

The primary neurolathyrogen in the chickling pea is β-N-oxalyl-L-α-diamino propionic

Figure 6-23 Perennial sweet pea pink and white varieties (*Lathyrus latifolius*).

acid. It apparently interferes with glutamate receptors in the brain, causing an influx of calcium into the cells with resulting cell death.[241] Additionally, the neurolathyrogen may cause free radical damage to the brain.[242] This is supported by finding that the antioxidant ascorbic acid when given to guinea pigs prevented the neurotoxic effects of the peas.[242] The toxin can be destroyed by boiling the peas or by allowing them to ferment thoroughly.[243,244] The primary lathyrogen in the annual sweet pea is the amino acid, β-amino proprionitrile, that causes defective cross-linking of collagen and elastin molecules.[229]

Sorghums, especially those containing low levels of cyanogenic glycosides, appear to contain lathyrogens similar in effect to those found in pea seeds. The lathyrogens may either affect the developing fetus causing skeletal and brain degeneration or may cause loss of the myelin sheath surrounding peripheral nerves with resulting loss of nerve function. The neuronal degeneration and demyelinization of the peripheral nerves is thought to result from the conversion of the cyanide to β-cyanoalanine and then to T-glutamyl β-cyanoalanine, a known lathyrogen that interferes with neuro-transmitter activity.[231] The neuronal degeneration in the brain may also result when cyanogens cause the depletion of hydroxycobalamin.[240] Animals may slowly recover if they are removed from the Sudan grass before neuronal degeneration is severe and if cystitis is not complicated by an ascending infection of the kidneys.

Caltrop, Carpetweed
Kallstroemia hirsutissima – Hairy caltrop
K. parviflora – Warty caltrop
Zygophyllaceae (Caltrop family)

Habitat

Caltrop or carpetweeds are common in the southwestern states, especially in Texas and Arizona, preferring sandy soils in overgrazed or waste areas. As the name implies the plants can form dense carpets if allowed to proliferate.

Description

These multibranched prostrate annuals have up to 3 feet (1 meter) long stems arising radially from a central root. The leaves are opposite, evenly pinnate with four to six pairs of leaflets and conspicuously hairy. Single, orange to yellow, five-petalled flowers are produced from the leaf axil on peduncles longer than the leaf (Figure 6-24). The fruits contain 8 to 12 one-seeded segments.

Kallstroemia spp. are similar in appearance to puncture vine or goathead (*Tribulus terrestris*) and are both members of the caltrop family. They differ in that *T. terrestris* has flower stalks shorter than the leaves, and the fruits separate into five, three to five-seeded, hard, spiny segments that look like the head of a horned goat (see Figure 4-14A). The toxicity of this plant is discussed in Chapter 4.

Figure 6-24 Hairy caltrop (*Kallstroemia* spp.). (Courtesy of Mack Dimmit, Tuscon AZ).

Principal Toxin

The toxin in *Kallstroemia* spp. is not known. Cattle, sheep, goats, and rabbits are reportedly susceptible to poisoning when they eat large quantities of the caltrop.[245] Animals generally will not eat the plant unless hungry and little else is available for them to eat.

Clinical Signs

Initially affected cattle develop weakness of the hind legs and knuckling of the fetlock joints. Paralysis of the hind legs and convulsions occur before death.[246] Sheep may show similar signs but are reported to walk on their front knees. Removing animals from the caltrop before the animals are severely affected often results in their recovery. Those that die exhibit hemorrhages and congestion of the thoracic and abdominal organs.[245]

Hemp, Marijuana
Cannabis sativa
Cannabaceae (Hemp family)

Figure 6-25A Marijuana (*Cannabis sativa*) plant (Courtesy of Carol Salman, Fort Collins, Colorado.)

Figure 6-25B Marijuana showing typical leaf structure (*C. sativa*).

Habitat

Introduced originally from Asia, Cannabis is now found in most areas of the United States, and especially where it was originally grown for its fiber. It will grow in most soils except in desert areas and is often found growing along fence rows, weedy areas, and disturbed soils.

Description

Hemp is an annual that grows to 6 feet (2 meters) in height, with a central, multibranched stem (Figure 6-25A). The leaves, opposite near the base, becoming alternate near the top of the plant, are palmate, with three to seven lanceolate, serrated leaflets (Figure 6-25B). Male and female greenish white flowers are produced on separate plants. Male flowers are produced as loose, leafy panicles in the leaf axils, whereas the female flowers are dense panicles. The fruit consists of papery bracts surrounding a single smooth brown seed.

Principal Toxin

More than 60 cannabinoid compounds have been isolated from *C. sativa*, the most toxic of which on the nervous system is tetrahydrocannibinol.[247] The content of the alkaloids is highest in plants grown in warm climates. Poisoning in animals is rarely encountered because the plant is not very palatable. Cattle, horse, pigs, and dogs have been intoxicated after eating marijuana.[249] Dogs are most likely to be poisoned when fed human food containing *Cannabis*.[250] Pollen from the flowers is a cause of allergy in humans and dogs.[251]

Clinical Signs

Although a wide variety of signs of marijuana poisoning have been reported in humans, the most prevalent sign of poisoning in the dog is central nervous system depression.[251] Other animals may show hyperexcitability, vomiting, salivation, muscle tremors, and ataxia. Coma and death may result in severe cases.

Tobacco
Nicotiana tabacum – cultivated tobacco
N. trigonophylla – wild or desert tobacco
N. attenuata – wild or coyote tobacco
N. glauca – tree tobacco
Solanaceae (Nightshade family)

Habitat

With the exception of cultivated tobacco, which is grown predominantly in the southeastern United States, most wild tobaccos are found in drier areas of the southwest.

Description

Tobacco is an erect, herbaceous, branching annual that grows 1 to 4 feet (0.3 to 1.2 meters) in height. The stems and leaves are hairy and sticky. Leaves are alternate, petioled, lanceolate and vary in size depending on the species. Flowers have five parts and are white, fragrant, and tubular. *N. trigonophylla* flowers open at night (Figure 6-26A); *N. attenuata* is a day bloomer.

Tree tobacco (*N. glauca*) is an evergreen, shrub, or small tree growing 6 to 20 feet (2 to 6 meters) in height. It is has slender, loosely branching stems, with alternate, bluish green, hairless leaves. The leaves and stems have a covering of whitish powder that rubs off easily. The tubular, yellow flowers, about 2 inches (5 cm) long, are produced on leafless branches at the end of the stems (Figure 6-26B). The plant reproduces via the large number of kidney-shaped seeds produced in the egg-shaped, pendulous, brown seed pods.

Figure 6-26A Desert tobacco (*Nicotiana trigonophylla*).

Figure 6-26B Tree tobacco (*N. glauca*).

Principal Toxins

Nicotine, a potent pyridine alkaloid, is probably present in all species of *Nicotiana* and in all parts of the plant. Animals usually find the plants unpalatable but will eat tobacco when other food is scarce. The minimum lethal dose of desert tobacco to cattle is 2 percent body weight of green plant.[252] Horses have been poisoned by eating dried tobacco leaves. Similarly dogs have been poisoned by eating tobacco.[253,254] Nicotine acts primarily on the autonomic nervous system, mimicking the action of acetylcholine at the autonomic ganglia and myoneural junction. The lethal dose of nicotine to animals is 100 to 300 mg total dose.[255]

The alkaloid anabasine and not nicotine is responsible for the skeletal deformities that develop in the fetus of pregnant animals eating tobacco in the first trimester of pregnancy.[256-258] These teratogenic effects of the *Nicotiana* spp. and other plants are discussed in Chapter 8.

Clinical Signs

In small doses nicotine is a stimulant that causes excitement, rapid heart rate, salivation, vomiting, colic, and diarrhea. Ruminants may develop bloat. At higher doses nicotine causes blockade at the neuromuscular junction with resulting muscle weakness, staggering, collapse of the front legs, and a rapid, weak, and irregular heart rate. Difficulty in breathing due to respiratory paralysis, blindness, prostration, coma, and death develop rapidly when large quantities of tobacco are consumed. Some animals may be found dead without having been observed with neurologic signs.[259]

Treatment

There is no specific treatment for nicotine poisoning other than to try and prevent further intestinal absorption by administering an oral adsorbent such as activated charcoal. Cathartics may help in the rapid removal of the toxic plant material from the intestinal tract. Additional supportive therapy in the form of intravenous fluids is necessary in severely poisoned animals.

Miscellaneous Plants Associated with Neurologic Signs

The plants listed in Table 6-4 have been suspected of causing poisoning characterized by neurologic signs. They are not discussed in detail because either poisoning from them rarely occurs in animals in North America or there is minimal current corroborating documentation of their toxicity to animals.

Table 6-4 Plants Suspected of Causing Neurologic Poisoning		
COMMON NAMES	SCIENTIFIC NAMES	TOXICITY PLANT PART
African Rue	*Peganum harmala*	Leaves and seeds
Coonties, Florida arrow root	*Zamia* spp.	Root and leaves
Cycad sago	*Cycas* spp.	Fruits
Matrimony vine	*Lycium halimifolium*	Entire plant
Sleepy grass	*Stipa robusta*	Entire plant
Morning glory	*Ipomoea tricolor*	Seeds
Bunch flower	*Melanthium hybridum*	Entire plant
Water parsnip	*Berula pusilla*	Entire plant

Texas Plants Poisonous to Livestock. Agricultural Extension Service B-1028.
Texas A & M University. College Station, Texas; Morton JF: *Plants Poisonous to People.*
South Eastern Printing Co. 1982; Kingsbury JM, Ed: *Poisonous Plants of the United States and Canada.* Englewood Cliffs, NJ: Prentice-Hall; 1964.

REFERENCES

Locoweeds

1. Marsh CD, Clawson AB, Eggleston WW: The locoweed disease. *US Department of Agriculture Bulletin* 1919;1054:1-31.

2. Kingsbury JM, Ed: *Poisonous Plants of the United States and Canada.* Englewood Cliffs, NJ: Prentice-Hall; 1964:305-313.

3. Williams MC, Barneby RC: The occurrence of nitro-toxins in North American Astragalus (*Fabaceae*). *Brittonia* 1977;29:310-326.

4. Mathews FP: Locoism in domestic animals. *Texas Agricultural Experiment Station Bulletin* 1932;456:5-27.

5. Mathews FP: The toxicity of red-stemmed peavine (*Astragalus emoryanus*) for cattle, sheep, and goats. *J Am Vet Med Assoc* 1940;97:125-134.

6. Nielson DB: The economic impact of poisonous plants on the range livestock industry in the 17 western states. *J Range Manage* 1978;31:325-327.

7. James LF, Nielson DB: Locoweeds: assessment of the problem on western range lands. *In* The Ecology and Economic Impact of Poisonous Plants on Livestock Production. James LF, Ralphs MH, Nielson DB, Eds. Boulder, CO: Westview Press; 1988:171-180.

8. James LF: The effects of *Astragalus tenellus* in sheep. *J Range Manage* 1971;24:161.

9. Williams MC: Nitro compounds in foreign species of *Astragalus*. Weed Sci 1981;29:261-269.

10. James LF, Hartley WJ, Van Kampen KR: Syndromes of *Astragalus* poisoning in livestock. *J Am Vet Med Assoc* 1981;178:146-150.

11. Van Kampen KR, James LF: Pathology of locoweed poisoning in sheep. *Pathol Vet* 1969;6:413-423.

12. Van Kampen KR, James LF: Pathology of locoweed poisoning in sheep: sequential development of cytoplasmic vacuolization in tissues. *Pathol Vet* 1970;7:503-508.

13. Williams CB, Barneby RC: The occurrence of nitro-toxins in North American *Astragalus* (Fabaceae). *Brittonia* 1977;29:310-326.

14. Cheeke PR: Neurotoxins, cardiac/pulmonary toxins and nephrotoxins. *In* Natural Toxicants in Feeds, Forages and Poisonous Plants. Cheeke PR, Ed. Danville, IL: Interstate Publishers; 1998:365-409.

15. Fraps GS, Wender SH: Studies on toxic substances of locoweeds, *Astragalus earlei* and others. *Texas Agricultural Experimental Station Bulletin* 1944;650:1-23.

16. Huxtable CR, Dorling PR: Poisoning of livestock by *Swainsona* spp: current status. *Aust Vet J* 1982;59:50-53.

17. Dorling PR, Huxtable CR, Colegate SM: Isolation of swainsonine from Swainsona canescens: Historical aspects. *In* Swainsonine and Related Glycosidase Inhibitors. James LF, Elbien AD, Molyneux RJ, Warren CD, Eds. Ames, IA: Iowa State University Press; 1989:83-90.

18. Hartley WJ: A comparative study of darling pea (*Swainsona* spp) poisoning in Australia with locoweed (*Astragalus* and *Oxytropis* spp) poisoning in North America. *In* Effects of Poisonous Plants on Livestock. Keeler RF, Van Kampen KR, James LF, Eds. New York: Academic Press;1978:363-369.

19. Molyneux RJ, James LF: Loco intoxication: indolizidine alkaloids in spotted locoweed (*Astragalus lentiginosus*). *Science* 1982;216:190-191.

20. Molyneux RJ, James LF, Panter HE, Ralphs MH: The occurrence and detection of

swainsonine in locoweeds. *In* Swainsonine and Related Glycosidase Inhibitors. James LF, Elbein LD, Molyneux RJ, Warren CD, Eds. Ames, IA: Iowa State University Press; 1989:100-117.

21. Smith GS, Allred KW, Kiehl DE: Swainsonine content in New Mexican locoweeds. *Proceed Western Section, Am Soc Anim Sci* 1992;43:405-407.

22. James LF, Van Kampen KR, Staker G: Locoweed (*Astragalus lentiginosus*) poisoning in cattle and horses. *J Am Vet Med Assoc* 1969;155:525-530.

23. Oehme F, Barbie WE, Hulbert LC: *Astragalus mollissimus* (Locoweed) toxicosis of horses in western Kansas. *J Am Vet Med Assoc* 1968;152:271-278.

24. James LF, Hartley WJ: Effects of milk from animals fed locoweed on kittens, calves and lambs. *Am J Vet Res* 1977;38:1263-1265.

25. Wolfe GJ, Lance WR: Locoweed poisoning in a northern New Mexico elk herd. *J Range Manage* 1984;37:59-63.

26. Ralphs MH, Panter KE, James LF: Feed preferences and habituation of sheep poisoned by locoweed. *J Anim Sci* 1990;68:1354-1362.

27. Ralphs MH, Panter KE, James LF: Grazing behavior and forage preference of sheep with chronic locoweed toxicosis suggest no addiction. *J Range Manage* 1991;44:208-209.

28. Ralphs MH, Molyneux RJ: Livestock grazing locoweed and the influence of swainsonine on locoweed palatability and habituation. *In* Swainsonine and Related Glycosidase Inhibitors. James LF, Elbien AD, Molyneux RJ, Warren CD, Eds. Ames, IA: Iowa State University Press; 1989:39-49.

29. Dorling PR, Huxtable CR, Colegate SM: Inhibition of lysosomal alpha-mannosidase by swainsonine, an indolizidine alkaloid isolated from *Swainsona canescens*. *Biochem J* 1980;191:649-651.

30. Tulsiani DR, Broquist HP, James LF, *et al.*: The similar effects of swainsonine and locoweed on tissue glycosidases and oligosaccharides of the pig indicate that the alkaloid is the principle toxin responsible for locoism. *Arch Biochem Biophys* 1984;232:76-85.

31. Stegelmeier BL, Molyneux RJ, Elbein AD, James LF: The lesions of locoweed (*Astragalus mollissimus*), swainsonine, and castanospermine in rats. *Vet Pathol* 1995, 32:289-298.

32. Hartley WJ, Baker DC, James LF: Comparative pathologic aspects of locoweed and swainsona poisoning in livestock. *In* Swainsonine and Related Glycosidase Inhibitors. James LF, Elbien AD, Molyneux RJ, Warren CD, Eds. Ames, IA: Iowa State University Press; 1989:50-56.

33. Dorling PR, Huxtable CR, Vogel P: Lysosomal storage disease in *Swainsona* spp. toxicosis: an induced manosidosis. *Neuropathol Appl Neurobiol* 1978;4:285-295.

34. Jolly RD, Hartley WJ: Storage diseases of domestic animals. *Aust Vet J* 1977;53:1-8.

35. Leipold HW, Smith JE, Jolly RD, Eldridge FE: Mannosidosis of Angus calves. *J Am Vet Med Assoc* 1979;175:457-459.

36. Driemeier D, Colodal EM, Gimeno EJ, Barros SS: Lysosomal storage disease caused by *Sida carpinifolia* poisoning in goats. *Vet Pathol* 2000, 37:153-159.

37. De Balogh KKIM, Dimande AP, van der Lugt JJ *et al*: Ipomoea carnea: the cause of a lysosomal storage disease in goats in Mozambique. *In:* Toxic Plants and Other Natural Toxicants. Garland T, Barr C, Eds. New York, CAB International: 1998:428-434.

38. Molyneux RJ, James LF, Ralphs MH *et al.*: Polyhydroxy alkaloid glycosidase inhibitors from poisonous plants of global distribution: analysis and identification. *In* Plant-Associated Toxins. Colegate SM, Dorling PR, Eds. Wallingford, England: CAB

International; 1994:107-112.

39. James LF, Van Kampen KR: Acute and residual lesions of locoweed poisoning in cattle and horses. *J Am Vet Med Assoc* 1971;158:614-618.

40. Kirkpatrick JG, Burrows GE: Locoism in horses. *Vet Hum Toxicol* 1990;32:168-169.

41. Balls LD, James LF: Effect of locoweed (*Astragalus* spp.) on reproductive performance of ewes. *J Am Vet Med Assoc* 1973;162:291-292.

42. Hartley WJ, James LF: Fetal and maternal lesions in pregnant ewes ingesting locoweed (*Astragalus lentiginosus*). *Am J Vet Res* 1975;36:825-826.

43. Hartley WJ, James LF: Microscopic lesions in fetuses of ewes ingesting locoweed (*Astragalus lentiginosus*). *Am J Vet Res* 1973;34;209-211.

44. James LF, Keeler RF, Binns W: Sequence in the abortive and teratogenic effects of locoweed fed to sheep. *Am J Vet Res* 1969;30:377-380.

45. James LF, Shupe JL, Binns W, Keeler RF: Abortive and teratogenic effects of locoweed on cattle and sheep. *Am J Vet Res* 1967;28:1379-1388.

46. Panter KE, Ralphs MH, James LF, Stegelmeier BL, Molyneux RJ: Effects of locoweed (*Oxytropis sericea*) on reproduction in cows with a history of locoweed consumption. *Vet Hum Toxicol* 1999, 41:282-286.

47. Panter KE, Bunch TD, James LF, Sisson DV: Ultrasonographic imaging to monitor fetal and placental developments in ewes fed locoweed (*Astragalus lentiginosus*). *Am J Vet Res* 1987;48;686-690.

48. James LF. Lesions in neonatal lambs resulting from maternal ingestion of locoweed. *Cornell Vet* 1971;61:667-670.

49. McIlwraith CW, James LF: Limb deformities in foals associated with ingestion of locoweed by mares. *J Am Vet Med Assoc* 1982;181:255-258.

50. Bunch TD, Panter KE, James LF: Ultrasound studies of the effects of certain poisonous plants on uterine function and fetal development in livestock. *J Anim Sci* 1992;70:1639-1643.

51. James LF, Van Kampen K: Effects of locoweed intoxication on the genital tract of the ram. *Am J Vet Res* 1971;32:1253-1256.

52. James LF, Panter KE: Locoweed poisoning in livestock. *In* Swainsonine and Related Glycosidase Inhibitors. James LF, Elbien AD, Molyneux RJ, Warren CD, Eds. Ames, IA: Iowa State University Press; 1989:23-38.

53. Tulsiani DRP, Skudlarek MD, Orgebin-Crist MC: Swainsonine induces the production of hybrid glycoproteins and accumulation of oligosaccharides in male reproductive tissues of the rat. *Biol Reprod* 1990;43:130-138.

54. Panter KE, James LF, Stegelmeier BL, Ralphs MH, Pfister JA: Locoweeds: effects on reproduction in livestock. *J Nat Toxins* 1999, 8:53-62.

55. James LF, Hartley WF, Van Kampen KR, Nielsen D: Relationship between ingestion of locoweed *Oxytropis sericea* and congestive right-side heart failure in cattle. *Am J Vet Res* 1983;2:254-259.

56. James LF, Hartley WJ, Nielsen D, *et al.:* Locoweed (*Oxytropis sericea*) poisoning and congestive heart failure in cattle. *J Am Vet Med Assoc* 1986;189:1549-1556.

57. Panter KE, James LF, Nielsen D, *et al.:* The relationship of *Oxytropis sericea* (green and dry) and *Astragalus lentiginosus* with high mountain disease. *Vet Hum Toxicol* 1988;30:318-323.

58. James LF, Panter KE, Broquist HP, Hartley WJ: Swainsonine-induced high mountain disease in calves. *Vet Hum Toxicol* 1991; 33:217-219.

59. Knight AP, Greathouse G: Locoweed in Northern Colorado: its effects on beef

calves. Locoweed and Broom Snakeweed Research Update. *Albuquerque NM: Ag Exp Station/Cooperative Extension Service.* New Mexico State University; 1996.

60. Sharma RP, James LF, Molyneux RJ: Effect of repeated locoweed feeding on peripheral lymphocytic function and plasma proteins of sheep. *Am J Vet Res* 1984;10:2090-2093.

61. Stegelmeier BL, Ralphs MR, Gardner DR, *et al.:* Serum alpha-mannosidase activity and the clinicopathologic alterations of locoweed (*Astragalus mollissimus*) intoxication in range cattle. *J Vet Diagn Invest* 1994;6:473-479.

62. Stegelmeier BL, James LF, Panter KE, Molyneux RJ: Serum swainsonine concentration and alpha-mannosidase activity in cattle and sheep ingesting *Oxytropis sericea* and *Astragalus lentiginosus* (locoweeds). *Am J Vet Res* 1995;56:149-154.

63. James LF, Binns W: Blood changes associated with locoweed poisoning. *Am J Vet Res* 1967;28:1107-1110.

64. James LF, Van Kampen KR, Johnson AE: Physiopathologic changes in locoweed poisoning of livestock. *Am J Vet Res* 1970;31:663-672.

65. Van Kampen KR, James LF: Ophthalmic lesions in locoweed poisoning of cattle, sheep, and horses. *Am J Vet Res* 1971;32:1293-1295.

66. Van Kampen KR: Sequential development of the lesions in locoweed poisoning. *Clin Toxicol* 1972;5:575-580.

67. Jones MZ, Cunningham JG, Dade AW, *et al.:* Caprine beta-mannosidosis: clinical and pathological features. *J Neuropath Exp Neurol* 1983;42:268-285.

68. Abbitt M, Jones MZ, Kasari TR, *et al.:* Beta-mannosidosis in 12 Salers calves. *J Am Vet Med Assoc* 1991;198:109-113.

69. Staley EE: An approach to treatment of locoism in horses. *VM/SAC* 1978;1205-1206.

70. Ralphs MH: Cattle grazing white locoweed: Influence of grazing pressure and palatability associated with phenological growth stage. *J Range Manage* 1987;40:330-332.

71. Ralphs MH, Graham D, James LF: Social facilitation influences cattle to graze locoweed. *J Range Manage* 1994;47:123-126.

72. Ralphs MH, Graham D, Molyneux RJ, James LF: Seasonal grazing of locoweeds by cattle in northeastern New Mexico. *J Range Manage* 1993;46:416-420.

73. Ralphs MH, Mickelsen LV, Turner DL: Cattle grazing white locoweed: diet selection patterns of native and introduced cattle. *J Range Manage* 1987;40:333-335.

74. Ralphs MH, James LF, Pfister JA: Utilization of white locoweed (*Oxytropis sericea* Nutt.) by range cattle. *J Range Manage* 1986;39:344-347.

75. Taylor CA, Ralphs MH: Reducing livestock losses from poisonous plants through grazing management. *J Range Manage* 1992;45:9-12.

76. Ralphs MH, Graham D, James LF: Cattle grazing white locoweed in New Mexico: influence of grazing pressure and phenological growth stage. *J Range Manage* 1994;47:270-274.

77. Ralphs MH, Panter KE, James LF: Grazing behavior and forage preference of sheep with chronic locoweed toxicosis suggest no addiction. *J Range Manage* 1991;44;208-209.

78. Ralphs MH, Cronin EH: Locoweed seed in soil: density, longevity, germination, and viability. *Weed Sci* 1987;35:792-795.

79. Ralphs MH, Ueckert DN: Herbicide control of locoweeds: a review. *Weed Technology* 1988;2:460-465.

80. Ralphs MH: Conditioned food aversion: training livestock to avoid eating poisonous plants. *J Range Manage* 1992;45:46-51.

81. James LF,Van Kampen KR: Effect of protein and mineral supplementation on potential locoweed (*Astragalus* spp.) poisoning in sheep. *J Am Vet Med Assoc* 1974;164:1042-1043.

82. Bachman SE, Gaylean ML, Smith GS, *et al.*: Early aspects of locoweed toxicosis and evaluation of a mineral supplement or clinoptilolite as dietary treatments. *J Anim Sci* 1992;70:3125-3132.

83. Beath OA, Draize JA, Eppson HF: Three poisonous vetches. *Wyoming Agricultural Experimental Station Bulletin* 1932;189.

84. Williams MC: Toxicological investigations on Toano, Wasatch, and stinking milkvetches. *J Range Manage* 1989;42:366-368.

85. Cronin EH, Williams MC: The poisonous timber milkvetches. *US Department of Agriculture Handbook* 1974;459:1-15.

86. Williams MC, Norris FA: Distribution of miserotoxin in varieties of *Astragalus miser Dougl.ex Hook*. *Weeds Sci* 1969:236-238.

87. Majak W, Williams JR, Van Ryswyk AL, Brooke BM: The effect of Columbia milkvetch toxicity. *J Range Manage* 1976;29:281-283.

88. Stermitz FR, Norris FA, Williams MC: Miserotoxin, a new naturally occurring nitro compound. *J Am Chem Soc* 1969;91:4599-4600.

89. Stermitz FR, Lowry WT, Norris FA, *et al.*: Aliphatic nitro-compounds from *Astragalus* species. *Phytochemistry* 1972;11:1117-1124.

90. Williams MC, Stermitz FR, Thomas RD: Nitrocompounds in *Astragalus* species. Phytochemistry 1975;14:2306-2308.

91. Stermitz FR, Yost GS: Analysis and characterization of nitrocompounds from Astragalus species. *In* Effects of Poisonous Plants on Livestock. Keeler RF, Van Kampen KR, James L, Eds. New York: Academic Press; 1978:371-378.

92. Pass MA, Majak W, Muir AD, Yost GS: Absorption of 3-nitro-propanol and 3-nitropropionic acid from the digestive system in sheep. *Toxicology* 1984;23:1-7.

93. Williams MC, James LF: Livestock poisoning from nitro-bearing Astragalus. *In* Effects of poisonous plants on livestock. Keeler RF, Van Kampen KR, James LF, Eds. New York: Academic Press; 1978:379-389.

94. Williams MC, James LF: Poisoning in sheep from emory milkvetch and nitro-compounds. *J Range Manage* 1976;29:165-167.

95. Williams MC, James LF, Bond BO: Emory milkvetch (*Astragalus emoryanus* var. *emoryanus*) poisoning in chicks, sheep, and cattle. *Am J Vet Res* 1989;40:403-406.

96. James LF, Hartley WI, *et al.*: Field and experimental studies in cattle and sheep poisoned by nitro-bearing *Astragalus* or their toxins. *Am J Vet Res* 1980;41:377-382.

97. Williams MC: Nitro compounds in Indigofera species. *Agronomy J* 1981;73:434-436.

98. Gustine DL, Moyer BG, Wangness PJ, Shenk JS: Ruminal metabolism of 3-nitropopanol-D-glucopyranoses from crown vetch. *J Anim Sci* 1977;44:1107-1111.

99. Gustine DL: Aliphatic nitrocompounds in crown vetch: A review. *Crop Sci* 1979;19:197-203.

100. Morton JF: Creeping indigo (*Indigofera spicata Forsk*) (*Fabaceae*). A hazard to herbivores in Florida. *Econ Bot* 1989;43:314-327.

101. Alston TA, Mela L, Bright HJ: 3-Nitropropionate, the toxic substance of Indigofera, is a suicide-inactivator of succinate dehydrogenase. *Proc Natl Acad Sci U S A* 1977;74:3767-3771.

102. Hegarty MP, Kelly WR, McEwan D, *et al.*: Hepatotoxicity to dogs of horse meat contaminated with indospicine. *Aust Vet J* 1988;65:337-340.

103. Shenk JS, Wangsness PJ, Leach RM, *et al.*: Relationship between b-nitropropionic acid content of crown vetch and toxicity in non-ruminant animals. *J Anim Sci* 1976;42:616-621.

104. Kemp A, Guerink JH, Malestein A: Nitrate poisoning in cattle 2. Changes in nitrate in rumen fluid and methemoglobin formation in blood after high nitrate intake. *Neth J Agric Sci* 1977;25:51-62.

105. Williams MC, Van Kampen KR, Norris FR: Timber milkvetch poisoning in chickens, rabbits and cattle. *Am J Vet Res* 1969;30:2185-2190.

106. Williams MC, Ralphs MH: Effect of herbicides on miserotoxin concentration in Wasatch milkvetch (*Astragalus miser* var. *oblongifolius*). *Weed Sci* 1987;33:746-748.

Sagebrush

107. Beath OA, Eppson HF, Gilbert CS, Bradley WB: Poisonous plants and livestock poisoning. *University of Wyoming Agricultural Experimental Station Bulletin* 1939;231.

108. Flemming CE: Poisonous range plants. *Nevada Agricultural Experimental Station Annual Report* 1919; 39:19-20.

109. Sampson AW, Malmsten HE: Stock poisoning plants of California. *California Agricultural Experimental Station Bulletin* 1942:593.

110. Arnold WN: Vincent van Gogh and the thujone connection. *J Am Med Assoc* 1988;260:3042-3044.

111. Kingsbury JM: *Poisonous Plants of the United States and Canada.* Englewood Cliffs, NJ: Prentice-Hall; 1964:393.

112. Kelsey RG, Stephens JF, Shafizadeh F: The chemical constituents of sagebrush foliage and their isolation. *J Range Manage* 1982;35:617-622.

113. Buttkus HA, Bose RJ, Shearer DAP: Terpenes in the essential oils of sagebrush (*Artemisia tridentata*). *J Agric Food Chem* 1977;25:288-291.

114. Cedarleaf JD, Welch BL, Brotherson JD: Seasonal variation in monoterpenoids in big sagebrush (*Artemisia tridentata*). *J Range Manage* 1983;36:492-494.

115. Welch BL, McArthur ED: Variation of monoterpenoid content among subspecies and accessions of *Artemisia tridentata* grown in a uniform garden. *J Range Manage* 1981;34:380-384.

116. Johnson AE, James LF, Spillet J: The abortifacient and toxic effects of big sagebrush (*Artemisia tridentata*) and juniper (*Juniperus osteosperma*) on domestic sheep. *J Range Manage* 1976;29:278-280.

Coyotillo

117. Charlton RM, Pierce KR, Storts RW, Bridges CH: A neuropathy in goats caused by experimental Coyotillo (*Karwinskia humboldtiana*) poisoning. *Pathol Vet* 1970;7:385-447.

118. Charlton KM, Claborn LD, Pierce KR: A neuropathy in goats caused by experimental poisoning: Coyotillo (*Karwinskia humboldtiana*) poisoning: Clinical and neurophysiologic studies. *Am J Vet Res* 1971;32:1381-1389.

119. Calderon-GonzalezR, Rizzi-Hernandez H: Buckthorn polyneuropathy. *N Engl J Med* 1967;277:69-71.

120. Munoz-Martinez EJ, Cueva J, Joseph-Nathan P: Denervation caused by tullidora (*Karwinskia humboldtiana*). *Neuropathol Appl Neurobiol* 1983;9:121-124.

121. Kingsbury JM: *Poisonous Plants of the United States and Canada.* Englewood Cliffs, NJ: Prentice-Hall; 1964;220-221.

122.   Dewan ML, Hensen JB, Dollahite JW, Bridges CH: Toxic myodegeneration in in goats produced by feeding mature fruits of the coyotillo plant (*Karwinskia humboldtiana*). *Am J Pathol* 1965;46:215-226.

Bracken Fern

123.   Evans WC: Bracken poisoning of farm animals. *Vet Rec* 1964;76:365-369.

124.   Kingsbury JM. *Pteridium aquilinum*. *In* Poisonous Plants of the United States and Canada. Kingsbury JM, Ed. Englewood Cliffs, NJ: Prentice-Hall; 1964:105-113.

125.   Cheeke PR: Carcinogens and metabolic inhibitors. *In* Natural Toxicants in Feeds, Forages, and Poisonous Plants. Cheeke PR, Ed. Danville, IL: Interstate Publishers; 1998:423-444.

126.   Evans WC, Widdop B, Harding JD: Experimental poisoning by bracken rhizomes in pigs. *Vet Res* 1972;90:471-475.

127.   Evans WC: Bracken thiaminase-mediated neurotoxic syndromes. *Bot J Linn Soc* 1976;73:113-131.

128.   Roberts HE, Evans ET, Evans WC: The production of "bracken staggers" in the horse, and its treatment with vitamin B1 therapy. *Vet Rec* 1949;61:549-550.

129.   Konishi K, Ichijo S: Experimentally induced equine bracken poisoning by thermostable anti-thiamine factor (SF factor) extracted from dried bracken. *J Japan Vet Med Assoc* 1984;37:730-734.

130.   Carpenter KJ, Phillipson AT, Thomson W: Experiments with dried bracken (*Pteris aquilina*). *Br Vet J* 1950;106:292-308.

131.   Bakker HJ, Dickinson J, Steele P, Nottle MC: Experimental induction of ovine polioencephalomalacia. *Vet Rec* 1980;107:364-366.

132.   Henderson JA, Evans EV, McIntosh RA: The antithiamine action of Equisetum. *J Am Vet Med Assoc* 1952; :375-378.

133.   Meyer P: Thiaminase activities and thiamine content of *Pteridium aquilinum*, *Equisetum ramosissimum*, *Malva parviflora*, *Pennisetum clandestinum*, and *Medicago sativa*. *Onderstepoort J Vet Res* 1989;56:145-146.

Horsetail

134.   Kingsbury JM: Horsetails. *In* Poisonous Plants of the United States and Canada. Kingsbury JM, Ed. Englewood Cliffs, NJ: Prentice-Hall; 1964:114-118.

135.   Meyer P: Thiaminase activities and thiamine content of *Pteridium aquilinum*, *Equisetum ramosissimum*, *Malva parviflora*, *Pennisetum clandestinum*, and *Medicago sativa*. *Onderstepoort J Vet Res* 1989;56:145-146.

136.   Henderson JA, Evans EV, McIntosh RA: The antithiamine action of Equisetum. *J Am Vet Med Assoc* 1952;120:375-378.

137.   Cheeke PR: Carcinogens and metabolic inhibitors. *In* Natural Toxicants in Feeds, Forages, and Poisonous Plants. Cheeke PR, Ed. Danville, IL: Interstate Publishers; 1998;423-444.

138.   Lott DG: The use of thiamine in mare's tail poisoning in horses. *Canada J Comp Med* 1951;15: 274-278.

Yellow Star Thistle, Russian Knapweed

139.   Maddox DM, Mayfield A, Poritz MW: Distribution of yellow star thistle (*Centaurea solstitialis*) and Russian knapweed (*Centaurea repens*). *Weed Sci* 1985;33:315-327.

140.   Gard GP, De Sarem WG, Ahrens PJ: Nigropallidal encephalomalacia in horses in New South Wales. *Aust Vet J* 1973;49:107-108.

141. Ajilvsgi G: *Wildflowers of Texas*. Fredericksburg, TX: Shearer Publishing; 1991;108-109.

142. Cheng CH, *et al.*: Toxic effects of solstitialin A 13-acetate and cyanaropicrin from *Centaurea solstitialis* L. (*Asteraceae*) in cell cultures of foetal rat brain. *Neuropharmacology* 1992;31:271-277.

143. Craig AM, Blythe LL, Roy DN, Spencer PS: Detection and isolation of neurotoxins from yellow star thistle (*Centaurea solstitialis*), the cause of nigropallidal encephalomalacia. *In* Plant Associated Toxins. Colegate SM, Dorling PR, Eds. Wallingford, UK: CAB International, Short Run Press;1994:257-262.

144. Roy DN, Peyton DH, Spencer PS: Isolation and identification of two potent neurotoxins, aspartic acid and glutamic acid, from yellow star thistle (*Centaurea solstitialis*). *Natural Toxins* 1995;3:174-180.

145. Cordy DR: Nigropallidal encephalomalacia (chewing disease) in horses associated with ingestion of yellow star thistle. *J Neuropathol Exp Neurol* 1954;13:330-342.

146. Young S, Brown WW, Klinger B: Nigropallidal encephalomalacia in horses caused by ingestion of weeds of the genus Centaurea. *J Am Vet Med Assoc* 1970;157:1602-1605.

147. Young S, Brown WW, Klinger B: Nigropallidal encephalomalacia in horses fed Russian knapweed—*Centaurea repens*. *Am J Vet Res* 1970;31:1393-1404.

148. Cordy DR: Centaurea species and equine nigropallidal encephalomalacia. *In* Effects of Poisonous Plants on Livestock. Keeler RK, Van Kampen KR, James LF, Eds. New York: Academic Press; 1970:327-336.

149. Robles M, Wang N, Kim R, Choi BH: Cytotoxic effects of repin, a principle *sesquiterpene lactone* of Russian knapweed. *J Neuro Sci* 1997;47:90-97.

150. Larsen KA, Young S: Nigropallidal encephalomalacia in horses in Colorado. *J Am Vet Med Assoc* 1970;156:626-628.

151. Stevens Kl, Riopelle RJ, Wong RY: Repin, a sesquiterpene lactone from *Acroptilon repens* possessing exceptional biological activity. *J Nat Prod* 1990;53:218-221.

152. Mettler FA, Stern GM: Observations on the toxic effects of yellow star thistle. *J Neuropathol Exp Neurology* 1963;22:164-169.

153. Fowler ME: Nigropallidal encephalomalacia in the horse. *J Am Vet Med Assoc* 1965;147:607-616.

154. Farrell RK, Sande RD, Lincoln SD: Nigropallidal encephalomalacia in a horse. *J Am Vet Med Assoc* 1971;158:1201-1204.

White Snake Root

155. Moseley EL: The cause of trembles in cattle, sheep, and horses and of milk sickness in people. *Med Rec* 1909;75:699-715.

156. Couch JF: The toxic constituent of richweed or white snakeroot (*Eupatorium urticaefolium*). *J Agric Res* 1927;35:547-576.

157. Couch JF: *Trembles (or Milk Sickness)*. Washington, DC: US Department of Agriculture; 1933;306.

158. Wolfe FA, Curtis RS, Kaupp BF: A monograph on trembles or milk sickness and white snakeroot. *North Carolina Agricultural Experimental Station Technical Bulletin* 1918;15.

159. Christensen WI. Milk sickness: a review of the literature. *Econ Bot* 1965;19:293-300.

160. Stotts R: White snakeroot toxicity in dairy cattle. *Vet Med* 1984;79:118-120.

161.   Thompson LJ: Depression and choke in a horse: probable white snakeroot toxicosis. *Vet Hum Toxicol* 1989;31:321-322.

162.   Sanders M: White snakeroot poisoning in a foal: a case report. *J Equine Vet Sci* 1983;3:128-131.

163.   Beier RC, Norman JO: The toxic factor in white snakeroot: identity, analysis and prevention. *Vet Hum Toxicol* 1990;32 Suppl:81-88.

164.   Beier RC, Norman JO, Reagor JC, *et al.*: Isolation of the major component in white snakeroot that is toxic after microsomal activation: possible explanation of sporadic toxicity of white snakeroot plants and extracts. *Natural Toxins* 1993;1:286-293.

165.   Beier RC, Norman JO, Irvin TR, Witzel DA: Microsomal activation of constituents of white snakeroot (*Eupatorium rugosum*) to form toxic products. *Am J Vet Res* 1987;48:583-585.

166.   Kingsbury JM: *Poisonous Plants of the United States and Canada*. Englewood Cliffs, NJ: Prentice-Hall; 1964;397-404.

167.   Hansen AA: Stock poisoning plants. *North Am Vet* 1928;March 32-36.

168.   Thompson LJ, White JL, Beasley VR, Buck WB: Diagnosing white snakeroot poisoning. *J Equine Vet Sci* 1983;3:184.

169.   White JL, Shivaprasad HL, Thompson LJ, Buck WB: White snakeroot (*Eupatorium rugosum*) poisoning. Clinical effects associated with cardiac and skeletal muscle lesions in experimental equine toxicoses. *In* Plant Toxicology. Seawright AA, Hegarty MP, James LF, Keeler RF, Eds. Melbourne, Australia: Dominion Press, Hedges and Bell; 1985:411-422.

170.   Smetzer DL, Coppock RW, Ely RW, *et al.*: Cardiac effects of white snakeroot intoxication in horses. *Equine Practice* 1983;5:26-32.

171.   Panter KE, James LF: Natural toxicants in milk: a review. *J Anim Sci* 1990;68:892-904.

172.   O'Sullivan BM: Crofton weed (*Eupatorium adenophorum*) toxicity in horses. *Aust Vet J* 1979;55:19-24.

173.   O'Sullivan BM, Gibson JA, McKenzie RA: Intoxication of horses by *Eupatorium adenophorium* and *E. riparium* in Australia. Plant Toxicology. *Proceedings of the Australian USA Poisonous Plants Symposium.* Yeerongpilly, Queensland, Australia, 1984.

174.   O'Sullivan BM: Investigations into crofton weed (*Eupatorium adenophorum*) toxicity in horses. *Aust Vet J* 1985;62:30-33.

175.   Gibson JA, McKenzie RA: Lung lesions in horses fed mist flower (*Eupatorium riparium*). *Aust Vet J* 1984;81:271.

176.   Nicholson SS: Tremorgenic syndromes in livestock. *Vet Clin North Am- Food Anim Pract* 1989;5:291-300.

177.   George LW: Grass staggers. *In* Large Animal Internal Medicine, edn 2. Smith BP, Ed. St. Louis: Mosby; 1996;1115-1120.

178.   Hansen AA: Two common weeds that cause death. *Purdue University Agricultural Experimental Station Circular* 1923;110:1-8.

179.   Olson CT, Keller WC, Gerken DF, Reed SM: Suspected tremetol poisoning in horses. *J Am Vet Med Assoc* 1984;185:1001-1003.

Jimmy Weed

180.   Marsh CD, Roe GC: The "alkali disease" of livestock in the Pecos valley. *US Department of Agriculture Bulletin* 1921;180:1-8.

181. Schmutz EM, Freeman BN, Reed RE: Jimmyweed (*Haplopappus heterophylus*) *In* Livestock Poisoning Plants of Arizona. Schmutz EM, Freeman BM, Reed RE, Eds. Tucson: University Arizona Press; 1974:42-45.

182. Sperry OE, Dollahite JW, Hoffman GO, Camp BJ: *Isocoma wrightii* (*Aplopappus heterophylus*). *In* Texas Plants Poisonous to Livestock. Sperry OE, Dollahite JW, Hoffman GO, Eds. Camp BJ. College Station, TX: Texas Agricultural Experimental Station and Texas A&M University; 1968:26-27.

Mescal Bean

183. Boughton IB, Hardy WT: Mescal bean (*Sophora secundiflora*) poisonous for livestock. *Texas Agricultural Experimental Station Bulletin* 1935:519.

184. Knauer KW, Reagor JC, Bailey EM: Mescal bean (*Sophora secundiflora*) toxicity in a dog. *Vet Hum Toxicol* 1995;37:237-239.

185. Izaddoost M, Harris BG, Gracy RW: Structure and toxicity of alkaloids and amino acids of *Sophora secundiflora*. *J Pharmaceut Sci* 1976;65:352-354.

186. Hatfield GM, Valdes LJ, Keller WJ, *et al.*: An investigation of *Sophora secundiflora* seeds (mescalbean). *Lloydia* 1977;40: 374-383.

187. Kingsbury JM: *Poisonous Plants of the United States and Canada.* Englewood Cliffs, NJ: Prentice-Hall; 1964:357-358.

Horse Chestnut, Buckeye

188. Kingsbury JM: *Poisonous Plants of the United States and Canada.* Englewood Cliffs, NJ: Prentice-Hall; 1964;218-220.

189. Williams MC, Olsen JD. Toxicity of three *Aesculus* spp. to chicks and hampsters. *Am J Vet Res* 1984;45:539-542.

190. Yoshikawa M, Murakami T, Matsuda H, *et al.*: Bioactive saponins and glycosides. III Horse chestnut. The structures, inhibitory effects on ethanol absorption, and hypoglycemic activity of escins Ia, Ib, IIa, IIb, IIIa from the seeds of *Aesculus hippocastanum* L. *Chem Pharmaceut Bull* 1996;44:1454-1464.

191. Cary CA: Poisonous action of red buckeye on horses, mules, cattle, hogs, and fish. *Alabama Agricultural Experimental Station Bulletin* 1922;218:1-40.

192. Vansell GH: Buckeye poisoning of the honey bee. *University of California Expimental Station Circular* 1926;301:1-12.

193. Burrows GE, Edwards WC, Tyrl RJ: Toxic plants in Oklahoma. oak, black locust, Kentucky coffee tree, and buckeye. *Okla Vet Med Assoc* 1981;33:37-39.

194. Casteel SW, Wagstaff DJ. *Aesculus glabra* intoxication in cattle. *Vet Hum Toxicol* 1992;34:55.

195. Magnusson RA, Whittier WD, Veit HP, *et al.*: Yellow buckeye (*Aesculus octandra* Marsh) toxicity in calves. *Bovine Practitioner* 1983;18:195-199.

196. Kornheiser KM: Buckeye poisoning in cattle. *Vet Med/Small Animal Clinician* 1983;78: 769-770.

197. Edwards AJ, Mount ME, Oehme FW: Buckeye toxicity in Angus calves. *Bovine Practitioner* 1980;1:18-20.

Kentucky Coffee Tree

198. Kingsbury JM: *Poisonous Plants of the United States and Canada.*, Englewood Cliffs, NJ: Prentice-Hall; 1964:322-323.

199. Evers RA, Link RP: *Poisonous Plants of the Midwest and Their Effects on Livestock.* Urbana-Champaign: University of Illinois; 1972:57-58.

Carolina Jessamine

200.   Kingsbury JM: *Poisonous Plants of the United States and Canada.* Englewood Cliffs, NJ: Prentice-Hall; 1964:260-261.

201.   Morton JF: *Plants Poisonous to People in Florida*, edn 2. Stuart, FL: Southeastern Printing; 1982:31-32.

Carolina Allspice

202.   Kingsbury JM: *Poisonous Plants of the United States and Canada.* Englewood Cliffs, NJ: Prentice-Hall; 1964:124

203.   Bradley RE, Jones TJ: Strychnine-like toxicity of Calycanthus. *Southeastern Vet* 1963;14: 40,71,73.

204.   Miller JF, Kates AH, Davis DE, McCormack J: *Poisonous Plants of the southern United States.* Cooperative Extension Service Circular, University of Georgia, Athens, Georgia 1974.

Golden Chain Tree

205.   Kingsbury JM: *Poisonous Plants of the United States and Canada.* Englewood Cliffs, NJ: Prentice-Hall; 1964:325.

206.   Leyland A: Laburnum (*Cytisus laburnum*) poisoning in two dogs. *Vet Rec* 1981;109:287.

207.   Keeler RF, Baker DC: Myopathy in cattle induced by alkaloid extracts from *Thermopsis montanum, Laburnum anagyroides*, and a *Lupinus* sp. *J Comp Pathol* 1990;103:169-182.

Corydalis, Fitweed

208.   Oertli EH, Rowe LD, Slovering SL, *et al.:* Phototoxic effect of *Thamnosma texana* (Dutchman's breeches) in sheep. *Am J Vet Res* 1983;44:1126-1129.

209.   Flemming CE, Miller MR, Vawter LR: The fitweed (*Capnoides caseana*). A poisonous range plant of the northern Sierra Nevada Mountains. *University of Nevada Bulletin* 1931;121:8-29.

210.   Kingsbury JM: Poisonous Plants of the United States and Canada., Englewood Cliffs, NJ: Prentice-Hall; 1964:153-157.

211.   Smith RA, Lewis D: Apparent *Corydalis aurea* intoxication in cattle. *Vet Hum Toxicol* 1990;3263-3264.

212.   Schmutz EM, Freeman BN, Reed RE: *Livestock Poisoning Plants of Arizona.* Tucson: University of Arizona Press; 1974:78-79.

213.   Evers RA, Link RP: *Poisonous Plants of the Midwest and Their Effects on Livestock.* Urbana-Champaign: University of Illinois; 1972:49-50.

Lobelia, Cardinal Flower

214.   Kingsbury JM: *Poisonous Plants of the United States and Canada.* Englewood Cliffs, NJ: Prentice-Hall; 1964:390-392.

215.   Dollahite JW, Allen TJ: Poisoning of cattle, sheep, and goats with *Lobelia* and *Centaurium* species. *Southwestern Vet* 1962;15:126-130.

216.   Sperry OE, Dollahite JW, Hoffman GO, Camp BJ: Texas Plants Poisonous to Livestock. College Station, TX: *Texas Agricultural Experimental Station*, B-1028.

Solanum, Horse Nettle

217.   Pienarr JG, Kellerman TS, Basson PA, *et al.:* Maldronsiekte in cattle. A neuropathy caused by *Solanum kwebense. Ondersterpoort J Vet Res* 1985;43:67-74.

218. Kellerman TS, Coetzer JAW, Naude TW: *Plant Poisonings and Mycotoxicoses of Livestock in Southern Africa.* Capetown, South Africa: Oxford University Press; 1988:64-67.

219. Molyneux RJ, James LF, Ralphs MH, *et al.:* Polyhydroxyalkaloid glycosidase inhibitors from poisonous plants of global distribution: analysis and identification. *In* Plant Associated Toxins. Agricultural, Phytochemical, and Ecological Aspects. Colegate SM, Dorling PR, Eds. Wallingford, UK: CAB International; 1994:107-112.

220. Riet-Correa F, Mendez MDC, Schield AL, *et al.:* Intoxication by *Solanum fastigiatum* var. *fastigiatum* as a cause of cerebellar degeneration in cattle. *Cornell Vet* 1983;73:240-256.

221. Bourke CA: Cerebellar degeneration in goats grazing *Solanum cinereum* (*Narrawa* burr). *Aust Vet J* 1997;75:363-365.

222. Molyneux RJ, Pan YT, Goldman A, *et al.:* Calystegins, a novel class of alkaloid glycosidase inhibitors. *Arch Biochem Biophys* 1993;304:81-88.

223. Menzies JS, Bridges CH, Bailey EM: A neurologic disease of cattle associated with *Solanum dimidiatum. Southwest Vet* 1979;32:45-49.

Sorghum

224. Roman WM, Adams LG, Bullard TL, Dollahite JW: Cystitis syndrome of the equine. *Southwest Vet* 1966;19:95-99.

225. Adams LG, Dollahite JW, Romane WM, *et al.:* Cystitis and ataxia associated with sorghum ingestion in horses. *J Am Vet Med Assoc* 1969;155:518-524.

226. McKenzie RA, McMicking LI: Ataxia and urinary incontinence in cattle grazing sorghum. *Aust Vet J* 1977;53:496-497.

227. Knight PR: Equine cystitis and ataxia associated with grazing pastures dominated by sorghum species. *Aust Vet J* 1968;44:257.

228. Bradley GA, Metcalf HC, Reggiardo C, *et al.:* Neuroaxonal degeneration in sheep grazing Sorghum pastures. *J Vet Diagn Invst* 1995;7:229-236.

229. Cheeke PR. *Natural Toxicants in Feeds, Forages, and Poisonous Plants*, edn 2. Danville, IL: Interstate Publishers; 1997:193-197.

230. Montgomery RD: Cyanogens. *In* Toxic Constituents of Plant Foodstuffs. Liener IE, Ed. New York: Academic Press; 1980:143-160.

231. Van Kampen KR: Sudan grass and sorghum poisoning of horses: a possible lathyrogenic disease. *J Am Vet Med Assoc* 1970;156:629-630.

232. Pritchard JT, Voss JL: Fetal ankylosis in horses associated with hybrid Sudan pasture. *J Am Vet Med Assoc* 1967;150:871-873.

233. Seaman JT, Smeal MG, Wright JC: The possible association of a sorghum (*Sorghum sudanense*) hybrid as a cause of developmental defects in calves. *Aust Vet J* 1981,57:351-352.

234. Padmanaban G: Lathyrogens. *In* Toxic Constituents of Plant Foodstuffs, edn 2. Liener IE, Ed. New York: Academic Press; 1980:239-263.

235. Spencer PS, Schaumburg HH: Lathyrism: a neurotoxic disease. *Neurobehavioral Toxicol Teratol* 1983;5:625-629

236. Haque A, Hossain M, Wouters G, Lambein F: Epidemiological study of lathyrism in northwestern districts of Bangladesh. *Neuroepidemiology* 1996;15:83-91.

237. Haimanot RT, Kidane Y, Kassina A, *et al.:* The epidemiology of lathyrism in

north and central Ethiopia. *Ethiopian Med J* 1993;31:15-24.

238.   Misra UK, Sharma VP, Singh VP: Clinical aspects of neurolathyrism in Unnao, India. *Paraplegia* 1993;31:249-254.

239.   Smith ADM, Duckett S, Waters AH: Neuropatholgical changes in chronic cyanide intoxication. *Nature* 1963;4902:179-181.

240.   McKenzie RA, McMicking LI: Ataxia and urinary incontinence in cattle grazing sorghum. *Aust Vet J* 1977;53:496-497.

241.   Ross SM, Roy DN, Spencer PS. b-N-oxalylamino-L-alanine action on glutamate receptors. *J Neurochem* 1989;53:710-715.

242.   Willis CL, Meldrum BS, Nunn PB, *et al.*: Neuroprotective effect of free radical scavengers on b-N-oxalylamino-L-alanine (BOAA) induced neuronal damage in rat hippocampus. *Neurosci Lett* 1994;182:159-162.

243.   Rotter RG, Marquardt RR, Low C, Briggs CJ: Influence of autoclaving on the effects of Lathyrus sativus fed to chicks. *Can J Anim Sci* 1990;70:739-741.

244.   Kuo YH, Bau HM, Quemener B, *et al.*: Solid state fermentation of *Lathyrus sativus* seeds using *Aspergillus oryzae* and *Rhizopus oligosporus* sp. T-3 to eliminate the neurotoxin b-OADP without loss of nutritional value. *J Sci Food Agric* 1995;69:81-89.

Caltrop

245.   Sperry OE, Dollahite JW, Hoffman GO, Camp BJ: Texas plants poisonous to livestock. *Agricultural Extension Service B-1028. College Station, TX:* Texas A & M University; 1972:28-29.

246.   Mathews FP: The toxicity of *Kallstroemia hirsutissima* (carpet weed) for cattle sheep and goats. *J Am Vet Med Assoc* 1944;10:152-155.

Marijuana

247.   Keeler RF, Tu AT: *Plant and Fungal Toxins*, Vol 1. New York: Marcel Dekker; 1983:473-508.

248.   Hulbert LC, Oehme FW: *Plants Poisonous to Livestock*. Manhattan, KS: Kansas State University; 1968:34.

249.   Driemeier D: Marijuana (*Cannabis sativa*) toxicosis in cattle *J. Vet Hum Toxicol* 1997, 39: 351-352.

250.   Godbold JC, Hawkins BJ, Woodward MG: Acute oral marijuana poisoning in the dog. *J Am Vet Med Assoc* 1979;175:1101-1102.

251.   Evans AG: Allergic inhalant dermatitis attributable to marijuana exposure in a dog. *J Am Vet Med Assoc* 1989;195:1588-1590.

Tobacco

252.   Kingsbury JM: *Poisonous Plants of the United States and Canada.* Englewood Cliffs, NJ: Prentice-Hall; 1964:284-287.

253.   Kaplan B: Acute nicotine poisoning in a dog. *Vet Med/Small Animal Clinician* 1968;63:1033.

254.   Vig MM: Nicotine poisoning in a dog. *Vet Hum Toxicol* 1990;32:573-575.

255.   Van Gelder GA: *Clinical and Diagnostic Veterinary Toxicology,* edn 2. Dubuque, IA: Kendall/Hunt Publishing; 1976:185-186.

256.   Crowe MW, Swerczek TW: Congenital arthrogryposis in offspring of sows fed tobacco (*Nicotiana tabacum*). *Am J Vet Res* 1974;35:1071-1073.

257. Keeler RF, Shupe JL, Crowe MW, *et al.: Nicotiana glauca*-induced congenital deformities in calves: clinical and pathologic aspects. *Am J Vet Res* 1981;42:1231-1234.

258. Keeler RF, Crowe MW: Teratogenicity and toxicity of wild tree tobacco, *Nicotiana glauca* in sheep. *Cornell Vet* 1984;74:50-59.

259. Plumlee KH, Holstege DM, Blanchard PC, *et al.: Nicotiana glauca* toxicosis in cattle. *J Vet Diagn Invest* 1993;5:498-499.

CHAPTER 7

Plants Causing Kidney Failure

In North America a variety of indigenous and exotic plants have been associated with kidney disease in animals. Most notable are the devastating livestock losses in the western alkaline desert areas of North America where plants such as greasewood (*Sarcobatus vermiculatus*), and halogeton (*Halogeton glomeratus*) that contain oxalates, are abundant. Interestingly, the most toxic of the plants that have established themselves in these semiarid areas is halogeton (*Halogeton glomeratus*), a plant introduced from Asia and highly adaptable to dry alkaline soils. This plant alone has caused heavy death loss in sheep in particular because of its extremely high oxalate content.[1-3] In addition to oxalate containing plants, kidney failure is also commonly associated with the consumption of plants such as oak trees (*Quercus* spp.) that contain high levels of tannins.

Oxalate Poisoning

Oxalate poisoning in animals generally occurs when quantities of oxalate-containing plants are grazed by livestock that are not accustomed to eating the plants. The most important oxalate-containing plants in North America include halogeton (*H. glomeratus*), and greasewood (*S. vermiculatus*).[1,2] In Australia and Africa oxalate poisoning has been associated with livestock grazing *Oxalis* spp.[4] Other plants known to contain significant quantities of oxalate are listed in Table 7-1. Plants normally accumulate oxalates in the form of soluble potassium and sodium oxalates. The more toxic potassium acid oxalate predominates in plants with a very acid cell sap (pH 2) (*Oxalis* spp.); sodium oxalate occurs in plants with a cell sap pH of 6 (*Halogeton, Sarcobatus* spp.).[5,6]

Oxalate poisoning most often occurs when unadapted sheep or cattle are allowed to graze large amounts of *Halogeton* or *Sarcobatus* as they pass through or are pastured overnight on rangeland containing large stands of these plants. Under normal range conditions sheep are most frequently poisoned by oxalate-containing plants. Ruminants in general tolerate relatively more oxalate in their diet than other animals because they are able to detoxify oxalate in the rumen thereby preventing the absorption of the soluble oxalates. When large quantities of soluble potassium and sodium oxalates are eaten that overwhelm the rumen's ability to metabolize the oxalates, they are absorbed into the bloodstream and form insoluble calcium and magnesium oxalates. It is these insoluble salts that precipitate in the kidneys and cause kidney failure.[2,7] Factors that predispose an animal to oxalate intoxication include the amount and rate at which the oxalate plant is eaten and the quantity of other feed diluting the oxalate in the rumen. Prior adaptation of rumen microflora to oxalates allows the animal to consume more oxalate because the increased number of oxalate-degrading bacteria in the rumen more effectively metabolize the oxalate.[8] Ruminants allowed to graze small quantities of oxalate-containing plants are able to increase their tolerance for oxalate 30 percent or more over a few days.[2,8,9] Once adapted to oxalate, sheep and cattle can make effective use of range forages containing oxalate that would otherwise be toxic.

Toxic Effects of Oxalates

Once absorbed from the gastrointestinal tract, soluble oxalates rapidly combine with

Table 7-1
Plants Containing Oxalates

SCIENTIFIC NAME	COMMON NAME
Amaranthus spp.	Red-rooted pigweed
Bassia hyssopifolia	Five hooked bassia
Beta vulgaris	Sugar beet
Chenopodium spp.	Lambs-Quarter
Halogeton glomeratus	Halogeton
Kochia scoparia	Kochia, summer cypress
Oxalis spp.	"Shamrock," soursob, sorrel
Portulaca oleraceae	Purslane
Phytolacca americana	Poke berry
Rumex spp.	Sorrel, dock
Rheum rhaponticum	Rhubarb
Salsola spp.	Russian thistle, tumbleweed
Sarcobatus vermiculatus	Greasewood

Grasses

Cenchrus ciliaris	Buffel grass
Panicum spp.	Elephant grass
Pennisetum clandestinum	Kikuyu grass
Setaria sphacelata	Setaria grass

**Common House and Garden Plants
Containing Oxalates** [35]

Arisaema spp.	Jack in the pulpit
Alocasia spp.	Elephant's ear
Anthurium spp.	Anthurium, flamingo flower
Arum spp.	Lords and Ladies, Cuckoo-pint
Calla palustris	Wild calla, water arum
Caladium spp.	Caladium
Dieffenbachia sequine	Dumb cane
Epiprenum spp.	Pothos, variegated philodendron, taro vine
Monstera spp.	Monstera, cutleaf philodendron, bread fruit vine
Philodendron spp.	Philodendron
Spathephyllum spp.	Peace lily
Schefflera spp.	Umbrella tree
Zantedeshia aethiopica	Calla lily

serum calcium and magnesium, causing a sudden decrease in available serum calcium and magnesium.[2,9,10] In the acute phase of oxalate poisoning the sudden decrease in soluble serum calcium (hypocalcemia) impairs normal cell membrane function, causing animals to develop muscle tremors and weakness, leading to collapse and eventually death. Oxalates also interfere with cellular energy metabolism that contributes to the acute death of affected animals. In chronic oxalate poisoning, insoluble calcium oxalate filtered by the kidneys causes severe damage to the kidney tubules (oxalate nephrosis). If animals do not die from the acute effects of the low blood calcium levels and impaired cellular energy metabolism, death results from kidney failure.

Clinical Signs

Within a few hours of consuming toxic levels of oxalate, sheep and cattle develop muscle tremors, tetany, weakness, reluctance to move, depression, and recumbency resulting from hypocalcemia and hypomagnesemia.[9-12] Coma and death may result within 12 hours. Animals that survive the acute effects of oxalate poisoning frequently succumb to kidney failure. As animals become uremic (increased serum creatinine and urea nitrogen levels), they develop severe depression, stop eating, and after a few days become comatose and die.

Horses may over time develop chronic calcium deficiency while grazing certain tropical grasses containing soluble oxalates. (Table 7-1) Oxalates in the grasses combine with calcium to form relatively insoluble calcium oxalate, thereby reducing calcium absorption and altering the calcium:phosphorus ratio. This causes mobilization of bone calcium through the action of parathyroid hormone to compensate for the low blood calcium levels. Over time the horse's bones loose sufficient calcium so that they become soft and misshapen (nutritional secondary hyperparathyroidism).[13-15] Sheep and cattle can be similarly affected, but are better able to metabolize oxalates in the rumen thereby reducing their effect on dietary calcium.[16]

Diagnosis

A diagnosis of oxalate poisoning can be made on the basis of the type of plants being eaten, the clinical signs, hypocalcemia, and the presence of oxalate crystals in the urine. However, oxalate crystals in the urine are usually not present in acute oxalate poisoning. Necropsy lesions seen in oxalate poisoning will depend on the severity of the poisoning. In acute poisoning, the kidneys may be edematous and dark red in color, whereas in chronic poisoning the kidneys may be pale and smaller than normal. Perirenal edema is a characteristic feature of oxalate poisoning in pigs and cattle consuming pigweed (*Amaranthus retroflexus*).[17-19] The rumen may often become hemorrhagic due to the presence of large quantities of oxalate in the rumen epithelial lining. The demonstration of calcium oxalate crystals in the kidneys and rumen epithelium histologically is diagnostic of oxalate poisoning.[10,11]

Treatment

Treatment with intravenous calcium gluconate, although theoretically appropriate for correcting hypocalcemia, is not effective in reversing the effects of the oxalate on cellular energy metabolism. Irreversible oxalate nephrosis and the effects of oxalates on cellular energy metabolism are more detrimental to the animal than hypocalcemia. A theoretical approach to treating acute oxalate poisoning would be to administer intravenous calcium gluconate, magnesium sulfate, glucose, and a balanced electrolyte solution to maintain kidney perfusion. Giving limewater ($Ca[OH]_2$) orally will help to prevent absorption of further soluble oxalate.

Prevention of Oxalate Poisoning

Livestock should not be grazed on rangeland on which oxalate-containing plants predominate without precaution, especially if the animals are hungry and have not been adapted to oxalate in their diet. Livestock should be introduced to oxalate plants for at least 4 days by incrementally increasing the time they are allowed to graze the plants. Overstocking and overgrazing will potentiate oxalate poisoning if there is not other vegetation for animals to eat. Cattle and sheep driven through or held overnight in pastures rich in oxalate-containing plants are prone to poisoning, and such circumstances should be avoided. Supplementary dicalcium phosphate in the diet before and during high-risk oxalate exposure is an effective means of reducing losses. High levels of dietary calcium bind oxalate in the rumen as insoluble, nonabsorbable calcium oxalate. Calcium may be provided to the animals in a salt mix (75 lb salt, 25 lb dicalcium phosphate) or in pelleted alfalfa at a 5 percent concentration and fed at the rate of 0.5 lb per sheep per day. Livestock diets can also be supplemented with hay to help reduce the total intake of oxalate-containing plants.

Five Hook Bassia, Smother Weed
Bassia hyssopifolia
Chenopodiaceae (Goosefoot family)

Habitat
Introduced from Eurasia, bassia has adapted well to the alkaline soils of western North America and is a weed of waste ground.

Description
Bassia is an erect (2 to 5 feet), annual, much-branched plant with alternate, sessile leaves ranging to 2 inches in length. The leaves and stem are hairy. The flowers are borne in pannicles or short spikes and sometimes singly in leaf axils. The greenish flowers are small, without petals, and the five sepals are armed with a stout, spreading, hooked spines that distinguish it from the similar appearance of Kochina weed. The plant spreads by seed only.

Principal Toxin
Although not widespread in this country as yet, *Bassia* has the potential for accumulating greater quantities of oxalates than *H. glomeratus*.[12]

Halogeton, Bavilla
Halogeton glomeratus
Chenopodiaceae (Goosefoot family)

Habitat
Introduced from Russia, halogeton prefers the arid alkaline soils or clays of western North America. The plant continues to spread covering large areas of rangeland especially in flood plains and along rivers and roadsides.

Description
Halogeton is an annual, multibranched herb with branches spreading horizontally before curving upward to 2 feet in height (Figure 7-1A). Seedlings are usually prostrate with four main branches in the form of a cross. Mature plants have red stems with fleshy, blue-green leaves terminating in a solitary hair (Figure 7-1B). The small, inconspicuous flowers appear in the leaf axil. The fruits are bracted and are often mistaken for the flowers. The single seed is surrounded by five reddish to yellow-green bracts.

Principal Toxin
Soluble sodium oxalate may comprise 30 to 40 percent of the dry matter content of the plant. Poisoning in sheep occurs when 0.3 to 0.5 percent of the animal's body weight of plant is consumed over a short period. A lethal dose of halogeton for an

Figure 7-1A Halogeton (*Halogeton glomeratus*)

adult sheep is about 1.5 lb of green plant. In animals deprived of their normal feed, the toxic dose of halogeton is about one-third to one-quarter of that needed to induce poisoning in normally fed animals.

Halogeton remains toxic when dried and is found to be quite palatable by sheep and cattle that commonly utilize halogeton as a winter forage when grazing western rangelands. They are able to do this once they are adapted to eating halogeton, and there are some other forages available.

Figure 7-1B Halogeton showing leaves with terminal hair (*H. glomeratus*).

Greasewood, Chico
Sarcobatus vermiculatus
Chenopodiaceae (Goosefoot family)

Habitat

Greasewood often forms the dominant shrubby vegetation of the arid alkaline soils of western North America. It commonly grows in the same habitat with *Halogeton glomeratus*.

Figure 7-2A Greasewood (*Sarcobatus vermiculatus*).

Description

Greasewood is an erect, deciduous, branched shrub often growing 4 to 6 feet (1 to 2 meters) in height (Figure 7-2A) The stem and branches turn gray with maturity and have 1 to 3 inches (2.5 to 7.5 cm) woody spines. The leaves are alternate, bright green, fleshy, loosely round in cross section, and up to 1.25 inches (3 cm) long (Figure 7-2B). The flowers are unisexual with the plants having both sexes of flower on the same plant. The female flowers are inconspicuous in the axils and the male flowers occur as terminal spikes (catkins). The fruits are winged and conical in shape.

Principal Toxin

Sodium oxalate is the principal toxin and may comprise from 10 to 22 percent of the dry matter content of the plant. The leaves contain the highest concentration of oxalate. Toxicity occurs when 1.5 to 5.0 percent of an animal's weight of the plant is ingested over a short period of time.

Sheep and cattle will often browse greasewood in the winter months, and do well on it as long as they are adapted to eating oxalate containing plants.

Figure 7-2B Greasewood leaves (*S. vermiculatus*).

Habitat

Various species of sorrel are found throughout North America, growing at altitudes up to 9000 feet (2,743 meters). The plants are common in moist pastures, forests, along roadsides, and as weeds of cultivated soils.

Figure 7-3 Oxalis, shamrock, wood sorrel (*Oxalis* spp.)

Description

Wood sorrels are herbaceous annuals or perennials that grow horizontally but occasionally erect with characteristic alternate or basal leaves that are palmately three-foliate and heart-shaped (Figure 7-3). The flowers are usually yellow, with perfect stipules, and regular umbel-like or dichotomous cymes. The fruit is a capsule.

Principal Toxins

The principal toxins are soluble potassium and sodium oxalates. *Oxalis* spp. poisoning is rarely a problem in North America, but it has been associated with severe poisoning of livestock in Australia.[3]

Pigweed, Amaranth
Amaranthus retroflexus
Amaranthaceae (Pigweed family)

Habitat

A variety of pigweeds grow throughout North America; the most common and typical of the genus is red-rooted pigweed (*A. retroflexus*). It is a common weed of most soils, found in cultivated and disturbed soils along roadsides and waste areas. It is common in and around corrals and animal enclosures.

Other common species of pigweed include prostrate pigweed (*A. blitoides*), tumble pigweed (*A. albus*), and Palmer amaranth (*A. palmeri*). All have similar characteristics to the red-rooted pigweed.

Description

Plants are stout, erect, rapidly growing annuals, reaching 3 to 4 feet (1 to 1.2 meters) tall depending on the growing conditions. The stems are usually much branched and hairy and may be red to purple in color. The leaves are alternate, petiolate, ovate to lanceolate, and acute at apex. The flowers are monoecious in densely crowded spikes or pannicles 3 to 8 inches (8 to 20 cm) long (Figure 7-4A). The flowers are greenish with long, spine-tipped bracts. The taproot is usually bright red in color (red-rooted pigweed; Figure 7-4B). Numerous, shiny black seeds are produced that ensure successful proliferation of the pigweed.

Figure 7-4A Pigweed, amaranth flower spike (*Amaranthus retroflexus*).

Principal Toxin

Amaranthus spp. may accumulate significant quantities of oxalate that causes a syndrome in pigs and cattle characterized by perirenal edema.[17,18] Renal tubular nephrosis probably causes death of the animal. However, renal nephrosis is often present without the presence of oxalate crystals, suggesting that *Amaranthus* spp. may contain other as yet undefined toxic substances.

As annual weeds, pigweeds also accumulate nitrates and can be a source of nitrate poisoning (see Chapter 1).

Figure 7-4B Red-rooted pigweed (*A. retroflexus*).

Habitat

Purslane is a very common weed of gardens, cultivated ground and waste places.

Figure 7-5A Purslane showing prostrate stems (*Portulacca oleracea*)

Figure 7-5B Purslane flowers

Description

Purslane is an annual, succulent plant with prostate stems, that are pink to red in color (Figure 7-5A). Leaves are fleshy and flat, obovate in shape, and either alternate or clustered on the stem. Flowers are yellow, solitary, and open for only a short time (Figure 7-5B). Fruit is a circumscissile capsule.

Principal Toxin

Oxalates are the principal toxins and account for the sour taste to the plant. As much as 9 percent dry matter of the plant may consist of oxalate. Few cases of poisoning in livestock have been reported.

Purslane is eaten by some people as a vegetable, and does not pose a toxicity problem because, in cooking the plant, the oxalates are leached out by the steaming or boiling process.

Dock, Curly Leaf Dock, Sorrel
Rumex crispus
Polygonaceae (Buckwheat family)

Habitat

Dock is a common plant of gravelly soils of pastures, plains, and roadsides throughout North America. There are a variety of *Rumex* spp. with similar characteristics and habitats.

Description

Dock is a perennial weed from a stout tap root. Flower stems may reach 6 to 7 feet (2 meters) in height without axillary branches. Leaves are 4 to 12 inches (10 to 30 cm) long, oblong to linear-lanceolate with crisped, wavy margins (Figure 7-6A). An ocrea is present at the base of each petiole. The flower is a compound raceme or panicle having many small perfect or unisexual flowers. The perianth is six-parted with the inner three segments becoming the wings of the fruit. The inflorescence turns a dark brown when dry (Figure 7-6B).

Principal Toxin

Oxalates are the primary toxins in *Rumex* spp. Although not a common source of oxalate poisoning, *Rumex* spp. will cause poisoning if eaten in excess by livestock.[11,18,19]

Figure 7-6A Dock, sorrel (*Rumex crispus*).

Figure 7-6B Dock with curly leaves and ripe seed head (*Rumex crispus*) .

Rhubarb
Rheum rhabarbarum (R. rhaponticum)
Polygonaceae (Buckwheat family)

Habitat

Rhubarb is a domestic plant found throughout North America, although it may escape and be found great distances from any dwelling. Moist, fertile soils are preferred habitats.

Description

Rhubarb is a stout herbaceous, perennial plant with large, fleshy roots and large leaves that are heart shaped, with entire margins and long reddish petioles. Flowering stems, 1 to 2 meters long, support a many-flowered panicle of greenish or whitish flowers. The fruit is a winged achene that turns a dark brown color when mature.

Principal Toxin

The leaves contain large amounts of soluble oxalates. The red leaf stalk (petiole) contains much less oxalate than the leaves. Feeding the leaves to livestock may induce severe oxalate nephrosis. Human poisoning is almost always associated with eating rhubarb leaves used as a vegetable or salad. Human poisoning from eating rhubarb stems is very unlikely as they contain minimal amounts of oxalate, and a person would have to eat literally several pounds of rhubarb to be affected. Cooking them with some calcium carbonate to produce insoluble calcium oxalate can further reduce the toxicity of the rhubarb stems.

Anthraquinone compounds in the leaves that have cathartic properties are the likely cause of the abdominal pain accompanied with vomiting encountered in people who have eaten rhubarb leaves

Oak
Quercus spp.
Fagaceae (Oak family)

Habitat

Some 60 species of oak grow in North America in a wide variety of habitats ranging from moist, rich soils of hard wood forests to drier mountainous areas.

Description

Ranging from large trees to shrubs, oaks have alternate, simple, toothed, or lobed dark green glossy leaves. The leaves may be deciduous or persistent depending on the species of oak. The plants are monoecious with the staminate flowers occurring in long catkins and the pistallate flowers occurring singly or in small clusters. The fruit, an acorn, is a nut partially enveloped by an involucre of scales (Figure 7-7). Two common species of oak growing in western North America commonly associated with livestock poisoning are scrub oak and shinnery oak.

Q. gambelii (Gambels oak, scrub oak) is a shrub or small tree reaching heights of 15 to 20 feet (4.5 to 6 meters). It grows in dense stands in the dry foothills and mountain slopes up to altitudes of 9000 feet (2,743 meters).

Figure 7-7 Oak leaves and acorn (*Quercus* spp.) .

Q. havardii (*Shinnery oak*) is a shrub that seldom attains heights over 4 feet (1.2 meters). It is confined more to the lower elevations and sandy soils of southwestern North America.

Principal Toxin

The principal toxin is gallotannin, a polyhydroxphenolic combination of tannic and gallic acid.[20-22] The tannins found in the leaves, bark, and acorns of oaks produce poisoning through their effect on the intestinal tract and kidneys.[21-23] Gallotannins are hydrolyzed in the rumen to smaller molecular weight compounds including gallic acid, pyrogallol, and resorcinol.[24] These compounds react with cell proteins to denature them, with resulting cell death. Most severe lesions occur in the kidneys, liver, and digestive tract. In small quantities the rumen microflora detoxify the tannins, and only when large amounts of tannic acid are eaten and bypass the rumen does poisoning occur.[25] Goats and wild ruminants are apparently better able to detoxify tannic acid than other livestock because they have a tannin-binding protein in their saliva that neutralizes tannic acid.[26,27] Goats have been used effectively to browse on oaks thereby reducing the spread of the oak and increasing the grazing capacity of the range. Oaks at any stage of growth are poisonous, but they are particularly toxic when the leaf and flower buds are just opening in the spring. Consumption of oak buds can be markedly increased in a heavy, late spring snowstorm , when cattle browse the oak that protrudes above the snow. As the leaves mature they become less toxic. Ripe acorns are less toxic than when green. Cattle sheep, horses, and pigs are susceptible to oak poisoning.[21,28-30]

Clinical Signs

Signs of oak poisoning will vary according to the quantity of oak consumed. Initially affected animals stop eating, become depressed, and develop intestinal stasis.[22-24] Excessive thirst and frequent urination may be observed. The feces are hard and dark initially, but a black tarry diarrhea often occurs later in the course of poisoning. Teeth grinding and a hunched back are often indicative of abdominal pain. Severe liver and kidney damage is detectable by marked elevations in serum liver enzymes, creatinine, and urea nitrogen.[30] Icterus, red-colored urine, and dehydration are further signs encountered in oak poisoning. Animals may live for 5 to 7 days after the onset of clinical signs.

Necropsy

A mucoid, hemorrhagic gastroenteritis is a common finding in oak poisoning.[23,29,31] Hemorrhages on various organs and excessive amounts of fluid in the peritoneal and pleural spaces are often present. The kidneys are usually pale swollen and covered with small hemorrhages. Histologically kidney tubular necrosis, and liver necrosis are characteristic of oak poisoning. [23,29,31]

Treatment

Animals should be removed from the oak and given supportive care in the form of fresh water and good quality hay. Oral administration of a calcium hydroxide solution is helpful in neutralizing residual tannic acid in the rumen.[32] Intravenous fluids should be given to rehydrate severely affected animals and maintain kidney function. Animals that continue to eat have a much better prognosis. Cattle that survive oak poisoning appear to have compensatory weight gains and appear to do well.[33]

Grain or pelleted rations containing 10 to 15 percent calcium hydroxide are beneficial in preventing oak poisoning if cattle have to graze pastures overgrown with oak brush.[32] Goats are effective biological controls and may be used to browse oak for range management purposes because they are unaffected by the tannins in the oak.[34]

Acorn Calf Syndrome

The acorn calf syndrome is not related to oak poisoning attributed to gallotannins but is encountered in calves born to cows on a low plain of nutrition and which have consumed quantities of acorns.[36] Acorn calves are born with laxity of the joints, shortened legs (dwarfism), deformed hooves, and either a domed skull or long narrow head. Compared to normal calves, the acorn calves are stunted and grow poorly. The acorn calf syndrome has also been reported in cows that graze heavily on lupine during mid pregnancy.[37] The toxic principal responsible for this congenital syndrome has not been determined. Protein malnutrition and the presence of a teratogen may be involved in the development of the acorn calf syndrome.

REFERENCES

1. Cook CW, Stodart LA: The Halogeton problem in Utah. *Utah Agricultural Experimental Station Bulletin* 1953, 364:44.

2. James LF: Oxalate poisoning in livestock. *In* Effects of Poisonous Plants on Livestock. Keeler RF, Van Kampen KR, James LF, Eds. New York: Academic Press; 1978:139–145.

3. Seawright AA: Oxalates. *In* Animal Health in Australia, Vol 2. Chemical and plant poisons. Australian Government Publication 1982, pp 74–76.

4. James LF, Butcher JE: Halogeton poisoning of sheep: effect of high level oxalate intake. *J Anim Sci* 1982, 35:1233–1238.

5. Cheeke PR. Plants and toxins affecting the gastrointestinal tract and liver. *In:* Natural Toxicants in Feeds, Forages, and Poisonous Plants, edn 2. Cheeke PR., Ed.Danville, IL: Interstate Publishers; 1998:329–332.

6. Van Kampen KR, James LF: Acute Halogeton poisoning of sheep: pathogenesis of lesions. *Am J Vet Res* 1969, 30:1779–1783.

7. Lincoln SD, Black B: Halogeton poisoning in range cattle. *J Am Vet Med Assoc* 1980, 176:717–718.

8. Fassett DW: Oxalates. *Toxicants Occurring Naturally in Foods.* Washington, DC: National Academy of Sciences; 1973:346–362.

9. Allison MJ, Dawson KW, Mayberry WR, Foss JG: *Oxalobacter formigenes* gen. nov. sp. nov: oxalate-degrading anaerobes that inhabit the gastrointestinal tract. *Arch Micrbiol* 1985, 141:1–7.

10. Osweiler GD, Buck WB, Bicknell EJ: Production of perirenal edema in swine with *Amaranthus retroflexus*. *Am J Vet Res* 1969, 30:557-566.

11. Stuart BP, Nicholson SS, Smith JB. Perirenal edema and toxic nephrosis in cattle associated with ingestion of pigweed. *J Am Vet Med Assoc* 1975, 167:949–950.

12. James LF, Williams MC, Bleak AT: Toxicity of *Bassia hyssopifolia* to sheep. *J Range Mange* 1976, 29:284-285.

13. Jones RJ, Seawright AA, Little DA: Oxalate poisoning in animals grazing the tropical grass *Setaria sphacelata*. *J Austr Inst Agric Sci* 1970, 36:41-43.

14. Blaney BJ, Gartner RJW, McKenzie RA: The effects of oxalate in tropical grasses on calcium, phosphorus and magnesium availability in cattle. *J Agric Sci* 1981a, 97:507-514.

15. McKenzie RA, Bell AM, Storie GJ, *et al:* Acute oxalate poisoning of sheep by buffel grass (*Cenchrus ciliaris*). *Austr Vet J* 1988, 65:26.

16. Allison MJ, Cook HM, Dawson KA: Selection of oxalate degrading rumen bacteria in continuous cultures. *J Anim Sci* 1981, 53:810-816.

17. Everist LS: *Poisonous Plants of Australia.* Sydney, Australia: Angus and Robertson Publishers 1974:394–397.

18. Dickie CW, Hamann MH, Carroll WD, Chow F: Oxalate (*Rumex venosus*) poisoning in cattle. *J Am Vet Med Assoc* 1978, 173:73–74.

19. Casteel SW, *et al.:* *Amaranthus retroflexus* (redroot pigweed) poisoning in cattle. *J Am Vet Med Assoc* 1994, 204:1068–1070.

20. Dollahite JW, Pigeon RF, Camp BJ: The toxicity of gallic acid, pyrogallol, tannic acid, and *Quercus havardii* in the rabbit. *Am J Vet Res* 1962, 23:1264–1267.

21. Pigeon RF, Camp BJ, Dollahite JW: Oral toxicity and polyhydroxyphenol moiety of tannin isolated from *Quercus havardii* (Shin oak). *Am J Vet Res* 1962, 23:1268–1270.

22. Sandusky GE, Fosnaugh CJ, Smith JB, Mohan R: Oak poisoning of cattle in Ohio. *J Am Vet Med Assoc* 1977, 171:627–629.

23. Panciera RJ: Oak poisoning in cattle. *In:* Effects of Poisonous Plants on Livestock. Keeler RF, Van Kampen KR, and James LF, Eds. New York: Academic Press; 1978:499–506.

24. Anderson GA, Mount ME, Veins AA, Zima EL: Fatal acorn poisoning in a horse: pathologic findings and diagnostic considerations. *J Am Vet Med Assoc* 1983, 182: 1105–1110.

25. Dollahite JW, Householder GT, Camp BJ: Effects of calcium hydroxide on the toxicity of post oak (*Quercus stellata*) in calves. *J Am Vet Med Assoc* 1966, 148: 908–912.

26. Nastis AS, Malechek IC: Digestion and utilization of nutrients in oak browsed by goats. *J Anim Sci* 1981, 83:283–290.

27. Austin PJ, Suchar LA, Robbins CT, Hagerman AE: Tannin-binding proteins in saliva of deer and their absence in saliva of sheep and cattle. *J Chem Ecol* 1989, 15:1335–1347.

28. Panciera RJ, Martin T, Burrows GE, *et al.:* Acute oxalate poisoning attributable to ingestion of curly dock (*Rumex crispus*) in sheep. *J Am Vet Med Assoc* 1990, 196:1981–1984.

29. Duncan CS: Oak leaf poisoning in two horses. *Cornell Vet* 1961, 51:159–162.

30. Spier SJ, Smith BP, Seawright AA, *et al.:* Oak toxicosis in cattle in northern California: clinical and pathological findings. *J Am Vet Med Assoc* 1987, 191:958–964.

31. Tor ER, Francis TM, Holstege DM, Galey FD: GC/MS determination of pyrogallol and gallic acid in biological matrices as diagnostic indicators of oak exposure. *J Agric Food Chem* 1996, 44:1275–1279.

32. Zhu J, Filippich LJ: Acute intra-abomasal toxicity of tannic acid in sheep. *Vet Hum Toxicol* 1995, 37:50–54.

33. Ostrowski SR, Smith BP, Spier SJ, *et al.:* Compensatory weight gain in steers recovered from oak bud poisoning. *J Am Vet Med Assoc* 1989, 195:481–484.

34. Davis GG, Bartel LE, Cook CW: Control of gambel oak sprouts by goats. *J Range Manage* 1975, 28:216–218.

35. Spoerke DG, Smolinske SC: *In* Toxicity of Houseplants. Spoerke DG, Smolinske SC, Eds. Boca Raton, FL: CRC Press; 1990:31–32.

36. Radostits OM, Gay CC, Blood DC, Hinchcliff KW, Eds. *Veterinary Medicine*, edn. 9. WB Sanders, Philadelphia. p 1810-2000.

37. Hawkins CD. "Acorn" calves and retained placental following grazing on sand plain lupins (*Lupinus coswentinii*). *In* Plant associated toxins. Colegate SM, Dorling PR, Eds. CAB International. 351-356, 1994.

CHAPTER 8

Plants Associated with Congenital Defects and Reproductive Failure

Considerable livestock losses occur annually as the result of toxic plants that cause embryonic death, abortion, and fetal abnormalities.[1-3] Plant toxins also interfere with reproduction through their effects on male fertility, especially by affecting spermatogenesis. The plant species that have been historically associated with abortion and infertility in livestock are listed in Table 8-1. Those plant toxins capable of causing fetal death or deformity are referred to as teratogens. To be a teratogen, a plant toxin must readily cross the placenta at a high enough dose and be present at a specific time in gestation to exert its effect on the developing fetus.[3,4] In addition, susceptibility to a plant teratogen depends on the animal species, because not all species are equally susceptible to teratogens. In general, the fetus is most susceptible to teratogens during the first trimester of pregnancy. The western false hellebore (*Veratrum californicum*) is a classical example of a teratogen because it only induces the cyclops deformity in lambs if the pregnant ewe eats sufficient quantity of the plant during the 13-14th days of gestation.[5] Other than plant teratogens, viruses, nutritional imbalances and deficiencies, chemicals (nitrates), and radiation may also be teratogenic and must be considered when determining the cause of congenital deformities. The plants that have been confirmed or suspected of being teratogens in animals are listed in Table 8-2.

Teratogenic Plants

Milk Vetches and Locoweeds

In general, the milk vetches or locoweeds (*Astragalus and Oxytropis* spp.) have the greatest economic impact on animal reproduction compared to any other group of plants. Most of the 370 or more species of milk vetch and locoweed thrive in the more arid and alkaline areas of western North America. Relatively few of these species, however, are known to be toxic and those that are more commonly associated with livestock poisoning are discussed under locoism in Chapter 6. Swainsonine, one of a group of indolizidine alkaloids found in the poisonous milk vetches and locoweeds, is the principle toxin responsible for causing

Table 8-1 Plants Associated with Livestock Abortion	
SCIENTIFIC NAME	COMMON NAME
Agave lechequilla	Lechuguilla
Astragalus spp.	Milk vetch, locoweeds
Brassica spp.	Rape
Conium spp.	Poison/spotted hemlock
Cupressus spp.	Cyprus
Festuca spp.	Fescue
Gutierrezia sarothrae	Broomweed, snakeweed
Halogeton spp.	Halogeton
Indigofera glomeratus	Juniper
Juniperus spp.	Juniper
Medicao sativa	Alfalfa
Oxtropis spp.	Locoweeds
Phytolacca americana	Poke weed
Pinus ponderosa	Ponderosa pine
Solidago spp.	Goldenrods
Tanacetum spp.	Tansy
Trifolium spp.	Clovers
Veratrum spp.	False hellebore, skunk cabbage

Table 8-2	
Known and Suspected Teratogenic Plants	
SCIENTIFIC NAME	COMMON NAME
Astragalus spp.	Milk vetch, locoweed
Blighia sapida	Akee
Colchicum autumnale	Autumn crocus
Conium maculatum	European or spotted hemlock
Cycadaceae spp.	Cyads
Datura stramonium	Jimson weed
Indigofera spicata	Creeping indigo
Lathyrus spp.	Wild pea
Leucaena leucocephala	Mimosa
Lupinus spp.	Lupine
Nicotiana glauca	Wild tree tobacco
Nicotiana tabacum	Tobacco
Oxytropis spp.	Locoweed
Papaveraceae	Poppies
Senecio spp.	Groundsel
Veratrum californicum	Western false hellebore
Vinca rosea	Periwinkle

reproductive problems in animals that eat them.[6] The alkaloid affects almost all organs because it inhibits normal cellular energy metabolism. Consequently swainsonine has a profound effect on all aspects of the reproductive system in both females and males.

Cattle and sheep, and to a lesser extent horses, are the most susceptible to the reproductive and teratogenic effects of chronic locoweed poisoning.[7-10] Unlike most teratogens, swainsonine may exert its effects on the dam and the fetus at any time during gestation, causing a variety of problems. These reproductive problems are most likely due to the combined effects of swainsonine on the pituitary gland affecting gonadotrophin production, the ovary affecting estrogen and progesterone levels, the uterus and placenta, and directly on the fetus.[11] Abortion, infertility, fetal deformity, and disturbances in placental circulation that results in massive accumulation of fluid in the uterus (hydrops). Some calves and lambs may be born weak and do not thrive. Abortions can occur at anytime during gestation.[9] Animals that have aborted tend to cycle normally and will conceive again provided they are kept from eating more locoweed, and there is not a chronic secondary uterine infection resulting from the abortion.

Common fetal deformities encountered in livestock eating locoweeds include twisted and deformed limbs resulting from contracted flexor tendons of the legs, and abnormal development of the bones and joints. These congenital defects are identical to those produced by lupines, poison hemlock and members of the family Solanaceae such as wild and cultivated tobacco. Lambs born to ewes eating locoweed during the 60th to 90th day of pregnancy may develop enlarged hearts and thyroid glands. Using ultrasound technology, it is possible to detect the effects of swainsonine on the formation of cotyledons of the placenta, heart enlargement and irregularity in the heart beat in fetal lamb.[12] Initially the fetal heart rate is accelerated but then it slows and eventually stops, causing fetal death and abortion.

Locoweeds also affect reproduction by decreasing sperm production in rams and bulls.[13] Swainsonine affects the sperm producing cells of the testicles causing the formation of abnormal sperm with reduced motility, thereby severely affecting the reproductive capacity of the ram or bull eating locoweeds.[13,14] Testicular function is also affected by the action of swainsonine on the pituitary gland that alters normal gonadotrophin levels.[15] These changes are not permanent, and normal spermatogenesis returns after a period of 60-90 days once the animal has stopped eating milk vetch or locoweed.[13]

In those areas of North America where milk vetch and locoweeds are prevalent, poisoning from these plants should always be considered as a cause for reproductive failure and congenital abnormalities in livestock. It is however difficult to confirm a diagnosis of locoweed poisoning as the cause for a reproductive problem as the effects of the plants are likely to be encountered long after the plants were consumed.

Freshly aborted fetuses and the placenta when examined histologically may show the characteristic cellular vacuolation of swainsonine poisoning. The alkaloid itself can be detected in the serum of animals if they have been eating locoweed within 2 days of sampling because the serum half-life of swainsonine is only about 20 hours. Further discussion of the effects swainsonine, and the descriptions of the milk vetches and locoweeds themselves can be found in Chapter 6.

Lupine, Blue Bonnet
Lupinus spp.
Fabaceae (Legume family)

Habitat
Approximately 100 species of lupine occur in North America, ranging from the dry hills and plains to moist mountain valleys up to 11,000 feet (3,352 meters) altitude.

Description
Lupines are perennial herbaceous plants growing up to 3 feet (1 meter) in height, with alternate, palmately compound leaves, each with 5 to 17 leaflets. The leaflets are oblanceolate smooth or hairy, especially on the underside. The showy inflorescence is a terminal raceme of compact white, blue-purple, red or yellow pea-shaped flowers (Figures 8-1A and 8-1B). The fruit is a mul-tiseeded leguminous pod.

Principal Toxin
Most species of lupine are not toxic and are used extensively in some parts of the world as a high-protein food source for human and animal consumption. Those species that have been associated with poisoning contain quinolizidine and piperidine alkaloids in all parts of the plant. (Table 8-3) The principal teratogenic quinolizidine alkaloid in lupines is anagyrine.[16,17] Drying the plants in hay does not reduce their toxicity. The mechanism by which anagyrine and

Figure 8-1A Lupine (*Lupinus* spp.).

other alkaloids causes fetal deformity is poorly understood. Restricted fetal movement due to either general or localized uterine contraction as has been demonstrated with the coniine, the teratogenic alkaloid in poison hemlock (*Conium maculatum*), is most likely the reason the fetal calf develops the skeletal deformities seen in crooked calf disease.[16,18-20]

Table 8-3 Lupine Species Known to Contain Anagyrine[3]	
SCIENTIFIC NAME	COMMON NAME
Lupinus. argenteus	Silvery lupine
L. caudatus	Kellog's spurred lupine
L. erectus	Tall silvery lupine
L. evermanii	Evermans lupine
L. laxiflorus	Douglas spurred lupine
L. leucophyllus	Wooly leafed lupine
L. nootkatensis	Nootka lupine
L. serecius	Silky lupine

Figure 8-1B Lupine showing typical palmate leaf and flowers (*L. leucophylus*).

Crooked Calf Disease

Crooked calf disease is a well recognized syndrome in the western United States characterized by skeletal deformities in calves born to cows that consume 0.5 to 1.0 kg/day of toxic lupine species between days 40 and 70 of gestation.[16,20,21,24] Affected calves may show varying degrees of limb deformity (arthrogryposis), vertebral column malformation (scoliosis, kyphosis, torticollis), and cleft palate.[20,21] The front limbs are usually most severely affected with deformities occurring in the elbow, knee (carpus), and fetlock joints (Figure 8-2). Unlike commonly encountered contracted tendons that are similar in appearance, crooked calf disease involves permanent bone and joint structure changes referred to as arthrogryposis. Congenital limb deformities and cleft palate are also caused by inherited genetic disorders such as occur in the Charolais breed and by viruses such as the Akabane virus that is yet to become a problem in North America.[25]

Other plants capable of causing skeletal deformities in animals include poison hemlock (*Conium maculatum*) (see Chapter 1), tobacco stalks (*Nicotiana tabacum*), wild tree tobacco (*N. glauca*), and locoweeds (*Astragalus* and *Oxytropis* spp.). Mountain thermopsis (*Thermopsis montana*), a common wild flower of the Rocky Mountains, also contains anagyrine and is potentially teratogenic.[26] It has been suggested that there is a risk to pregnant women who drink the milk of cows consuming lupines and other teratogenic plants because the toxic alkaloids are secreted in milk.[27]

Nervous Syndrome

Lupines have been associated with three different syndromes of poisoning in livestock. In North America, lupine poisoning is most commonly associated with a terato-

Figure 8-2 Crooked legged calf (arthrogryposis) due to lupine teratogenicity.

genic syndrome that most frequently affects cattle, and is commonly referred to as "crooked calf disease."[20,21] Occasionally lupines have caused an acute fatal neurologic disease in sheep, and rarely in cattle and horses. The toxins responsible for the neurotoxicity are a variety of alkaloids other than anagyrine. Sheep ingesting from 0.25 to 0.5 percent of their body weight of seeds from certain lupines found in the western United States develop an acute neurologic disease characterized by muscle tremors, noisy labored breathing, convulsions, coma, and death. Species of lupine that have been incriminated in this neurologic syndrome include *L. leucophyllus*, *L. argenteus*, *L. leucopsis*, and *L. siriccus*[23]. A third syndrome, lupinosis, is associated with livestock grazing lupine pods and stalks infected with a fungus *Phomopsis leptostromiformis* (*Diaporthe toxica*) that produces a mycotoxin (phomopsin) capable of causing severe liver, kidney and muscle disease.[28-31] The fungus persists in the lupine stubble after harvesting and causes severe liver disease and poor growth in sheep and occasionally cattle that graze the stubble.[29-31] Lupinosis is especially important in those parts of the world where lupines are grown for the purpose of harvesting their seeds for animal consumption.[28]

Poison Hemlock, Spotted Hemlock
Conium maculatum

Poison hemlock is a widespread weed throughout most of North America, and has toxicologic significance because of its profound effects on the nervous system and its teratogenic properties. The alkaloid coniine, found in all parts of the plant, is a potent neurotoxin and a teratogen (see Chapter 1). Cattle and pigs seem to be the most susceptible to the teratogenic effects of coniine, while horses and sheep appear to be unaffected.[18-20,32] Sublethal doses of coniine ingested by cows between the 50th and 70th days of gestation cause congenital malformations in the calf (crooked calf disease) that are indistinguishable from the teratogenic effects of tobacco (*Nicotiana* spp.)[33] and lupine (*Lupinus* spp.) poisoning.[18] Calves with crooked legs due to deformity of the bones of the carpal and hock joints (arthrogryposis) may cause difficulty in calving, as they are unable to pass through the birth canal of the cow normally. In addition the vertebrae may show varying degrees of malformation and malalignment, and a cleft palate may be present.[34]

As with other congenital deformities, it is often difficult to determine the teratogen involved, because the deformity is not evident until the fetus is born many months after the plant teratogen was eaten by the dam. A presumptive diagnosis is usually possible by establishing the presence of poison hemlock in the pasture that the dam happened to be in during early pregnancy.

Plants Causing Abortion

Western False Hellebore, Skunk Cabbage, Corn Lily
Veratrum tenuipetalum (V. californiuom)
V. viride (Green false hellebore)
Liliaceae (Lily family)

Habitat

Western false hellebore is more common in moist mountain meadows and valleys above 8500 feet (2,590 meters), where it can form dense stands. Green false hellebore (*V. viride*) is more common at lower altitudes in moist meadows and forested areas.

Description

Both species are very similar, course, erect, 4 to 8 feet (2 to 3 meters) tall, with short perennial rootstalks. The leaves are smooth, alternate, parallel veined, broadly oval to lanceolate, up to 12 inches (3 cm) long, 6 inches (15 cm) wide, in three ranks and sheathed at the base (Figure 8-3A). The inflorescence is a panicle of very numerous, small, greenish white, star-shaped flowers, the lower ones often staminate and the upper ones perfect (Figure 8-3B). The flowers of *V. viride* are distinctly green. The fruit is three-chambered with several seeds.

Principal Toxin

Over 50 complex alkaloids have been identified from *Veratrum* spp., some of which have been used as hypotensive drugs.[35] The plant is most toxic when it first emerges in the early spring, becoming less toxic and more unpalatable as it matures. The roots are more toxic than the leaves.

Figure 8-3A False hellebore (*Veratrum viride*).

Figure 8-3B Western false hellebore or skunk cabbage (*V. tenuipetalum*).

Several of the alkaloids including cyclopamine, jervine, and cyclopasine are teratogenic[36] *Veratrum* is the classical teratogen, causing pregnant ewes that consume the plant on the 14th day of gestation to produce a lamb that has a single eye located in the center of its head (cyclopia).[5] If *Veratrum* is eaten later in gestation (30-35th. day) other defects including shortened legs and tracheal agenisis may develop.[37] Embryonic death in lambs without development of cyclopia may also occur.[37,38] *Veratrum* poisoning occurs mostly in sheep, but cattle, goats, and llamas are also susceptible to the teratogenic effects of cyclopamine.[5]

Clinical Signs

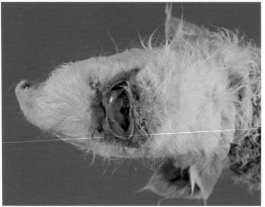

Figure 8-4 Cyclops lamb.

Veratrum tenuipetalum is primarily important because of its teratogenic effects in sheep, but if consumed in quantity over a short period of time can produce acute poisoning. Sheep eating 6 to 12 oz of the plant may within a few hours show signs of excessive salivation, vomiting, fast irregular heart rate, muscular tremors, incoordination, and coma in severe cases. There is no specific treatment, but the affected sheep should be kept quiet and given symptomatic treatment as needed until they recover.

Pregnant ewes consuming *Veratrum* spp. on the 14th day of gestation may develop characteristic cyclopia in up to 25 percent of the lambs born.[5,35,39] Affected lambs are born with varying degrees of facial deformity including cyclopia (single or double centrally located eye), protruding and twisted mandible, shortened upper jaw and proboscis- like structure located above the eye (Figure 8-4). The lambs have been referred to as "monkey faced" because of their appearance. Ewes eating *V. tenuipetalum* between 30 and 33 days of gestation produced lambs with different deformities that included cleft palate, hairlip, shortened legs, and tracheal stenosis.[37-39] Most deformed lambs are either born dead or die shortly after birth. Prolonged gestation periods are also associated with *V. tenuipetalum* poisoning.

The teratogenic and other effects of *Veratrum* poisoning can be avoided by keeping sheep off of pastures containing the plants, especially during the first trimester of pregnancy. If this is not possible, delaying the breeding season until after the first killing frost is a way of avoiding problems because the plants die off and lose their toxicity.

Tree Tobacco
Nicotiana glauca
Solanaceae (Nightshade family)

Other Poisonous Species of Tobacco

N. trigonophylla	desert tobacco
N. attenuata	wild tobacco
N. tabacum	cultivated tobacco

Habitat

Tree tobacco is a common weedy shrub of waste areas, roadsides, dry hillsides, washes and canyons at lower altitudes.

Description

Tree tobacco is a perennial shrub or small tree growing to 20 feet (5 to 6 meters) in height. Stems are loosely branched, with leaves that are entire, ovate, hairless, and bluish green. Flowers are pale yellow, tubular, 2 inches (4 to 5 cm) long, in loose pendant clusters at the ends of branches (see Figures 6-26A and 6-26B). Blooms are produced throughout the year.

Principal Toxin

Many alkaloids are present in the various species of Nicotiana. Two distinctly different clinical syndromes can occur with Nicotiana poisoning depending upon the predominant active alkaloid present in the plant. Nicotine and anabasine are the 2 most active alkaloids with respect to animal poisoning. Nicotine is poisonous to all animals, although ruminants are more tolerant of the alkaloid effects than are simple-stomached animals. Nicotine is readily absorbed through the digestive and respiratory tracts and has a rapid effect on the nervous system, frequently causing muscle tremors, excitement, ataxia, rapid heart and respiratory rates, and coma. Death results from respiratory paralysis.

The alkaloid, anabasine, is teratogenic to pigs, lambs, and calves.[40-42] Quantitatively, anabasine forms 99 percent of the alkaloid content of *N. glauca* and will cause severe skeletal deformities in lambs and calves born to their mothers that consumed the plant during the 30th to 60th day of gestation.[40,42]

Clinical Signs

Pregnant cows, ewes, and sows that consume the green plants of wild or domestic tobacco may occasionally show signs of nicotine poisoning, but it is the anabasine content of the plants that more commonly prevails to produce fetal deformities.[3,40] Anabasine, when consumed between days 30 and 60 of pregnancy, causes severe bony deformity (arthrogryposis) of the limbs and vertebrae.[3,40] Calves, lambs, and piglets are born with varying degrees of crooked legs due to malformation of the bones of the carpal, fetlock, and pastern joints.[3,40-42] Abnormal spinal curvature (scoliosis) and twisted necks (torticollis) may also occur. These same defects may also be produced by lupine (*Lupinus* spp.)[16,24] certain locoweeds (*Astragalus* and *Oxytropis* spp.), and poison hemlock (*Conium maculatum*).[18]

Broom Snakeweed, Turpentine Weed, Slinkweed
Gutierrezia sarothrae
G. microcephala Threadleaf snakeweed
Asteraceae (Sunflower family)

Habitat

Snakeweed is commonly found in the dry plains and hills at 4000 to 10,000 feet (1,219 to 3,048 meters) of altitude, where it can form dense stands over thousands of acres. The predominance of snakeweed in rangelands is generally indicative of over-grazing and poor range management.

Figure 8-5A Broom snakeweed invading range-land (*Gutierrezia sarothrae*).

Figure 8-5B Broom snakeweed detail of flower and leaves (*Gutierrezia sarothrae*).

Description

Snakeweed is a herbaceous perennial small shrub with a woody base, growing up to 24 inches (70 cm) in height (Figure 8-5A). The stems are branching, and the leaves are linear, alternate, and hairless. The yellow flowers are small, numerous, and in clusters at the ends of branches (Figure 8-5B). A given flower head will have no more than three to eight small ray flowers and three to eight disc flowers. The corollas have five lobes and the pappus is composed of several to many oblong scales. Snakeweed is considered an evergreen shrub in south western North America, and is deciduous in colder areas.

Threadleaf snakeweed (*G. microcephala*) is quite similar in appearance but has smaller flowers, each with 1 to 3 disk flowers and 4 to 5 ray flowers.

Principal Toxin

Several potentially toxic compounds are present in snakeweed including steroids, ter-penoids, saponins, and flavones.[43] Triterpene Saponins are thought to be the com-pounds responsible for the toxicity of broom snakeweed.[44-47] However, the toxicity of the plant appears to be quite variable, with animal susceptibility being dependent on the quantity of snakeweed consumed. Cattle on a good plane of nutrition can consume up to 30 percent snakeweed without apparent detrimental effect.[48,49] The green and to a lesser extent the dried plant may cause abortions in cattle, sheep, and goats at any stage of gestation.[44,45]

Clinical Signs

A variety of clinical signs have been attributed to snakeweed poisoning including poor appetite, weight loss, initial diarrhea followed by constipation, jaundice, and abortions.[50] Calves near term may be born weak and die shortly after birth. Retention of the placenta is a common sequel to abortion. Swelling of the vulva and udder edema has also been associated with snakeweed poisoning. Severe cases may have uremia and blood in the urine indicative of kidney degeneration. At postmortem examination liver, kidney, and gastrointestinal lesions may be present. Diffuse toxic hepatitis with hydropic degeneration of hepatocytes and acute tubular nephrosis characterize the pathologic changes of snakeweed poisoning.[51]

Western Yellow or Ponderosa Pine
Pinus ponderosa
Pinacea (Pine tree family)

Habitat

Ponderosa pines are common large trees of western North America preferring mountain slopes, valleys, and mesas, generally on south-facing slopes.

Description

Growing to 100 feet (30 meters) in height, ponderosa pines are broad and round-topped trees with thick, orange-brown bark that separates into scales or plates. Young trees have a dark brown or black bark. The dark green needles are 5 to 10 inches (12.5 to 25 cm) long in threes or twos and threes on the same tree and clustered near the end of branches. The cones are ovoid in shape, 3 to 6 inches (7.5 to 15 cm) long, each scale tipped with a recurved prickle (Figure 8-6).

Figure 8-6 Ponderosa pine needles and cone (*Pinus ponderosa*).

Principal Toxin

A variety of agents have been associated with the toxicity of pine needles including phytoestrogens, mycotoxins, resins, and lignols.[51-57] However, the principal compound in ponderosa pine needles responsible for causing premature parturition or abortion in cattle is isocupressic acid.[58] Junipers also contain significant levels of isocupressic acid. Other compounds present in the pine needles and bark of the tree including abietic and dehydroabietic acids (diterpene abietane acids) are toxic to cattle. Individually or in combination these compounds are possibly responsible for the kidney and brain degeneration described in cattle that have eaten pine needles.[60] Pregnant cattle and bison are the most likely species to abort from pine needles, while sheep, goats, and mares do not abort after eating pine needles.[61]

Other species of pine tree and junipers (*Juniperus* spp.) have been suspected of causing abortion, but in North America the ponderosa pine is most frequently incriminated.[62] Pine needles from the new branch tips and the bark are the most toxic.[63]

The means by which pine needles cause abortion or premature parturition is thought to be due to isocupressic acid and possibly other compounds causing a marked decrease uterine blood flow as a result of vasoconstriction.[47,64,65] This results in a change in the equilibrium of the hormones that maintain pregnancy and premature parturition results.[66] Specifically progesterone levels progressively decline as a result of necrosis of the corpus luteum.[67] The viability of the calf will depend on how close it was to term before premature delivery.

Pine Needle Abortion in Cattle

Cattle consuming pine needles in the last trimester of pregnancy will undergo premature parturition from 2 to 21 days later.[62-64] Cattle will eat pine needles when stressed or when normal forage is scarce such as during winter snowstorms. The quantity of pine needle necessary to produce premature parturition experimentally is in the range of 2.2 to 2.7 kg/day for at least 3 days.[63] Greater quantities may be readily consumed if cows in late pregnancy are without food during a winter storm. Affected cows develop edematous swelling of the vulva and udder and a mucoid vaginal discharge before premature parturition or abortion. Weak uterine contractions and uterine inertia may result in poor cervical and vaginal dilation that can result in difficulty in delivery of the calf.[60,67] Retention of the fetal membranes (placenta) and secondary uterine infections commonly occur in the cows that abort, which have serious consequences for the future fertility of the cow if it does not die from septic metritis. Some aborted fetuses may be autolyzed in utero several days before abortion. Calves that are born alive are often weak and their ability to survive is compromised by the fact that their mother often produces little or no colostrum or milk.

At postmortem examination of cows that have died from the complications of pine needle-induced premature parturition, septic necrosis of the uterus and placenta and necrosis of the corpora lutea are commonly present.[67,68] Histologically however, kidney and skeletal muscle degeneration and patchy neuronal degeneration in the brain also contribute to the syndrome of pine needle abortion.[60]

Phytoestrogens

Phytoestrogens are substances found in some plants that have estrogen-like properties and compete for estrogen receptors on cells. Consequently, phytoestrogens can mimic the effects of estrogen and cause infertility. Phytoestrogens (isoflavones, coumestans) are found in legumes such as alfalfa (*Medicago sativa*), burr medic (*Medicago* spp.), red clover (*Trifolium pratense*), and subterranean clover (*Trifolium subterran*).[69,70] Silage made from red clover has been incriminated in an estrogenic syndrome.[71] Similar estrogenic compounds are found in soybeans.[72]

Sheep in particular that graze subterranean clover pastures over a period of years develop lowered fertility that is referred to as "clover disease." A similar disease has been reported in cattle grazing burr medic that contained significant quantities of phytoestrogens.[73] This decreased flock fertility has been problematic in Australia, causing economic losses.[69,74,75] Phytoestrogens alter cellular structure and mucous consistency in the cervix, thus lowering chances of fertilization.[76] Affected sheep do not exhibit normal seasonal breeding cycles and develop cystic ovaries and irregular estrus cycles. Wethers exhibit teat enlargement when chronically exposed to phytoestrogens. Fertility returns once ewes are removed from clover pastures. Ram fertility does not appear to be affected adversely by grazing estrogenic pastures.[74] Selection of different clover species or new hybrids with low estrogen activity has reduced infertility problems associated with grazing clover pastures.

Nitrate Poisoning

Nitrate poisoning is primarily a problem of cattle and occasionally other ruminants resulting from the consumption of plants and water containing high levels of nitrates. Nitrate, is readily reduced by rumen organisms to nitrite, that once absorbed, converts hemoglobin of red blood cells to methemoglobin. As the level of methemo-globin increases, the oxygen carrying capacity of the blood decreases, with the animal eventually dying from lack of oxygen. Aside from the lethal effect of nitrate poisoning discussed in Chapter 1, pregnant cows may also abort from the effects of the nitrite. The fetus is particularly susceptible to decreases in oxygen crossing the placenta as a result of methemoglobinemia and the effects of nitrite on the maternal and fetal circulation.[77,78] Fetal death and abortion may occur at any stage of gestation as a result of the combined effects of decreased placental oxygen transport and the limited ability of the fetus to metabolize nitrite.[79-81] Abortions may also result from the decrease in progesterone production induced by chronic nitrate poisoning interfering with luteal production of progesterone.[82]

Plants or hay containing 1 percent or more nitrate (10,000 ppm) dry matter are potentially toxic and should be fed with caution. Forages containing more than 1 percent nitrate should only be fed if the total nitrate intake can be reduced to less than 1 percent by diluting the nitrate-forage with nitrate-free forages. Further description of plant nitrate poisoning and the plants commonly associated with nitrate poisoning are discussed in Chapter 1.

REFERENCES

1. Keeler RF: Known and suspected teratogenic hazards in range plants. *Clin Toxicol* 1972, 5:529–565.

2. Shupe JL, James LF: Teratogenic plants. *Vet Hum Toxicol* 1972, 25:415–421.

3. Keeler RF: Teratogens in plants. *J Anim Sci* 1984, 58:1029–1039.

4. Wilson JG: Current status of teratology–general principles and mechanisms derived from animal studies. *In:* Handbook of Teratology, vol. 1. Wilson JG, Fraser FC, Eds. New York: Plenum Press, 1977:48–52.

5. Binns WM, Keeler FR, Balls LD: Congenital deformities in lambs, calves and goats resulting from maternal ingestion of *Veratrum californicum*: hair lip, cleft palate, aplasia and hypoplasia of metacarpal and metatarsal bones. *Clin Toxicol* 1972, 5:245–261.

6. Molyneux RJ, James LF, Panter HE, Ralphs MH: The occurrence and detection of swainsonine in locoweeds. *In:* Swainsonine and Related Glycosidase Inhibitors. James LF, Elbein LD, Molyneux RJ, Warren CD, Eds. Ames, IA: Iowa State University Press, 1989:100–117.

7. James LF, Keeler RF, Binns W: Sequence in the abortive and teratogenic effects of locoweed fed to sheep. *Am J Vet Res* 1969, 30:377–383.

8. James LF: Effect of locoweed on fetal development: preliminary study in sheep. *Am J Vet Res* 1972, 33:835–840.

9. James LF, Shupe JL, Binns W, Keeler RF: Abortive and teratogenic effects of locoweed on cattle and sheep. *Am J Vet Res* 1967, 28:1379–1386.

10. McIlwraith CE, James LF: Limb deformities in foals associated with ingestion of locoweed by mares. *J Am Vet Med Assoc* 1982, 181:255–258.

11. Hartley WJ, James LF: Fetal and maternal lesions in pregnant ewes ingesting locoweed (*Astragalus lentiginosus*). *Am J Vet Res* 1975, 36:825–826.

12. Bunch TD, Panter KE, James LF: Ultrasound studies of the effects of certain poisonous plants on uterine function and fetal development in livestock. *J Anim Sci* 1992, 70:1639–1643.

13. Panter KE, Hartley WJ: Transient testicular degeneration in rams fed locoweed (*Astragalus lentiginosus*). *Vet Hum Toxicol* 1989, 30:318–323.

14. James LF, Van Kampen KR: The effects of locoweed intoxication on the genital tract of rams. *Am J Vet Res* 1971, 32:1253–1256.

15. Van Kampen KR: Sequential development of the lesions in locoweed poisoning. *Clin Toxicol* 1972, 5:575–580.

16. Keeler RF: Lupine alkaloids from teratogenic and nonteratogenic lupines. Crooked calf disease incidence with alkaloid distribution by gas chromatography. *Teratology* 1983, 7:23–30.

17. Davis AM: The occurrence of anagyrine in a collection of Western American lupines. *J Range Manage* 1982, 35:81–84.

18. Keeler RF: Teratogenic effects of *Conium maculatum* and Conium alkaloids and analogs. *Clin Toxicol* 1978, 12:49–64.

19. Keeler RF, Balls LD, Shupe JL, Crowe W: Teratogenicity and toxicity of coniine in cows, ewes and mares. *Cornell Vet* 1980, 70:19–26.

20. Keeler RF: Coniine, a teratogenic principle from *Conium maculatum* producing congenital malformations in calves. *Clin Toxicol* 1974, 7:195–206.

21. Shupe JL, James LF, Binns W: Observations on crooked calf disease. *J Am Vet Med*

Assoc 1967, 151:191–197.

22. Shupe JL, Binns W, James LF, Keeler RF: Lupine, a cause of crooked calf disease. *J Am Vet Med Assoc* 1967, 151:198–203.

23. Cheeke PR, Shull LR: Mycotoxins. *In* Natural Toxicants in Feeds and Poisonous Plants. Cheeke PR, Shull LR, Eds. Westport, CT: AVI Publishing; 1985:453–455.

24. Keeler RF: Alkaloid teratogens from *Lupinus, Conium, Veratrum* and related genera. *In:* Effects of Poisonous Plants on Livestock. Keeler RF, VanKampen KR, James L, Eds.. New York: Academic Press; 1978:397–408.

25. Van Huffel X, De Moore A: Congenital multiple arthrogryposis of the forelimbs in calves. *Compendium Food Animal* 1987, 9:333–339.

26. Chase RL, Keeler RF: Mountain thermopsis toxicity in cattle. *Utah Sci* 1983, 44:28–31.

27. Kilgore WW, Crosby DG, Craigmill AL, Poppen NK: Toxic plants as possible human teratogens: circumstantial evidence points to lupine toxin in goats milk as a cause of human birth defects. *Calif Agric* 1981, 35, 6.

28. Van Warmelo KT, Marasas WFO, Adelaar TF, *et al.:* Experimental evidence that lupinosis of sheep is a mycotoxicosis caused by the fungus *Phomopsis leptostromiformis* (kuhn) Bubak. *J South Afr Vet Med Assoc* 1970, 41:235–247.

29. Allen JG, Steele P, Masters HG, Lambe WJ: A lupinosis-associated myopathy in sheep and the effectiveness of treatments to prevent it. *Aust Vet J* 1992, 69:75–81.

30. Allen JG, Randal AG: The clinical biochemistry of experimentally produced lupinosis in the sheep. *Aust Vet J* 1993, 70:283–288.

31. Allen JG: An evaluation of lupinosis on cattle in western Australia. *Aust Vet J* 1981, 57:212–215.

32. Edmonds LD, Selby LA, Case AA: Poisoning and congenital malformations associated with consumption of poison hemlock by sows. *J Am Vet Med Assoc* 1972, 160:1319–1324.

33. Keeler RF, Shupe JL, Crowe MW, *et al.:* Nicotiana glauca-induced congenital deformities in calves, clinical and pathologic aspects. *Am J Vet Res* 1981, 42:1231–1234.

34. Panter KE, Keeler RF, Buck WB: Induction of cleft palate in newborn pigs by maternal ingestion of poison hemlock (*Conium maculatum*). *Am J Vet Res* 1985, 46:1368–1371.

35. Cheeke PR, Shull LR: Mycotoxins. *In* Natural Toxicants in Feeds and Poisonous Plants. Cheeke PR, Shull LR, Eds. Westport, CT: AVI Publishing; 1985:135–139.

36. Keeler RF, Binns W: Teratogenic compounds of *Veratrum californicum. Teratology* 1968, 1:5.

37. Keeler RF, Young S, Smart R: Congenital tracheal stenosis in lambs induced by maternal ingestion of *Veratrum californicum. Teratology* 1985, 31:83–88.

38. Keeler RF: Alkaloid teratogens from *Lupinus, Conium Veratrum* and related general. *In* Effects of Poisonous Plants on Livestock. Keeler RF, Van Kampen KR, James LF, Eds. New York: Academic Press; 1978:397–408.

39. Binns W, Shupe JL, Keeler RF, James LF: Chronologic evaluation of teratogenicity in sheep fed *Veratrum californicum. J Am Vet Med Assoc* 1965, 147:839–842.

40. Keeler RF, Crowe MW: Teratogenicity and toxicity of wild tree tobacco *Nicotiana glauca* in sheep. *Cornell Vet* 1984, 1:50–59.

41. Crowe MW, Swerczek TW: Congenital arthrogryposis in offspring of sows fed tobacco (*Nicotiana tabacum*). *Am J Vet Res* 1974, 35:1071–1073.

42. Crow MW: Tobacco—a cause of congenital arthrogryposis. *In* Effects of Poisonous

Plants on Livestock. Keeler RF, Van Kampen KR, James LF, Eds. New York: Academic Press; 1978:419-428.

43. Roitman JH, James LF, Panter KE: Constituents of broom snakeweed (*Gutierrezia sarothrae*); an abortifacient rangeland plant. *In* Plant Associated Toxins. Colegate SM, Dorling PR, Eds. CAB International 345-350, 1994.

44. Dollahite JW, Anthony WW: Poisoning of cattle with *Gutierrezia microcephala*, a perennial broomweed. *J Am Vet Med Assoc* 1957, 130:525–530.

45. Dollahite JW, Shaver T, Camp BJ: Injected saponins as abortifacients. *Am J Vet Res* 1962, 23:1261–1263.

46. Molyneux RJ, Stevens KL, James LF: Chemistry of toxic range plants, volatile constituents of broomweed (*Gutierrezia sarothrae*). *J Agric Food Chem* 1980, 28:1332–1333.

47. Gardner DR, James LF, Panter KE, *et al.:* Ponderosa pine and broom snakeweed: poisonous plants that affect livestock. *J Natural Toxins* 1999, 8:27–34.

48. Martinez JH, Ross TT, Becker KA, Smith GS: Ingested dry snakeweed did not impair reproduction in ewes and heifers during late gestation. *Proc West Sect Am Soc Anim Sci* 1993, 44:322–325.

49. Williams JL, Campos D, Ross TT, *et al.:* Snakeweed (*Gutierrezia* spp.) toxicosis in beef heifers. *Proc West Sect Am Soc Anim Sci* 1992, 43:67.

50. Kingsbury JM: *Poisonous Plants of the United States and Canada.* Englewood Cliffs, NJ: Prentice-Hall; 1964:406–408.

51. Jones TC, Hunt RD, King NW: *Veterinary Pathology.* Philadelphia Williams & Wilkins; 1997:728.

52. Call JW, James LF: Pine needle abortion in cattle. *In* Effects of Poisonous Plants on Livestock. Keeler RF, Van Kampen KR, James LF, Eds. New York: Academic Press; 1978:587–590.

53. McDonald MA: Pine needle abortion in range beef cattle. *J Range Manage* 1952, 5:150–155.

54. Anderson CK, Lozano EA: Embryonic effects of pine needles and pine needle extracts. *Cornell Vet* 1979, 69:169–175.

55. Chow FH, Hamar DW, Udall RW: Mycotoxic effect on fetal development: pine needle abortion in mice. *J Reprod Fert* 1974, 40:203–204.

56. Cogswell C, Kamstra LD: Toxic extracts in ponderosa pine needle that produce abortion in mice. *J Range Manage* 1980, 33:46–48.

57. Kubik YM, Jackson LL: Embryo resorptions in mice induced by diterpene resin acids of Pinus ponderosa needles. *Cornell Vet* 1981, 33:34–42.

58. Manners GD, Penn DD, Lurd L, James LF: Chemistry of toxic range plants. Water soluble lignols of ponderosa pine needles. *J Agric Food Chem* 1982, 30:401–404.

59. Gardner DR, Molyneux RJ, James LF, *et al.:* Ponderosa pine needle-induced abortion in beef cattle: identification of isocuppressic acid as the principle active compound. *J Agric Food Chem* 1994, 42:756–761.

60. Stegelmeier BL, Gardner DR, James LF, *et al.:* The toxic and abortifacient effects of ponderosa pine. *Vet Pathol* 1996, 33:22–28.

61. Short RE, *et al.:* Effects of feeding ponderosa pine needles during pregnancy: comparative studies with buffalo, cattle, goats, and sheep. *J Anim Sci* 1992, 70:3498–3504.

62. James LF, Short RE, Panter KE, *et al.:* Pine needle abortion in cattle: a review and report. *Cornell Vet* 1989, 79:39–52.

63. Panter KE, James LF, Molyneux RJ: Ponderosa pine needle-induced parturition in

cattle. *J Anim Sci* 1992, 70:1604–1608.

64. Christenson LK, Short RE, Rosazza JP, Ford SP: Specific effects of blood plasma from beef cows fed pine needles during late pregnancy on increasing tone of caruncular arteries in vitro. *J Anim Sci* 1992, 70:525–530.

65. Christenson LK, Short RE, Farley DB, Ford SP: Effects of ingestion of pine needles (*Pinus ponderosa*) by late-pregnant beef cows on potential sensitive Ca2+ channel activity of caruncular arteries. *J Reprod Ferti* 1993, 98:301–306.

66. Short Re, James LF, Staigmiller RB, Panter KE: Pine needle abortion in cattle: associated changes in serum cortisol, estradiol and progesterone. *Cornell Vet* 1989, 79:53–60.

67. Jensen R, Pier AC,Kaltenbach CC, Murdoch WJ, *et al.:* Evaluation of histopathologic and physiologic changes in cows having premature births after consuming ponderosa pine needles. *Am J Vet Res* 1989, 50:285–289.

68. Stuart LD, James LF, Panter KE, *et al.:* Pine needle abortion in cattle: pathological observations. *Cornell Vet* 1989, 79:61–69.

69. Cox RI: Plant estrogens affecting livestock in Australia. *In* Effects of Poisonous Plants on Livestock. Keeler RF, VanKampen KR, James LF, Eds. New York: Academic Press; 1978:451–464.

70. Adams NR: Detection of the effects of estrogens on sheep and cattle. *J Anim Sci* 1995, 73:1509–1515.

71. Kallela K: The estrogenic effect of silage fodder. *Nord Vet Med* 1980, 32:180–184.

72. Eldridge AC: Determination of isoflavones in soybean flours, protein concentrates and isolates. *J Agric Food Chem* 1982, 30:353.

73. Donaldson LE: Clover disease in two Mississippi cattle herds. *J Am Vet Med Assoc* 1983, 182:412–413.

74. Cheeke PR, Shull L: Isofavones and coumestans. *In* Natural Toxicants in Feeds, Forages, and Poisonous Plants, edn 2. Cheeke PR, Ed. Danville, IL: IPP Interstate Publishers; 1998:297–302.

75. Neil HG, Lightfoot RJ, Fels HE: Effect of legume species on ewe fertility in South Western Australia. *Aust Soc Anim Prod* 1974, 10:136.

76. Lightfoot RJ, Adams NR: Changes in cervical histology in ewes following prolonged grazing on oestrogenic subterranean clover. *J Comp Pathol* 1979, 89:367–373.

77. Van't Klooster AT, Malestein A, Kemp A, *et al.:* Nitrate intoxication in cattle. *Proceedings X11th World Congress on Diseases of Cattle*. The Netherlands, 1982:398–403.

78. Van't Klooster AT, Taverne MAM, Malestein A, Akkersdijk EM: On the pathogenesis of abortion in acute nitrite toxicosis of pregnant dairy heifers. *Theriogenology* 1990, 33:1075–1089.

79. Malestein A, Guerink JH, Schuyt G, *et al.:* Nitrate poisoning in cattle. 4. The effect of nitrite dosing during parturition on the oxygen capacity of maternal blood, and the oxygen supply of the unborn calf. *Vet Q* 1980, 2:149–159.

80. Hibbs C M, Stencel EL, Hill RM: Nitrate toxicosis in cattle. *Vet Hum Toxicol* 1978, 20:1–2.

81. Vermunt J, Visser R: Nitrate toxicity in cattle. *N Z Vet J* 1987, 35:136–137.

82. Jainudeen MR, Hansel W, Davison K: Nitrate toxicity in dairy heifers. 3. Endocrine response to nitrate ingestion during pregnancy. *J Dairy Sci* 1965, 48:217–221.

CHAPTER 9

Plants Affecting the Musculoskeletal System

Lameness due to musculoskeletal disorders is relatively common in animals that consume poisonous plants. Many toxic plants cause lameness through muscular weakness induced by the debilitating effects of the toxins on other organs. For example, livestock with liver disease caused by eating tansy ragwort (*Senecio jacobea*) eventually have a severe weight loss that results in muscle weakness and lameness. Similarly, plants affecting the brain and peripheral nervous system can affect muscle function and cause secondary lameness. Relatively few toxic plants primarily effect the musculoskeletal system, but a few such as day blooming jessamine exert their primary effect on the muscles and bones[1] In this chapter only those plants that are a primary cause of lameness are discussed.

Plant-Induced Calcinosis

A variety of plants contain calcinogenic glycosides that may be converted to a vitamin D-like substance in animals. Chronic exposure to these vitamin D analogs result in excessive amounts of calcium being absorbed from the intestinal tract and deposited in the tissues. Over time these calcium deposits in muscles cause chronic lameness and weight loss. Affected cattle and horses may survive for several years before they are unable to walk and become permanently recumbent.

Plants that have been incriminated as a cause of calcinosis in animals include *Solanum malacoxylon*, *S. sodomeum* (sodom apple), *S. linnaeanum*, *Trisetum flavescens* (golden oat grass), *Cestrum diurnum* (day-blooming jessamine), and *Nierembergia veitchii*.[2-5] Of these plants, only *C. diurnum* is known to cause calcinosis in horses in North America.

Day-Blooming Jessamine
Cestrum diurnum
Solanaceae (Nightshade family)

Habitat

Introduced from the West Indies, *C. diurnum* has become widely distributed through Florida, Texas, California, and Hawaii.

Description

This plant is a shrub or small tree up to 15 feet (4 to 5 meters) high, with alternate elliptic leaves that have a dark green glossy upper surface. The fragrant white tubular flowers are born in small clusters on axillary peduncles. Multiple green berries that turn black when ripe are produced after flowering (Figure 9-1A).

Principal Toxin

Cestrum diurnum contains a toxin that has strong similarities to 1,25-dihydroxycholecalciferol, the active metabolite of vitamin D.[6,7] Consequently, all animals eating cestrum while on a diet adequate in calcium and phosphorus absorb excessive

Figure 9-1A Day-blooming jessamine (*Cestrum diurnum*) (Courtesy Dr. Julia F. Morton, Miami, Florida.).

amounts of calcium. Prolonged consumption of the plant results in the calcification of the elastic tissues of the arteries, tendons, and ligaments (calcinosis).[6,8,9] The calcium deposited in the muscles, tendons and ligaments may become evident within 2 weeks of the animal starting to eat the plant. Generalized increased density of the bones (osteopetrosis) may also be related to hypoparathyroidism and hypercalcitoninism induced by the plant toxin.[9]

Other members of the genus, *Cestrum nocturnum* (night-blooming jessamine) (Figure 9-1B), *C. aurantiacum* (Figure 9-1C), and *C. parqui* (green cestrum, willow-leafed jessamine) cause toxicity in livestock through the action of atropine-like alkaloids that are common in the family Solanaceae (see Chapter 3). These species of *Cestrum* have not been associated with calcinosis.

Clinical Signs

Chronic weight loss despite normal appetite, and generalized stiffness leading to severe lameness and prolonged periods of recumbency are characteristic. Lameness arises from pain in the ligaments and tendons where the calcium is deposited.[8,9] Heart murmurs may develop due to the calcification of the heart valves. Plasma calcium levels in animals with cestrum poisoning are consistently elevated in the range of 11 to 16 mg/dL.[8] Other blood parameters, including phosphorus levels, are generally normal. Radiographically, a marked increase in bone density (osteopetrosis) is apparent,

Figure 9-1B Night-blooming jessamine (*C. nocturnum*). (Courtesy of Dr. Gerald D. Carr. Botany Department, University of Hawaii).

with increased calcification of cartilage and increased metaphyseal and epiphyseal trabeculae.[9]

Severe calcification of the tendons, ligaments, and elastic arteries is often visible on postmortem examination. Mineralization of tissues is confined to those containing elastic tissue.[8,9] Chronic cases may show severe calcinosis of the aorta, pulmonary arteries, heart valves, and endocardium.[8]

Recovery from *C. diurnum* poisoning is rarely reported because animals are usually affected chronically. Recovery is likely in less severely affected animals if they are denied further access to the plant and are given a balanced ration. Care should always be taken to ensure horses are not placed in pastures or pens surrounded by or containing day-blooming jessamine.

Figure 9-1C Cestrum (*C. aurantiacum*)

Plants Causing Muscle Degeneration

A variety of plant toxins are capable of causing muscle degeneration (myodegeneration) in addition to their other effects on different organ systems.[1] The golden chain tree (*Laburnum anagyroides*), scotch broom (*Cytisus scoparius*), and coyotillo (*Karwinskia humboldtiana*) are plants containing toxins that affect the nervous system and cause myodegeneration. Lupinosis is a degenerative disease of muscle and liver of sheep that graze the stubble of cultivated lupines containing fungal toxins (phomopsins) produced by the fungus *Phomopsis leptostromiformis* (*Diaporthe toxica*).[10,11] Many of the toxic plants discussed in Chapter 6 that affect the nervous system also have secondary effects resulting in lameness.

Relatively few plants contain toxins that only cause degeneration of the musculature. A few plant toxins including gossypol found in cottonseed and tremetol in white snakeroot (*Eupatorium rugosum*) cause degeneration of the heart muscle. In North America, mountain thermopsis (*Thermopsis montana*) and coffee weed or sickle pod senna (*Cassia* spp.) are examples of plants capable of causing primary muscle degeneration.

Senna, Coffee Weed, Coffee Senna
Cassia occidentalis (Senna occidentalis)

Sickle Pod, Coffee Weed, Coffee Bean
Cassia obtusifolia (C. tora) (Senna obtusifolia)
Fabaceae (Legume family)

Less common species:

C. roemeriana	Twin-leaf senna
C. fasciculata	Showy partridge pea
C. lindheimeriana	Lindheimer senna
C. nictitans	Wild sensitive plant

Habitat

The *Cassia* spp. are common and troublesome weeds of the southeastern United States, Hawaii, Mexico, and most of the tropical world. As annuals they are opportunists, growing in waste areas, roadsides, fence lines, and are especially common as weeds of corn and soybean fields.

Description

The *Cassia* spp. are woody, erect, lightly branched annual and occasionally perennial shrubs, 6 to 8 feet (2 to 3 meters) tall. The leaves are alternate, pinnate, consisting of four to five pairs of leaflets widely spaced along a common stalk. Each leaflet is rounded at the base and pointed at the other end. A raised gland is present near the upper side of the base of each petiole. The flowers are yellow and produced in loose clusters in the leaf axils. Thick, dark brown, slightly flattened and curved seed pods with paler longitudinal stripes along the edges contain dark brown seeds.

As the name implies, sickle pod cassia has much longer curved pods, up to 8 inches (20 cm) in length, that are distinctly sickle-shaped and contain many seeds (Figure 9-2).

Principal Toxins

Figure 9-2 Sickle pod senna (*Cassia obtusifolia*).

Several compounds that bind strongly to cell membranes occur in *Cassia* spp., but the specific toxin responsible for muscle degeneration has not been identified.[12-14] The toxin induces acute muscle and liver degeneration that can be rapidly fatal in most animals.[15-20] The greatest concentration of the toxin appears to be in the seeds. *C. occidentalis* and *C. obtusifolia* are considered to be more toxic than other species, but all *Cassia* spp. should be considered toxic unless proven otherwise.

Cattle, sheep, goats, horses, pigs, rabbits, and chickens are susceptible to poisoning by *Cassia* spp.[13,19,20-26] All parts of the plant are toxic, although most poisoning occurs when animals eat the pods and beans, or are fed green-chop containing cassia.[27,28] Ground beans of coffee senna fed to cattle at the rate of 0.5 percent of their body weight induces severe muscle degeneration.[15] Roasting of the beans partially reduces their toxicity such that goats fed 2.5 g/kg body weight of roasted beans were unaffected, whereas unroasted beans at this dosage were fatal.[23,24,27] Cassia poisoning in cattle may occur when 0.4 to12.0 percent of body weight of the green plant is eaten.[15,17,28] At lower doses, *Cassia* spp. can cause diarrhea and decreased weight gain.[27] The plant is not very palatable and tends to reduce feed intake. As the amount of *Cassia* in the animal's diet increases muscle degeneration becomes a predominant characteristic of the poisoning and cause of the clinical signs. Experimentally, high doses of the plant (10 g/kg body weight daily for 3 days) induce acute liver degeneration and death before myodegeneration has time to develop.[13]

Clinical Signs

In cattle a moderate to severe diarrhea develops shortly after consumption of the plant.[30] Abdominal pain, straining (tenesmus), and diarrhea are thought to be due to the irritant effects of anthraquinones in *Cassia* spp. Affected animals remain afebrile. Depending on the amount of plant or seeds consumed, muscle degeneration begins after several days, causing weakness and recumbency.[20,21,31,32] The urine may be coffee colored due myoglobinuria from acute muscle degeneration.[13,14,30] The levels of serum enzymes creatine kinase and aspartate transaminase are usually markedly elevated, reflecting acute muscle degeneration. Renal failure may develop secondarily to the myoglobinuria. In severe cases hepatic failure may be the predominant organ failure leading to death of the animal.[14] In more chronic cases, cardiomyopathy and hyperkalemia from muscle degeneration cause cardiac irregularities and contribute to the death of the animal.[15,16,27] Respiratory difficulty develops as a result of the degeneration of the intercostal and diaphragm muscles.[22]

Horses may not exhibit the digestive and muscle degenerative signs of poisoning seen in cattle. Myoglobinuria may not develop in horses because they apparently succumb to liver degeneration sooner than to the degeneration of the musculature.[19] Poisoned horses are generally afebrile and severely ataxic and may die without showing other clinical signs. Serum liver enzymes may be elevated reflecting acute liver degeneration.

Gross lesions at postmortem examination consist primarily of pale skeletal muscles similar to those seen in white muscle disease associated with selenium and vitamin E deficiency. Skeletal muscle necrosis and renal tubular and hepatic centralobular necrosis are characteristic histologic findings that differentiate cassia poisoning from vitamin E and selenium deficiency.[15,16,20] Confirmation of the diagnosis should be based on access to and consumption of *Cassia* spp., along with the presence of degenerative lesions in the muscles, heart, and liver.[16,18,27]

Treatment

There is no specific treatment for cassia poisoning. Affected animals should be removed from the source of the plants as quickly as possible and fed a nutritious diet. Supportive care of the recumbent animal will help prevent further muscle degeneration due to pressure necrosis. Where warranted, intravenous fluid may be helpful in maintaining renal function if myoglobinuria is present. Recovery depends on the severity of muscle and liver degeneration that has resulted. Rarely does an animal recover once it has become recumbent.

It is important to differentiate white muscle disease due to selenium and vitamin E deficiency from cassia poisoning because the use of selenium and vitamin E in cassia poisoning is contraindicated. Increased myodegeneration and higher mortality occur when selenium and vitamin E are used to treat cassia poisoning.[15]

**Golden Banner
Mountain Thermopsis
False Lupine, Yellow Pea**
Thermopsis montana
T. rhombifolia
Fabaceae (Legume family)

Habitat

Golden banner is a common wildflower of the Rocky Mountains (*Thermopsis montana*) and the prairies (*T. rhombifolia*) from North Dakota to Oregon and Washington, and south to Nevada and Colorado.

Description

These plants are perennials, arising from a rhizomatous root system, with erect, branching stems that reach a height of 12 to 18 inches (30 to 46 cm). Leaves are alternate, with three leaflets, unlike lupine that have five or more. Flowers are bright yellow, pealike, and produced in dense racemes from the leaf axils (Figure 9-3A). The seed pods are densely haired, erect, and straight (*T. montana*) or curved (*T. rhombifolia*).

Principal Toxin

A variety of quinolizidine alkaloids including thermopsine, cytisine, N-methylcytisine, and anagyrine have been isolated from all parts of the plant. The

Figure 9-3A Golden banner, false lupine (*Thermopsis montana*).

specific toxin responsible for the myopathy has not been identified.[33-35] The quinolizidine alkaloid fraction, however, induced the same lesions as did the entire plant when fed to cattle.[34] Experimentally, a dose of 1 g/kg body weight of dried plant given orally, once daily for 2 to 4 days, consistently induced muscle degeneration.[3]

Clinical Signs

Cattle poisoned by *T. montana* initially show reluctance to move, walk with a stiff gait, and show muscle tremors when forced to move.[33-37] Severe depression, anorexia, arched back, and swollen eyelids have also been observed in affected animals. Depending on the quantity of plant consumed, animals become recumbent and eventually die. Severely affected animals may die acutely from respiratory arrest.

Levels of serum creatine kinase and aspartate transaminase are markedly elevated, reflecting acute muscle degeneration. Myoglobinuria has not been associated with poisoning caused by *Thermopsis* spp.

At postmortem examination, there are usually few gross signs. Microscopic muscle degeneration of the skeletal muscles is a characteristic finding, although there is usually no myocardial degeneration.[34, 35]

Figure 9-3B Scotch broom (*Cytisus scoparia*). Inset showing detail of the flowers.

Scotch broom (*Cytisus scoparius*), a bushy shrub that was introduced from Europe, has occasionally been suspected of animal poisoning, especially in the horse. Having escaped from cultivation, scotch broom has become a troublesome weed along the west coast of North America from British Columbia to southern California. It grows 6 to 8 feet (2 to 2.5 meters) in height and has evergreen or deciduous palmate three-lobed leaves with showy yellow pealike flowers (Figure 9-3B). Scotch broom contains several quinolizidine alkaloids including sparteine, cytisine, and isosparteine with the potential for similar toxic effects caused by golden banner (*Thermopsis montana*) and the golden chain tree (*Laburnum* anagyroides) (Figure 9-3C).

Figure 9-3C Golden chain tree (*Laburnum anagyroides*).

Black Walnut
Juglans nigra
Juglandaceae (Walnut family)

Habitat

About 15 species of walnuts are widely distributed throughout the world; 6 species are native to North America. Most are deciduous trees growing to 60 feet in height. The black walnut is most commonly found in cultivation, where it is extensively used for its wood, aromatic oils, and edible nuts.

Description

Black walnuts are large trees with rough dark brown bark, with pinnate leaves to 50 cm long and 11 to 23 leaflets (Figure 9-4). Male and female flowers are produced separately; the male flowers have 12-cm catkins, and the female flowers are 0.5 to 1 inch (1 to 2 cm) long with yellow-green stigmas. The fruits are ovoid, single hard-shelled nuts containing the edible fruit.

Principal Toxin

The toxin responsible for black walnut toxicosis in horses is not known. Juglone (5-hydroxy-1,4-naphthoquinone), present in the roots, bark, nuts, and pollen of the walnut tree, is possibly involved with poisoning in horses. Juglone is found in other members of the walnut family including English walnuts, butternuts, hickories, and pecans. Walnut trees are allelopathic, meaning they secrete chemical substances through their roots to inhibit the growth of other plants in the vicinity. Consequently, many plants will not grow under walnut trees.

Horses become poisoned if they are exposed to the wood shavings of black walnuts that are used for bedding.[38,39] Bedding containing as little as 20 percent of black walnut shavings can cause the development of laminitis in horses.[40] It is not necessary for horses to eat walnut shavings to develop laminitis. Pollen and the leaves in the autumn are also toxic to horses.[41]

The variability of laminitis, edema of the lower legs, colic, and other systemic signs associated with black walnut shavings is poorly understood. The clinical signs are not simply related to contact of the horse's skin with the walnut shavings. Purified juglone applied topically to the feet of horses causes mild dermatitis but does not cause laminitis.[39] However, horses experimentally treated with aqueous extracts of black walnut via nasogastric tube consistently develop acute laminitis, indicating that toxicity is due in part to the ingestion or inhalation of a toxic substance present in black walnut.[42-46] It has also been postulated that juglone or other substances act as haptens to induce toxicity.[39,41] Experimental evidence indicates that the toxin in black walnuts does not directly cause contraction of the digital vessels responsible for the laminitis, but it appears to enhance vasoconstriction of the blood vessels in the presence of catecholamines and corticosteroids.[44,45,47]

Fallen walnuts that have become moldy may contain the mycotoxin penitrem A, which is a neurotoxin capable of poisoning dogs and other animals.[48]

Clinical Signs

Naturally occurring black walnut toxicosis in horses is characterized by depression, edema of the lower legs, lameness, colic, and respiratory distress.[38,40] The severity of lameness depends on the duration and severity of laminitis. If affected horses are

Figure 9-4 Black walnut leaves and fruits (*Juglans nigra*).

removed from the source of the black walnut shavings in the early stages of laminitis and are treated for the laminitis, they recover without the severe consequences of hoof deformity and third phalanx rotation attributable to laminitis.

Until the toxin and conditions causing black walnut toxicosis are better understood, it is a wise precaution to avoid bedding horses with wood shavings containing shavings from walnuts. Similarly black walnut trees should not be voluntarily planted in horse pastures. Fallen, moldy walnuts should also be removed to prevent animals gaining access to them.

Wood shavings from bitterwood trees (*Quassia simarouba*), a tree indigenous to Central and South America, also contain irritant compounds that can cause blister-like lesions on the lips, nose, and around the eyes of horses that are bedded on the shavings. Horses that eat the shavings may also develop lesions around the anus.

Hoary Alyssum
Berteroa incana
Brassicaceae (Mustard family)

Habitat

Introduced from Europe, hoary alyssum has become established as a common weed from Nova Scotia to Washington, the northern midwestern states, and west to northern California. It is an invasive weed of disturbed soils, waste areas, and roadsides and can become troublesome in alfalfa fields.

Description

Hoary alyssum is an erect, branching, annual, growing to 3 feet (1 meter) in height. The plant is densely haired, giving it a grayish green appearance. The leaves are alternate, narrow, and lanceolate and have smooth edges. The flowers, which are produced at the ends of the branches, are white, with four deeply divided petals. The round, slightly flattened seed pods with a central septum, contain up to six brown seeds.

Principal Toxin

No specific toxin has been identified in hoary alyssum. Both green and dried plants are toxic to horses. Hay containing hoary alyssum may remain toxic for up to 9 months.[49] The quantity of plant that has to be ingested to cause poisoning has not been determined. In reported cases of hoary alyssum poisoning, the hay being fed contained up to 90 percent of the plant depending on the bale of hay.[49] The fact that the plant can contaminate alfalfa hay makes hoary alyssum poisoning of horses possible in any state to which the hay is transported.

Clinical Signs

Lameness due to laminitis and limb edema are associated with the consumption of hoary alyssum.[47,48] Depending on the quantity of plant consumed, horses may show signs ranging from stiffness, limb swelling, fever, diarrhea, laminitis, intravascular hemolysis, severe hypovolemic shock, and death secondary to endotoxemia.[51] Pregnant mares may undergo abortion or premature parturition.[48] Horses recover if they receive no further hoary alyssum and are treated symptomatically for laminitis and shock.

Flatweed, Cat's Ears
Hypochaeris radicata
Asteraceae (Sunflower family)

Habitat
Flatweed is a perennial native weed in Europe and now prevalent in many areas of Australia and North America, where it has become established in disturbed soils and overgrazed pastures.

Description
Resembling the common dandelion (*Taraxacum officinale*), flatweed has multiple, basally clustered, irregularly lobed, 3 to 12 inch (7.5 to 30 cm), hairy leaves. Multiple branched flower stalks up to 2 feet (0.5 meter) in height, each bearing a single yellow dandelion-like flower, are produced each season (Figure 9-5). Seeds are long-beaked, roughened, and tipped by a circle of bristles.

Figure 9-5 Flatweed, cat's ears (*Hypochaeris radicata*).

Principal Toxin
No specific toxin has been identified in flatweed. Horses preferentially grazing flatweed have been reported to develop a unique lameness syndrome described in Australia as Australian stringhalt.[52,53] A similar syndrome has been reported in horses grazing flatweed in California.[54] Attempts at experimentally reproducing the disease by feeding flatweed to horses have been unsuccessful, suggesting that other factors may be involved in the pathogenesis of the disease.[52,53]

The hypermetria and hyperflexion of the hind legs associated with hoary alyssum is similar to stringhalt, a well-recognized lameness of horses.[55]

Clinical Signs
Horses grazing flatweed over a period of several weeks may develop lameness characterized by high stepping and hyperflexion of the hind legs similar to that described for stringhalt.[54,55] Affected horses have difficult in stepping backward; others have such severe hyperflexion of the hind limbs when walking that the abdomen is kicked. Left laryngeal hemiplegia (roaring) associated with the Australian stringhalt syndrome has not been encountered in horses with this syndrome in North America.[52] Unless severely and chronically affected, horses tend to recover over a period of months once they are prevented from eating further flatweed.

Phytogenic Selenium Poisoning

Historically, selenium poisoning in animals has been associated with two disease syndromes referred to as alkali disease and blind staggers.[56,57] Both diseases were assumed to be associated with the chronic ingestion of forage and crop plants that had accumulated toxic levels of selenium from the soil in which they were growing. Records from 1860 indicate that cavalry horses in South Dakota suffered from chronic weight loss, lameness due to hoof deformity, and hair loss that was referred to as alkali disease.[58] In the 1930s in South Dakota and Wyoming, selenium poisoning (selenosis) was linked to animals grazing plants containing high levels of selenium. Since then, many reviews have stated the importance of chronic selenium poisoning in livestock raised on western rangelands of the United States.[59-63] However, the prevalence of confirmed selenosis in livestock is relatively low, and the economic significance of chronic selenium poisoning (alkali disease) is difficult to determine.[64]

Blind staggers is a disease of cattle and sheep characterized by aimless wandering, walking in circles, disregard for objects in their paths, loss of appetite, and blindness.[57] However, as discussed later, there is good evidence that blind staggers is not related to selenium poisoning, and is more likely associated with sulfate toxicity.

Selenium-rich soils are generally alkaline and exist in areas with low rainfall where there is minimal leaching of selenium from the soil. In North America, selenium is most abundant in the cretaceous shales and glacial deposits of the great plains.[65] (Figure 9-6). Plant uptake of selenium is variable and depends on the chemical form of selenium, soil pH, temperature, moisture, and the species and stage of plant growth.[66] Selenium in its inorganic form as selenide or selenite is minimally absorbed by plants,[67,68] whereas selenates, which occur in alkaline soils, are readily available to plants.[68] Total soil selenium content is, therefore, not a reliable predictor of selenium uptake by plants.

Some plants require selenium for normal growth and are capable of accumulating 10 times the amount of selenium that is present in the soil. Referred to as obligate accumulators, these plants may contain up to 10,000 ppm dry matter of selenium[69] (Table 9-1). Obligate accumulator plants with 25 ppm or more of selenium may cause acute poisoning but are generally distasteful and not eaten by animals unless

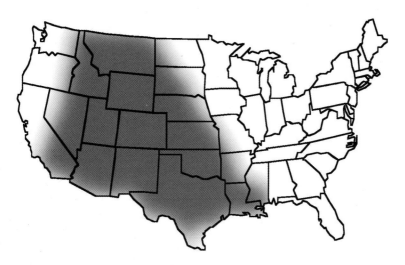

Figure 9-6 Seleniferous soils of North America.

Table 9-1
Plants That Accumulate Selenium

Obligate Selenium Accumulator Plants

SCIENTIFIC NAME	COMMON NAME
Astragalus (24 spp.)	Milk vetches
Conopsis spp.	Golden weeds
Xylorhiza spp.	Woody aster
Stanleya pinnata	Princes plume

Secondary Selenium Accumulators

SCIENTIFIC NAME	COMMON NAME
Acacia spp.	Acacia
Artemisia spp.	Sages
Aster spp.	Asters
Atriplex spp.	Saltbrush
Castilleja spp.	Paintbrush
Penstemon spp.	Beard tongue
Grindelia spp.	Gumweed

they are especially hungry. Recognition of selenium-accumulating plants or indicator plants, therefore, provides a means of identifying seleniferous soils and the potential for selenium poisoning of livestock kept in these areas. Other plants known as secondary or facultative accumulators will bind selenium in its organic forms if it is present in the soil but do not require selenium for growth. Crop plants and pasture grasses growing in seleniferous soils will accumulate toxic quantities of selenium. Phytogenic selenium poisoning is likely to occur after plants containing 5 to 40 ppm selenium have been consumed for several months.[69] High selenium content of water will compound a forage selenium toxicity problem by adding to the animal's total selenium intake.

Toxicity

Selenium has numerous complex effects on cellular function, many of which are poorly understood. It is well known that selenium inhibits cellular enzyme oxidation reduction reactions, especially those involving sulfur-containing amino acids.[70,71] This effect of selenium on sulfur alters the metabolism of sulfur-containing amino acids (methionine, cystine) thereby affecting cell division and growth. This causes degeneration and necrosis of the cells that form keratin (keratinocytes).[72,73] By replacing sulfur in the keratin molecule, the primary constituent of the hooves and hair, selenium weakens the keratin structurally at the site of selenium incorporation into its structure. Consequently the hair and hoof wall tend to fracture at this site when subjected to mechanical stresses.

Selenium poisoning in animals is variable and depends on the amount and rate of absorption of selenium from the intestinal tract. Horses appear to be more susceptible to chronic selenosis than are cattle and sheep.[74] Some animals such as pronghorn antelopes appear capable of consuming diets high in selenium (15 ppm) for long periods without ill effect.[75] Individual animal susceptibility, the chemical form of selenium present, and the bioavailability of selenium as a result of the interaction with other elements such as sulfur or arsenic present in the diet are also important in the pathogenesis of selenium poisoning.[57] Elevated levels of selenium in animals are also immunotoxic, possibly through the peroxidative damage of free radicals on lymphocytes.[76, 77]

Chronic Selenosis (Alkali Disease)

Chronic selenosis is a disease of horses, cattle, pigs, sheep, and poultry that consume forages or cereal crops grown in seleniferous soils and that have accumulated toxic levels of selenium. Plants or diets with 5 to 50 ppm selenium are most likely to cause chronic selenosis.[78-83] Animals that consume greater than 50 ppm, or are inadvertently injected with an overdose of selenium, develop acute selenium poisoning. The signs of acute selenium poisoning are quite different from those of chronic selenosis, and are characterized by sudden death from cardiac insufficiency, and pulmonary congestion and edema.[77,84] Congestion, edema, and necrosis of the lungs, liver, and kidneys are the major lesions seen at postmortem examination in acute selenium poisoning.

Clinical Signs

Although early reports of chronic selenium poisoning described a wide variety of clinical signs including lameness, hair loss, blindness, aimless wandering, and head pressing, the most distinctive lesions are those involving the keratin of the hoof and hair.[59, 72, 85] The long hairs of the tail and mane tend to break off at the same place giving the animal a "bob" tail and "roached" mane, respectively (Figure 9-7A). Lameness is due to abnormal hoof wall growth in all feet, which results in rapid uneven growth, circular ridges and subsequent cracking of the hoof wall (Figure 9-7B). Some horses may slough the hoof wall entirely. Cattle will show similar defective hoof growth but rarely loose the hoof wall. Sheep do not seem to develop as severe lesions as cattle and horses but show marked reduction in fertility when grazing on selenium-rich pastures.[84] Chronic selenium poisoning has also been associated with reduced reproductive performance, anemia, liver cirrhosis, heart atrophy, and degeneration of bones and joints in horses and cattle.[57,67]

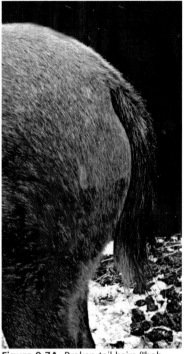

Figure 9-7A Broken tail hairs ("bob-tail") in a horse with chronic selenium poisoning.

Diagnosis

A diagnosis of selenium poisoning is best confirmed by submitting samples of hay, forages, water, serum, and liver for analysis. Western wheat grass accumulates selenium more readily than other common grasses and is therefore useful to submit for selenium analysis.[87] Forage selenium levels greater than 5 ppm should be considered potentially toxic.[88] Blood levels of 1 to 4 ppm are indicative of chronic selenium poisoning; serum levels up to 25 ppm have been reported in acute poisoning.[57, 89] Liver and kidney levels greater than 4 ppm are indicative of selenium toxicosis.[74,83,90] High tissue levels of selenium may take 6 to 12 months to return to normal after the animal has been removed from the source of the selenium.[57]

In chronic selenosis, levels of selenium in the tissues may be low, but hair and hoof samples retain high concentrations. Hair and hoof wall samples collected at the site where the hair is broken and where the hoof wall is cracked are useful in determining historical levels of selenium. Hoofwall containing 8 to 20 ppm of selenium is indicative of chronic selenosis.[90, 91] Similarly, hair samples containing in excess of 5 to 10 ppm selenium indicate excessive selenium levels capable of causing toxicity.[57]

Figure 9-7B Horizontal or circular hoof wall cracks resulting from chronic selenium poisoning.

Treatment

Successful treatment of selenium poisoning depends on early recognition of signs and the removal of livestock from the source of the excess selenium. Recovery from chronic selenium poisoning will occur in time if the animal is fed a diet low in selenium and high in sulfur-containing amino acids. Feeding a high-protein diet with adequate copper levels counteracts the effects of selenium on sulfur-containing amino acids. Feeding good-quality alfalfa hay, provided it is low in selenium, is helpful in providing adequate sulfur to counteract selenium in the diet. Alfalfa typically has low selenium levels (0.15-0.69 ppm), but can accumulate toxic levels (22 ppm) when growing in selenium-rich soils.[83] Adequate levels of copper in the diet has a protective effect against selenium poisoning.[92] Attention should be given to careful and regular trimming the deformed, overgrown hooves to avoid permanent lameness.

Selenium Deficiency

Deficiencies of selenium in livestock usually coincide with geographic areas that have high annual rainfall, which tend to leach selenium from acidic top soils[92] (Figure 9-7). In deficient soils, plants may contain less than 0.05 mg/kg of selenium, making it necessary to supplement livestock rations with selenium to prevent deficiency symptoms. Diets containing 0.1 mg/kg (0.1 ppm) of selenium are generally considered adequate for normal growth. Selenium is an essential trace mineral required by all animals for normal growth. It is an important antioxidant, preventing intracellular oxidation, cellular degeneration, and cell membrane destruction. A deficiency of selenium in the diet of animals results in muscle degeneration, a disease of livestock known as muscular dystrophy or white muscle disease.[92] As the name implies, the muscles of selenium-deficient animals undergo degeneration and characteristically develop pale or white areas, especially in the muscles of the limbs and heart.

Blind Staggers

Blind staggers in cattle and sheep, characterized by aimless wandering, circling, disregard for objects in their paths, loss of appetite, and blindness, was reported as a form of selenium poisoning.[57] The disease is characterized by front leg weakness, staggering gait, and eventual inability to stand. Weight loss accompanies poor appetite. Teeth grinding, an indication of abdominal pain, is common. Affected animals are often blind.

Recent review of the original micrographs used to document the cause of blind staggers indicates that the tissues were compromised by autolysis that led to misinterpretation.[93] The clinical signs and tissue findings more appropriately resemble those of sulfate toxicity.[93,94] Excessive consumption of sulfate (more than 2 percent of total diet) results in toxic levels of sulfide being absorbed from the rumen, which leads to polioencephalomalacia with clinical signs typical of blind staggers. This is substantiated by recent studies on sulfate toxicity in which cattle develop clinical signs of brain disease characterized by severe depression, uncoordination, and blindness.[94-96] Contributing to the history of the syndrome of blind staggers is the association of animals that eat two-grooved milk vetch (*Astragalus bisulcatus*), a known selenium accumulator. Sheep fed two-grooved milk vetch developed classical signs of blind staggers, suggesting high levels of selenium in the plant was the cause of the problem. However, two-grooved milk vetch also contains swainsonine, the alkaloid responsible for locoism, making the neurologic signs more likely due to the swainsonine than the selenium[97-99] Furthermore the microscopic cytoplasmic vacuolation found in the sheep's tissues were typical of those seen in locoweed poisoning and not selenium poisoning. In light of current evidence, it is appropriate to assume that selenium poisoning is not the cause of blind staggers and that sulfate toxicity is the cause of this disease syndrome. Locoweed poisoning may be a confounding issue in that clinical signs are similar to those of animals showing blind staggers.

The following plants are described in some detail as their recognition is helpful in identifying areas in which there is potential for selenium poisoning. Many of these plants are require selenium in the soil for them to flourish, thereby serving as indicator plants for selenium rich soils.

Two-Grooved Milk Vetch
Astragalus bisulcatus
Fabaceae (Legume family)

Habitat
Many of the vetches prefer the dry, alkaline seleniferous soils at lower and middle elevations especially in western North America.

Description
This leafy-stemmed, perennial plant often grows as a large clump (Figures 9-8A and 9-8B). The leaves are pinnate with numerous leaflets. Flowers are in dense spikelike racemes, which become reflexed in age. Flower color may vary from purple to pink or white. The fruit is a pod, from 1 cm to 1.5 cm(11 to 15 mm) long on a stipe, 3 to 4 mm in length. The pod is one-celled with two grooves running lengthwise on the upper surface of the pod, which gives the plants its common name.

Figure 9-8A Two-grooved milk vetch (*Astragalus bisulcatus*). **Figure 9-8B** Inset showing two distinctive grooves on seed pods.

Principal Toxins
Approximately 24 species of *Astragalus* are known to accumulate toxic levels of selenium[2] (Table 9-2). Livestock will rarely eat these plants except under starvation conditions apparently because of their distasteful seleniferous odor. These *Astragalus* spp. are useful indicator plants for soils high in selenium, and their presence can help identify pastures in which all plants may accumulate potentially toxic levels of sele-

nium. Two-grooved milk vetch is one of the few known vetches to contain both selenium and swainsonine, the alkaloid responsible for locoism.

Table 9-2 Selenium-Accumulating Astragalus Species
A. albulus
A. argillosus
A. beathii
A. bisulcatus
A. confertiflorus
A. crotalariae
A. diholcos
A. eastwoodae
A. ellisiae
A. grayi
A. haydenianus
A. moencoppensis
A. oocalycis
A. osterhouti
A. pattersonii
A. pectinatus
A. praelongus
A. preussii
A. racemosus
A. recedens
A. sabulosus
A. toanu

Rayless Goldenweed
(Oonopsis engelmannii)
Asteraceae (Sunflower family)

Habitat
Goldenweed prefers the dry, alkaline soils of the plains and foothills of central and southwest North America.

Description
Goldenweed is a perennial shrub with a woody root stock. The stems are from 4 to 12 inches (10 to 30 cm) tall, have brown bark below, and are glabrous. The leaves are 1 to 2.75 inches (3 to 7 cm) long and 1 to 3 mm wide, narrowly linear, rigidly erect, and glabrous. The heads are few to many, with bracts imbricated in three or more lengths, somewhat oblong to lanceolate. The color of the disc flowers is yellowish, and the pappus is brown. There are no ray flowers. *Oonopsis foliosa* var. *monocephal* is similar to this species, except that it has longer involucral bracts, oblong lanceolate leaves which are over 6 mm wide, and usually one head to a stem. *O. foliosa* differs from the preceding species in that it has ray flowers 8 to 15 mm long and has broadly lanceolate leaves, with one to several flower heads per stem.

Principal Toxin
Jimmy weed may accumulate toxic levels of selenium that can induce chronic selenium poisoning in livestock grazing it and other forages in the area.

Woody Aster
Xylorrhiza glabriuscula (Machaeranthera)
Asteraceae (Sunflower family)

Habitat
Woody aster requires the alkaline seleniferous soils of western North America, often growing at altitudes of 5000 to 6500 feet (1,524 to 1,981 meters).

Description
Woody aster is a low shrubby plant with a thick taproot and woody branching base. The leaves are 1 to 2 inches (2 to 5 cm) long and 2 to 6 mm wide. They are linear-oblanceolate to linear, hairy, and tipped with a callus point. The ray flowers are stiff and white (Figure 9-9). *X. venusta* (*Machaeranthera* spp., old name) resembles the preceding with the exception that the involucres are longer and the disc is somewhat wider.

Figure 9-9 Woody aster (*Xylorrhiza glabriuscula*)

Principal Toxin
Woody aster is an obligate selenium accumulator and may cause severe chronic selenium poisoning in livestock that are forced into grazing the plant when other forages are scarce.

Prince's Plume
Stanleya pinnata
Brassicaceae (Mustard family)

Habitat

Prince's plume requires the dry alkaline, selenium rich soils and shale rock formations of hills, valleys, and arroyo banks of western North America.

Description

The plant is a coarse, herbaceous perennial, ranging from 1.5 to 5 feet (0.5 to 2 meters) in height. The stems are stout and mostly unbranched. The leaves are entire to pinnately compound, from 2 to 8 inches (5 to 20 cm) long. Many small flowers with yellow petals with long claws are arranged in a plume-like inflorescence (Figure 9-10). The fruit is slender pod, nearly round in cross section, and with a stalk from 1 to 3 cm long.

Principal Toxin

Prince's plume is an obligate selenium-accumulating plant capable of concentrating high levels of selenium. It is relatively unpalatable and rarely eaten by livestock. However, the presence of prince's plume is indicative of high selenium content in the soil, and other grasses and plants growing in the same area may also accumulate selenium that could cause chronic selenium poisoning in livestock.

Figure 9-10 Prince's plume (*Stanleya pinnata*).

White Fall Aster, Rough White Aster
Aster falcatus
Asteraceae (Sunflower family)

Habitat

Rough white aster is a fall blooming plant of the prairies and foothills of central and western North America.

Description

White fall aster is a branched perennial plant with extensive root stalks. The leaves are 1 to 3 cm long, linear, and densely hairy. The inflorescence is a head with white ray flowers 3 to 4 mm long with many tawny pappus bristles (Figure 9-11).

Figure 9-11 White fall aster (*Aster falcatus*).

Principal Toxin

White fall aster is a secondary or facultative selenium accumulator when growing in alkaline, selenium-rich soils. It may, therefore, cause chronic selenium poisoning of livestock.

Broom Snakeweed
Gutierrezia sarothrae
Asteraceae (Sunflower family)

Habitat

Broom snakeweed is a common inhabitant of the dry plains and hills of western North America, usually growing at altitudes from 4000 to 10,000 feet (1,219 to 3,048 meters).

Description

This is a herbaceous perennial that is shrubby and woody at its base. The stems are branching; the leaves linear and glabrous. The heads are many, usually in clusters at the ends of the branches. A given head will have no more than three to eight ray flowers and three to eight disc flowers. The flowers are yellow, with the disc flowers usually perfect (see Figures 8-5A and 8-5B). The corollas have five lobes. The pappus is composed of several to many oblong scales.

Principal Toxin

The plant is a secondary selenium absorber. It has also been associated with abortion in cattle and sheep and may cause fatal liver disease (see Chapter 8).

Gumweed, Resin Weed
Grindelia spp.
Asteraceae (Sunflower family)

Habitat

Gumweed is common on the prairies, plains, roadsides and waste areas.

Description

Gumweeds are biennial or perennial herbaceous plants with leafy stems. The leaves are alternate, simple, and more or less resinous dotted. The heads are solitary, with the bracts being imbricated in several series with the tips often recurved and covered with a very gummy resinous material (Figure 9-12). The ray flowers, when present, are 8 to 10 mm long, lemon-yellow to bright yellow in color; the pappus awns are 2 to 3 mm long.

Figure 9-12 Gumweed (*Grindellia squarosa*).

Principal Toxin

Gumweeds will accumulate selenium if growing in alkaline, selenium-rich soils, and therefore may cause chronic selenium poisoning in livestock. Horses frequently like to eat the flower heads.

Saltbush
Atriplex spp.
A. canescens – four-wing saltbush
Chenopodiaceae (Goosefoot family)

Habitat
Saltbush is common on the dry plains and foothills of western North America.

Description
Saltbush is a perennial shrub growing to 6 feet (2 meters) in height that is more or less scaly and scurfy. The leaves are mostly alternate, often covered by a white powdery substance. The plants contain both male and female flowers, with staminate flowers in terminal panicles without bracts. They have a three- to five-parted perianth and three to five stamens. The pistillate flowers are subtended by two bracts, but are without a perianth. The ovary is one-celled with two stigmas. Four-wing saltbush has four distinct

Figure 9-13 Saltbush showing 4 wings (bracts) of the fruits (*Atriplex* spp.).

Principal Toxin
Saltbush will accumulate selenium when growing in selenium-rich soils and may cause chronic selenium poisoning of livestock under such growing conditions. Otherwise, the plant is considered a valuable range plant for grazing.

Saltbush Poisoning
Saltbush is also the name given to *Bacchaaris halimifolia*, a common shrub that grows in the coastal plains of Virginia, and from Florida to Texas. The plant is not a selenium accumulator but contains cardiotoxic glycosides. Cattle rarely eat saltbush unless other forages are scarce.

Indian Paintbrush
Castilleja spp.
Scrophulariaceae (Figwort family)

Habitat

Paintbrushes are widespread in the mountains and plains of western North America.

Description

These perennial and occasionally annual plants are herbaceous, but many are woody at the base. The leaves are alternate and sessile. The irregular flowers are arranged in terminal, bracted spikes. The bracts are usually petal-like ranging from scarlet to yellow in color (Figure 9-14). The calyx is tubular and four-lobed, more deeply cleft above and below than on the sides. The corolla is long and narrow, strongly two-lipped; the upper, called a galea (hood), is elongated and the lower lip is very short and three-toothed. The stamens number four, in pairs, and are enclosed by the galea. Paintbrush attaches to the roots of surrounding plants (sage species) in a symbiotic relationship.

Figure 9-14 Paintbrush (*Castilleja* spp.).

Principal Toxin

Indian paintbrush species are selenium accumulators if growing in high selenium soils.

Beard Tongue
Penstemon spp.
Scrophulariaceae (Figwort family)

Habitat

Many different species of *Penstemon* are widespread in the mountains and plains of North America. Their habitat varies according to species preferences.

Description

These are usually herbaceous perennial plants that are erect and tufted; however, many are low and creeping. The leaves are opposite with the upper one sessile and often clasping. The inflorescence is a panicle of flowers ranging from blue to red (Figure 9-15). The flowers are showy with a tubular corolla, which is bilabiate with the upper lip two-lobed and the lower lip three-lobed. There are four fertile stamens in two pairs with arched filaments. A fifth stamen, called a staminode, is represented by conspicuous sterile filament attached to the upper side of the corolla. It is widened and bearded at the apex, giving the plants their common name of beard tongue.

Figure 9-15 Penstemon (*Penstemon* spp.). Inset shows flower details.

Principal Toxin

Penstemon spp. will accumulate selenium if growing in selenium-rich soil. It is seldom a problem to livestock.

REFERENCES

1. Dolahite JW, Henson JB: Toxic plants as the etiologic agent of myopathies in animals. *Am J Vet Res* 1965;26:749-752.

2. Radostits OM, Blood DC, Gay CC, eds: *Veterinary Medicine*, edn 8. Philadelphia: Bailliere Tindall; 1994:1543-1544.

3. Krook L, Wasserman RH, Shively JN, *et al.*: Hypercalcemia and calcinosis in Florida horses: implication of the shrub, *Cestrum diurnum*, as the causative agent. *Cornell Vet* 1975;65:26-56.

4. Weisenberg M: Calcinogenic glycosides. *In* Toxicants of Plant Origin, Vol II Ed. Cheeke PR, Boca Raton, FL: CRC Press; 1989:201-238.

5. Morris KML: Plant induced calcinosis: a review. *Vet Hum Toxicol* 1982;24:34-48.

6. Kasali OB, Krook L, Pond WG, Wasserman RH: *Cestrum diurnum* intoxication in normal and hyperparathyroid pigs. *Cornell Vet* 1977;67:190-221.

7. Wasserman RH: The nature and mechanism of action of the calcinogenic principle of *Solanum malacoxylon* and *Cestrum diurnum*, and a comment on *Trisetum flavescens*. *In* Effects of Poisonous Plants on Livestock. Keeler RF, Van Kampen KR, James L, Eds. New York: Academic Press; 1978:545-553.

8. Krook L, Wasserman RH, McEntee K, *et al.*: *Cestrum diurnum* poisoning in Florida cattle. Cornell Vet 1975;65:557-575.

9. Krook L, Wasserman RH, Shively JH, *et al.*: Hypercalcemia and calcinosis in Florida horses: implication of the shrub *Cestrum diurnum*, as the causative agent. *Cornell Vet* 1975; 65:26-56.

10. Allen JG, Steele P, Masters HG, Lambe WJ: A lupinosis-associated myopathy in sheep and the effectiveness of treatments to prevent it. *Aust Vet J* 1992;69:75-81.

11. Williamson PM, Highet AS, Gams W, *et al.*: *Diaporthe toxica* sp. nov., the cause of lupinosis in sheep. *Mycological Res* 1994;98:1364-1368.

Cassia

12. Hebert CD, Flory W, Seger C, Blanchard RE: Preliminary isolation of a myodegenerative toxic principle from *Cassia occidentalis*. *Am J Vet Res* 1983;44:1370-1374.

13. Rowe LD, Corrier DE, Reagor JC, Jones LP: Experimentally induced *Cassia roemeriana* poisoning in cattle and goats. *Am J Vet Res* 1987;48: 992-997.

14. Rowe LD: Cassia-induced myopathy. *In* Handbook of Natural Toxins. Toxicology of Plant and Fungal Toxins. Keeler RF, Tu AT, Eds. New York: Marcel Dekker; 1991:335-351.

15. O'Hara PJ, Pierce KR, Reid WK: Degenerative myopathy associated with ingestion of *Cassia occidentalis*: clinical and pathologic features of the experimentally induced disease. *Am J Vet Res* 1969;30:2173-2180.

16. O'Hara PJ, Pierce KR: A toxic cardiomyopathy caused by *Cassia occidentalis II*. Morphological studies in poisoned rabbits. *Vet Pathol* 1974;11:97-109.

17. Mercer HD, Neal FC, Himes JA, *et al.*: *Cassia occidentalis* toxicosis in cattle. *J Am Vet Med Assoc* 1967;151:735-741.

18. Pierce KR, O'Hara PJ: Toxic myopathy in Texas cattle. *Southwest Vet* 1967;20:179-184.

19. Martin BW, Terry MK: Toxicity of *Cassia occidentalis* in the horse. *Vet Hum Toxicol* 1981; 23: 416-418.

20. Henson JB, Dolomite JW, Bridges CH, *et al.:* Myodegeneration in cattle grazing Cassia species. *J Am Vet Med Assoc* 1965;147:142-145.

21. Rogers RJ, Gibson J, Reichman KG: The toxicity of *Cassia occidentalis* for cattle. *Aust Vet J* 1979;55:408-412.

22. Colvin BM, Harrison LR, Sangster LT, Gosser HS: *Cassia occidentalis* toxicosis in pigs. *J Am Vet Med Assoc* 1986;189:423-426.

23. Suliman HB, Wasfi AI, Adam SEI: The toxicity of *Cassia occidentalis* in goats. *Vet Hum Toxicol* 1982;24:326-329.

24. Galal M, Adam SEI, Maglad MA, *et al.:* The effects of *Cassia occidentalis* on goats and sheep. *Acta Vet* 1985; 35: 163-174.

25. Graziano MJ, Flory W, Seger CL, *et al.:* Effects of *Cassia occidentalis* extract in the domestic chicken (*Gallus domestica*). *Am J Vet Res* 1983; 44:1238-1244.

26. O'Hara PJ, Pierce KR: A toxic cardiomyopathy caused by *Cassia occidentalis* II. Biochemical studies in poisoned rabbits. *Vet Pathol* 1974;11:110-124.

27. Suliman HB, Shommein AM: Toxic effect of the roasted and unroasted beans of *Cassia occidentalis* in goats. *Vet Hum Toxicol* 1986;28:6-11.

28. Nicholson SS, Thorton JT, Rimes AJ: Toxic myopathy in dairy cattle caused by *Cassia obtusifolia* in green-chop. *Bovine Practitioner* 1977;12:120-123.

29. Putnam MR, Boosinger T, Spano J, *et al.:* Evaluation of *Cassia obtusifolia* (sickle pod) seed consumption in Holstein calves. *Vet Hum Toxicol* 1988;30:316-318.

30. Burrows GE, Edwards WC, Tyrl RJ: Toxic plants of Oklahoma: coffeeweeds and sennas. *Oklahoma Vet Med Assoc* 34;1989:101-105.

31. Henson JB, Dolomite JW: Toxic myodegeneration in calves produced by experimental *Cassia occidentalis* intoxication. *Am J Vet Res* 1966;27:947-949.

32. Schmitz DG, Denton JH:. Senna bean toxicity in cattle. *Southwest Vet* 1977;30:165-170.

Thermopsis (Yellow Banner)

33. Keeler RF, Johnson AE, Chase RL: Toxicity of *Thermopsis montana* in cattle. *Cornell Vet* 1986;76:115-127.

34. Keeler RF, Baker DC: Myopathy in cattle induced by alkaloid extracts from *Thermopsis montana, Laburnum anagyroides,* and *Lupinus* spp. *J Comp Pathol* 1990;103:169-182.

35. Baker DC, Keeler RF: Myopathy in cattle caused by *Thermopsis montana.* In Handbook of Natural Toxins. Toxicology of Plant and Fungal Toxins. Keeler RF, Tu AT, Eds. New York: Marcel Dekker; 1991:61-69.

36. Spoerke DG, Murphy MM, Wruk KM, Rumack BH: Five cases of Thermopsis poisoning. *Clin Toxicol* 1988;26:397-406.

37. Baker DC, Keeler RF: *Thermopsis montana*-induced myopathy in calves. *J Am Vet Med Assoc* 1989;194:1269-1272.

Black Walnut

38. Uhlinger C: Black walnut toxicosis in ten horses. *J Am Vet Med Assoc* 1989;195:343-344.

39. True RG, Lowe JE: Induced juglone toxicosis in ponies and horses. *Am J Vet Res* 1980;41:944-945.

40. Ralston SL, Rich VA: Black walnut toxicosis in horses. *J Am Vet Med Assoc* 1983;183:1095.

41. MacDaniels LH: Perspective on the black walnut toxicity problem–apparent allergies to man and horse. *Cornell Vet* 1983;73:204-207.

42. Minnick PD, Brown CM, Braselton WE, *et al.:* The induction of equine laminitis with an aqueous extract of the heartwood of black walnut (*Juglans nigra*). *Vet Hum Toxicol* 1987;29:230-233.

43. McDaniels LH: Perspective on the black walnut toxicity problem. *Cornell Vet* 1987;29:230-233.

44. Galey FD, Beasley VR, Schaeffer D, Davis LE: Effect of an aqueous extract of black walnut (*Juglans nigra*) on isolated equine digital vessels. *Am J Vet Res* 1990;51:83-88.

45. Galey FD, Twardock AR, Goetz TE, *et al.:* Gamma scintigraphic analysis of the distribution of perfusion of blood in the equine foot during black walnut (*Juglans nigra*)-induced laminitis. *Am J Vet Res* 1990;51:688-695.

46. True RG, Lowe JE, Heissen J: Black walnut shavings as a cause of laminitis. *Proc Annu Meet Am Assoc Equine Pract* 1978;24:511-515.

47. Galey FD, Beasley VR, Twardock AR, Whiteley HE, *et al.:* Pathophysiologic effects an aqueous extract of black walnut (*Juglans nigra*) when administered via nasogastric tube to the horse. *In* Poisonous Plants Proceeding of the Third International Symposium. James LF, Keeler R, Bailey EM, *et al.,* Eds. Ames, IA: Iowa State University Press; 1992:630-635.

48. Richard JL, Arp LH, Bachetti P: The mycotoxin, penitrem A, as a cause of moldy walnut toxicosis in a dog. *California Vet* 1981;6:12.

Hoary Alyssum

49. Geor RJ, Becker RL, Kanara EW *et al.:* Toxicosis in horses after ingestion of hoary alyssum. *J Am Vet Med Assoc* 1992;201:63-67.

50. Ellison SP. Possible toxicity caused by hoary alyssum (*Berteroa incana*). *Vet Med/ Equine Practice* 1992;May:472-475.

51. Hovda LR, Rose ML: Hoary alyssum (*Berteroa incana*) toxicity in a herd of brood mares. *Vet Human Toxicol* 1993; 35,39-40.

Flatweed

52. Huntington PJ, Jeffcott LB, Friend SCE, *et al.:* Australian-stringhalt–epidemiological, clinical, and neurological investigations. *Equine Vet J* 1989;21:266-273.

53. Pemberton DH, Caple IW: Australian-stringhalt in horses. *Vet Annu* 1980;20:167-171.

54. Galey FD, Hullinger PJ, McCaskill J: Outbreaks of stringhalt in northern California. *Vet Hum Toxicol* 1991;33:176-177.

55. Stashak TS. Lameness. *In* Adams' Lameness in Horses, Stashak TS., Ed. Philadelphia: Lea & Febiger; 1987:723-725.

Selenium

56. Trelease SF, Beath OA: *Selenium: Its Geological Occurrence and Its Biological Effects in Relation to Botany, Chemistry, Agriculture, and Medicine.* Burlington, VT: Champlain Printers; 1949:1-12, 165-187.

57. Rosenfeld I, Beath OA: *Selenium: Geobotany, Biochemistry, Toxicity and Nutrition.* New York: Academic Press; 1964:1-7, 91-104, 141-163.

58. Durrell LW, Cross F: Selenium poisoning of livestock. *Colorado State College, Extension Service Bulletin* 1944;382-A.

59. Radostits OM, Blood DC, Gay CC, eds: Diseases caused by inorganic and farm chemicals–selenium poisoning. *In* Veterinary Medicine, edn 8. London: Bailiere Tindall; 1994:1484-1486.

60. James LF, Panter KE, Mayland HF, *et al.:* Selenium poisoning in livestock: a review

and progress. *In:* Jacobs LW, Ed. Selenium in Agriculture and the Environment. Madison, *Wisc: Am Soc Agron Soil Sci Soc of America;* 1989;23:123-131.

61. Whanger PD: Selenocompounds in plants and their effects on animals. *In* Toxicants of Plant Origin, Vol III. Proteins and Amino Acids. Cheeke PR, Ed. Boca Raton, FL: CRC Press. 1989;141-167.

62. James LF: Suspected phytogenic selenium poisoning in sheep. *J Am Vet Med Assoc* 1982;180:1478-1481.

63. Edmondson AJ, Norman BB, Suther D: Survey of state veterinarians and state veterinary diagnostic laboratories for selenium deficiency and toxicosis in animals. *J Am Vet Med Assoc* 1993; 202:865-872

64. Raisbeck MF, Dahl ER, Sanchez DA, *et al.:* Naturally occurring selenosis in Wyoming. *J Vet Diagn Invest* 1993;5:84-87.

65. Kubota J, Allaway WH: Geographic distribution of trace element problems. *In* Micronutrients in Agriculture. Proceedings Dinauer RC, Ed. Madison, Wisc. *Soil Science Society of America;* 1972:525-554.

66. Meyer RD, Burau RB: The geochemistry and biogeochemistry of selenium in relation to its deficiency and toxicity in animals. Selenium in the environment: essential nutrient, potential toxicant. *Proceedings University California, Division of Agriculture and Natural Resources.* 1995:38-44.

67. Ganje RJ, Whitehead EI: Selenium uptake by plants as affected by forms of selenium in the soil. *Proc South Dakota Acad Sci* 1958;37:85-88.

68. Olson OE, Whitehead EI, Moxon AI: Occurrence of soluble selenium in soils and its availability to plants. *Soil Sci* 1942;54:47-53.

69. Olson OE: Selenium in plants as a cause of livestock poisoning. *In* Effects of Poisonous Plants on Livestock. Keeler RF, Van Kampen KR, James LF, Eds. New York: Academic Press; 1978:121-133.

70. Shrift A: Metabolism of selenium by plants and microorganisms. *In* Organic Selenium Compounds: Their Chemistry and Biology. Klayman DJ, Gunther WHH, Eds. New York: John Wiley & Sons; 1973:693-726.

71. Underwood EJ, Ed: *Trace Elements in Human and Animal Nutrition.* New York: Academic Press; 1977:334-346.

72. O'Toole D, Raisbeck MF: Pathology of experimentally induced chronic selenosis (alkali disease) in yearling cattle. *J Vet Diagn Invest* 1995;7:364-373.

73. Raisbeck MF, O'Toole D: Morphologic studies of selenosis in herbivores. *In* Toxic Plants and Other Natural Toxicants. Garland T, Barr AC, Eds. New York: CAB International; 1998:380-388.

74. Crinion RA, O'Connor JP: Selenium intoxication in horses. *Irish Vet J* 1978;32: 81-86.

75. Raisbeck MF, O'Toole D, Schamber RA, *et al.:* Toxicologic evaluation of a high-selenium hay diet in captive pronghorn antelope (*Antilocapra americana*). *J Wildlife Dis* 1996;32:9-16.

76. Raisebeck MF, Schamber RA, Belden EL: Immunotoxic effects of selenium in mammals. *In* Toxic Plants and Other Natural Toxicants. Garland T, Barr AC, Eds. New York: CAB International; 1998:260-266.

77. Smyth JBA, Wang JH, Barlow RM, *et al.:* Experimental acute selenium intoxication in lambs. *J Comp Pathol* 1990;102:199-209.

78. Moxon AL: Alkali disease or selenium poisoning. *South Dakota Agricultural Experimental Station Technical Bulletin* 1937;311:1-91.

79. Flemming GA, Walsh T: Selenium occurrence in certain soils and its toxic effects on animals. *Proc Royal Irish Acad* 1957;58:151-167.

80. Knott SG, McCray CWR: Two naturally occurring outbreaks of selenosis in Queensland. *Aust Vet J* 1959;35:161-165.

81. Olsen OE, Embry LB: Chronic selenite toxicty in cattle. *Proc South Dakota Acad Sci* 1973;52:50-58.

82. Raisbeck MF, Dahl ER, Sanchez DA, *et al.:* Naturally occurring selenosis in Wyoming. *J Vet Diagn Invest* 1993;5:84-87.

83. Witte ST, Will LA, Olsen CR, Kinker JA: Chronic selenosis in horses fed locally produced alfalfa hay. *J Am Vet Med Assoc* 1993;202:406-409.

84. James LF, Smart RA, Shupe JL, *et al.:* Suspected phytogenic selenium poisoning in sheep. *J Am Vet Med Assoc* 1982;180:1478-1481.

85. Raisbeck MF, O'Toole D, Belden EL, Waggoner JW: Chronic selenosis in ruminants. *In* Toxic Plants and Other Natural Toxicants. Garland T, Barr AC, Eds. New York: CAB International; 1998:389-396.

86. Glenn MW, Jensen R, Griner LA: Sodium selenate toxicosis: pathology and pathogenesis of sodium selenate toxicosis in sheep. *Am J Vet Res* 1964;25:1486-1494.

87. Palmer IS: Water, soil, and plant selenium: analytical methodology. *In* Selenium in the Environment: Essential Nutrient, Potential Toxicant. Proceedings University California, Division of Agriculture and Natural Resources; 1995:20-35.

88. National Research Council. *Selenium in Nutrition*. Washington, DC: National Academy Press; 1983.

89. Traub-Dargatz, Knight AP, Hamar DW: Selenium toxicity in horses. *Compend Contin Ed Pract Vet* 1986;8:771-776.

90. Osweiler GD, Carson TL, Buck WB, Van Gelder GA: Selenium. *In* Osweiler GD, Carson TL, Buck WB, Van Gelder GA. Clinical and Diagnostic Veterinary Toxicology, edn. 3. Dubuque, Iowa, Kendall Hunt Publishing; 1985:132-142.

91. Hutline JD, Mount ME, Easley KJ, Oehme FW: Selenium toxicosis in the horse. *Equine Pract* 1979;1:57-60.

92. Radostits OM, Blood DC, Gay CC, Eds: Diseases caused by inorganic and farm chemicals–selenium poisoning. *In* Veterinary Medicine, edn. 8. London: Bailiere Tindall; 1994:1408-1425.

93. O'Toole D, Raisbeck MF, Case JC, Whitson TD: Selenium-induced "blind staggers" and related myths. *Commentary. Vet Pathol* 1996;33:104-116.

94. Gould DH, McAllister MM, Savage JC, Hamar DW: High sulfide concentrations in rumen fluid associate with nutritionally induced polioencephalomalacia in calves. *Am J Vet Res* 1991;52:1164-1169.

95. Raisbeck MF: Is polioencephalomalacia associated with high sulfate diets? *J Am Vet Med Assoc* 1982;180:1303-1305.

96. Jeffrey M, Duff JP, Higgins RJ, *et al.:* Polioencephalomalacia associated with the ingestion of ammonium sulfate by sheep and cattle. *Vet Rec* 1994;134:343-348.

97. Van Kampen KR, James LF: Manifestations of intoxication by selenium accumulating plants. *In* Effects of Poisonous Plants on Livestock. Keeler RF, Van Kampen KR, James LF, Eds. New York: Academic Press; 1978:135-138.

98. James LF, Van Kampen KR, Hartley WJ: *Astragalus bisulcatus*: a cause of selenium or locoweed poisoning. *Vet Hum Toxicol* 1983;25:86-89.

99. James LF, Hartly WJ, Van Kampen KR: Syndromes of Astragalus poisoning in livestock. *J Am Vet Med Assoc* 1981;178:146-150.

CHAPTER 10

Plants Affecting
the Mammary Gland

Plant Toxins in Milk

Ever since the deaths of Abraham Lincoln's mother and others were shown to be caused by drinking milk from cows that had eaten white snakeroot, concern has existed regarding the presence of plant toxins in milk.[1,2] The best known plant toxicants secreted in milk are tremetol and tremetone found in white snakeroot (*Eupatorium rugosum*) and rayless goldenrod (*Isocoma pluraflora*).[3,4] White snakeroot and rayless goldenrod poisoning has been reported in cattle, horses, sheep, pigs, dogs, and cats. It causes a syndrome of muscle tremors referred to as "trembles."[3,4] Milk sickness develops in humans who drink the milk from cows that have eaten these plants. Because most milk consumed in North America is now produced under intensive management conditions, there is little risk of "milk sickness." However, the risk of milk sickness exists where pastured milk cows or goats gain access to white snakeroot and rayless goldenrod.[5] Details of these plants and the disease they cause are discussed in Chapter 6. Other plants known to affect milk quality are listed in Table 10-1.

A variety of ingested plant compounds and their metabolites, once absorbed into the blood, are secreted by the mammary gland into the milk. There the toxins cause changes in milk flavor, and may pose a hazard to animals and people drinking the milk.[2] Many toxicants diffuse readily into the milk from the blood and reach concentrations similar to that in the blood. The extent to which plant compounds diffuse into milk depends either their water or fat solubility, and whether or not they are protein bound or have a pH that enhances their diffusion into the milk. Toxins that are water soluble and have a basic pH (plant alkaloids) may concentrate in the milk because normally it is more acidic (pH 6.5) than the blood. The quantities of toxin are generally very low because most of the toxins are more readily excreted through the liver and kidneys. The lactating mammary gland however, is one of the ways that the animal's body is able to eliminate toxins.

Plant alkaloids, especially the pyrrolizidine alkaloids found in the plant genera *Amsinckia*, *Crotolaria*, *Cynoglossum*, *Echium*, *Heliotropium*, and *Senecio*, are readily secreted in milk and can be hazardous to the young animal suckling its mother.[6-9] The quantity of pyrrolizidine alkaloids secreted in milk is generally very low and can be considered of minimal risk to humans. However, the chronic long-term consumption of milk containing these alkaloids may be carcinogenic and cause chronic liver disease.[9-11] Further details on the plants containing pyrrolizidine alkaloids are covered in Chapter 4.

The important indolizidine alkaloid swainsonine found in the *Astragalus*, *Oxytropis*, and *Swainsona* spp. is excreted in milk.[12,13] Swainsonine is the plant toxin primarily responsible for the syndrome known as locoism in animals that consume locoweeds (*Astragalus* and *Oxytropis* spp.) Young animals are likely to be exposed to more swainsonine because they learn to eat locoweeds from their mothers and suckle their mothers that are also eating the plants. Because swainsonine is rapidly cleared from the serum of cattle and sheep through the urine (serum half-life of about 20 hours), the milk is also likely cleared of the alkaloid in 2 to 3 days after the ingestion of locoweed has ceased.[14] With the resurgence of small acreage farms in western range-lands where locoweeds are abundant, there is a potential health risk to people who drink raw milk from milking cows or goats that are eating locoweed. Swainsonine and the locoweeds are discussed in detail in Chapter 6.

Other toxic alkaloids with the potential for being excreted in milk are found in lupines (*Lupinus* spp.). A woman whose child was born with skeletal deformities was found to have consumed the milk of goats that had been eating lupines.[15] A dog that was fed the goat's milk also produced deformed puppies. Some of the goats aborted fetuses with skeletal abnormalities.[15] Other plants with similar quinolizidine alkaloids include mountain thermopsis (*Thermopsis montanum*), Scotch broom (*Cytisus* spp.), and golden chain tree (*Laburnum* spp.)[16] Alkaloids found in poison hemlock (*Conium maculatum*), tobacco (*Nicotiana* spp.), and various other genera including *Cassia*, *Sedum*, *Prosopis*, and *Lobelia* may be passed in the milk and can pose a hazard to pregnant women drinking the milk.[17] The piperidine alkaloids from these plants are capable of causing congenital defects in the fetus.[18,19]

Other compounds found in plants that can be secreted into the milk of lactating animals eating the plants are the glucosinolates. Glucosinolates are found in the large mustard family and especially in the *Brassicaceae* that includes the mustards, cabbage, kale, rape, turnips, horseradish, radish, and watercress among others. The bitter-tasting compounds may flavor the milk of animals eating the plants. Hydrolysis of the glucosinolates produces compounds such as isothiocyanates, thiocyanates, and goitrin that have an antithyroid hormone effect and result in thyroid enlargement and decreased growth rates. Thyroid gland enlargement (goiter) may occur in young animals and human babies drinking the milk of animals that have eaten plants of the mustard family (*Brassicaceae*).[2,20] Goat kids and rabbits fed milk from goats eating meadowfoam meal (*Limnanthes alba*) have been reported to develop thyroid abnormalities.[20] Thyroid enlargement was similarly seen in lambs born to ewes fed meadowfoam meal.[21]

In Europe, poisoning in humans and animals has been associated with milk containing colchicine, a potent alkaloid found in the autumn crocus (*Colchicum autumnale*).[22]

Abnormal Milk Flavor

Objectionable flavors and odors are commonly encountered in milk and can result in its condemnation for human consumption. In many instances abnormal milk flavors are associated with bacterial spoilage and rancidity of the milk or absorption of odors from the cow's environment. At other times chemicals used to disinfect milking equipment or drugs used to treat mastitis and other diseases may accumulate in the milk and affect its flavor. The milk from cows that are severely ketotic due to metabolic disease may have a strong smell of ketones (acetone). Certain diets, especially those that are fermented such as silage or haylage, often cause a change in the flavor of milk. Brewer's grains, a by-product of the alcohol distillery industry, can cause a distinctive flavor in milk. Similarly, cull onions, cabbages, turnips, rape, and similar plants fed to lactating animals may induce an abnormal flavor to the milk. A sudden change of feed to a rich feed such as alfalfa or clover may cause a temporary change in milk flavor that some people find objectionable. Most grains, soybeans, sugar beets, potatoes, carrots, and grass hay do not affect the milk.

A variety of plants that individual cows and especially goats may ingest when they are on pasture and rangeland and that may affect the flavor of milk are listed in Table 10-1. A bitter taste to milk may result when a lactating cow or goat eats members of the buttercup family or the bitterweeds.[23]

Table 10-1
Plant Toxins That May Affect Milk Quality

COMMON NAME	SCIENTIFIC NAME	PRINCIPAL TOXIN
White snakeroot	*Eupatorium rugosum*	Acetylbenzofurans (tremetol)
Rayless golden rod	*Isocoma pluraflora*	Acetylbenzofurans (tremetol)
Groundsels, senecio	*Senecio* spp.	Pyrrolizidine alkaloids
Rattle pod	*Crotolaria* spp.	Pyrrolizidine alkaloids
Hound's tongue	*Cynoglossum* spp.	Pyrrolizidine alkaloids
Fiddleneck	*Amsinckia intermedia*	Pyrrolizidine alkaloids
Comfrey	*Symphytum* spp.	Pyrrolizidine alkaloids
Heliotrope	*Heliotropium* spp.	Pyrrolizidine alkaloids
Viper's bugloss	*Echium* spp.	Pyrrolizidine alkaloids
Mustards, rape, cabbage	*Brassica* spp.	Glucosinolates*
Horse radish	*Amoracia* spp.	Glucosinolates*
Radish	*Raphanus* spp.	Glucosinolates*
Water cress	*Nasturtium officinale*	Glucosinolates*
Poison hemlock	*Conium maculatum*	Piperidine alkaloids (coniine)
Tobacco	*Nicotiana* spp.	Piperidine alkaloids (coniine)
Locoweeds	*Astragalus, Oxytropis* spp.	Indolizidine alkaloids (swainsonine)
Lupine	*Lupinus* spp.	Quinolizidine alkaloids (anagyrine)
Bitterweeds	*Helenium, Hymenoxys* spp.	Sesquiterpene lactones*
Bracken fern	*Pteridium aquilinum*	Ptaquiloside
Buttercups	*Ranunculus* spp.	Protoanemonins*
Onions, garlic	*Allium* spp.	N-propyl disulphide*
Autumn crocus	*Colchicum* spp.	Alkaloids (colchicine)
Avocado	*Persea americana*	Unknown toxin
Sage	*Artemisia* spp.	Monoterpenes, diterpenes*
Marijuana	*Cannabis sativa*	Cannabinol

* These plants impart an abnormal flavor to milk.

Where lactating animals gain access to avocado leaves or fruits (*Persea americana*), they may develop a noninfectious mastitis, with a marked decrease in milk production. Cattle, horses, goats, and rabbits have been affected by this unique plant toxicity.[24] Goats fed as little as 31 g/kg body weight of avocado leaves showed dramatic reduction in milk production and developed hard swollen udders 24 hours after they had eaten the leaves. The milk was of a cheesy consistency and contained clots. The milk somatic cell counts also became markedly elevated. If no further avocado leaves were fed, the udder edema regressed and milk production returned, but not to the levels prior to feeding the avocado leaves.[25] Generalized necrosis of the mammary gland epithelium with sloughing of necrotic cells and minimal inflammation was the principle histologic finding.[25] Further discussion of the toxic effects of avocados is presented in Chapter 2.

REFERENCES

1. Drake D: A memoir on the disease called by the people "trembles" and the "sick stomach" or "milk sickness; as they have occurred in the counties of Fayette, Madison, Clark, and Green in the state of Ohio. *West J Med Surg* 1841, 3:161–226.

2. Panter KE, James LF: Natural toxicants in milk: a review. *J Anim Sci* 1990, 68:892–904.

3. Couch JF: Trembles (or milk sickness). *USDA Circ* 1933, 306.

4. Kingsbury JM: *Poisonous Plants of the United States and Canada.* Englewood Cliffs, NJ: Prentice-Hall; 1964:397–404.

5. Stotts R: White snakeroot toxicity in dairy cattle. *Vet Med Small Anim Clin* 1984, 1:118.

6. Dickinson JO, Cooke MP, King RR, Mohamed PA: Milk transfer of pyrrolizidine alkaloids in cattle. *J Am Vet Med Assoc* 1976, 169:1192.

7. Johnson AE: Changes in calves and rats consuming milk from cows fed chronic lethal doses of *Senecio jacobaea* (tansy ragwort). *Am J Vet Res* 1976;.37:107.

8. Miranda CL, Cheeke PR, Groeger DE, Buhler DR: Effect of consumption of milk from goats fed *Senecio jacobaea* on hepatic drug metabolizing enzyme activities in rats. *Toxicology* 1981, 8:343.

9. Molyneux RJ, James LF: Pyrrolizidine alkaloids in milk: thresholds of intoxication. *Vet Hum Toxicol* 1990, 32:94–103.

10. Dickinson JO, King RR: The transfer of pyrrolizidine alkaloids from *Senecio jacobaea* into the milk of lactating cows and goats. *In* Effects of Poisonous Plants on Livestock. Keeler RF, VanKampen KR, James LF, Eds. New York: Academic Press; 1978:201–208.

11. Groeger DE, Cheeke PR, Schmitz JA, Buhler DR: Effects of feeding milk from goats fed tansy ragwort (*Senecio jacobaea*) to rats and calves. *Am J Vet Res* 1982, 43:1631–1633.

12. James LF, Hartley WJ: Effects of milk from animals fed locoweed on kittens, calves and lambs. *Am J Vet Res* 1977, 38:1263–1266.

13. James LF, Molyneux RJ, Panter KE: The potential for the toxic principles of Astragalus and related plants to appear in meat and milk. *Vet Hum Toxicol* 1990, 32:104–110.

14. Stegelmeier BL, James LF, Panter KE, Ralphs MH, Gardner DR: The pathogenesis and toxicokinetics of locoweed (*Astragalus* and *Oxytropis* spp.) poisoning in livestock. *J Natural Toxins* 1999; 8:35-45.

15. Ortega JA, Lazerson J: Anagyrine-induced red cell aplasia vascular anomaly, and skeletal dysplasia. *J Pediatr* 1987, 3:87–91.

16. Keeler RF: Alkaloid teratogens from Lupinus, Conium, Veratrum and related genera. *In* Poisonous Plants of Livestock. Keeler RF, VanKampen KR, James LF, Eds. New York: Academic Press; 1978:397–408.

17. Panter KE, Keeler RF: Conium, Lupinus, Nicotiana alkaloids: fetal effects and the potential for residues in milk. *Vet Hum Toxicol* 1990, 32:89–93.

18. Keeler, RF, Crowe MW: Anabasine, a teratogen from the Nicotiana species. *In* Plant Toxicology. Seawright AA, Hegarty MP, James LF, Keeler RF, Eds. Queensland, Australia: Queensland Poisonous Plants Committee; 1985:324–333.

19. Kubik M, Refholec J, Zachoval Z: Outbreak of hemlock poisoning in cattle. *Veterinarstvi* 1980, 30:157–159.

20. White RD, Cheeke PR: Meadowfoam (*Limnanthes alba*) meal as a feed stuff for dairy goats and toxicologic activity of the milk. *Can J Anim Sci* 1983, 63:391–394.

21. Throckmorton JC, Cheeke PR, Patton NM, *et al.*: Evaluation of meadow foam (*Limnanthes alba*) meal as a feed stuff for broiler chickens and weanling rabbits. *Can J Anim Sci* 1981, 61:735–739.

22. Kingsbury JM: *Poisonous Plants of the United States and Canada.* Englewood Cliffs, NJ: Prentice-Hall; 1964:450.

23. Ivie GW, Witzel DA, Rushing DD: Toxicity and milk-bittering properties of tenulin, the major sesquiterpene lactone constituent of *Helenium amarum* (bitter sneezeweed). *J Agric Food Chem* 1975, 23:845–849.

24. Craigmill AL, Eide RN, Schultz TA, Hedrick K: Toxicity of avocado (*Persea americana* var.) leaves: review and preliminary report. *Vet Hum Toxicol* 1984, 26:381–383.

25. Craigmill AL, Seawright AA, Matilla T, Frost AJ: Pathological changes in the mammary gland and biochemical changes in milk of the goat following oral dosing with leaf of the avocado (*Persea americana*). *Aust Vet J* 1989, 66:206–211.

GLOSSARY

A

Achene - A dry indehiscent one-seeded fruit, attached to the pericarp at only one place; formed from a single carpel, the seed is distinct from the fruit, as in Asteracea.

Acuminate – Gradually tapering sides finished before arriving at the apex or tip.

Acute – Sharp-pointed.

Alternate – With a single leaf or other structure at each node.

Annual – Living one growing season.

Articulate – Jointed; breaking into distinct pieces without tearing at maturity.

Auriculate – Having ear-shaped appendages.

Awn – A bristle on the flowering glumes of grasses (oats).

Axil - The angle between the leaf or branch and the main axis.

B

Berry – Simple, fleshy indehiscent (not splitting open) fruit with one or more seeds (tomato, nightshade)

Biennial – Living two growing seasons.

Bipinnate – Twice pinnately compound.

Blade – The expanded part of a leaf or floral part.

Bract – A small, rudimentary or imperfectly developed leaf.

Bulb – A bud with fleshy bracts or scales, usually subterranean.

C

Calyx – The outer set of sterile, floral leaves called sepals.

Campanulate – Bell-shaped.

Canescent – Becoming gray or grayish.

Capitate – Arranged in a head, as the flowers in Compositae.

Capsule – A dry fruit of two or more carpels, usually dehiscent by valves.

Carpel – A portion of the ovary or female portion of the flower.

Catkin – Spike-like inflorescence, unisexual, usually with scaly bracts

Caulescent – Having a stem.

Cilia – Fine hairs or projections.

Ciliate – Having fine hairs or projections, usually as marginal hairs.

Circumscissile – Opening all around by a transverse split.

Compound – Composed of several parts or divisions.

Cordate – Heart-shaped.

Corolla – The inner set of sterile, usually colored, floral leaves; the petals considered collectively.

Corymb – A raceme with the lower flower stalks longer than those above, so that all the flowers are at the same level.

Crenate – With roundish teeth or lobes.

Cuneate – Wedge-shaped.

Cuspidate – Having a rigid point.

Cyme – An inflorescence; a convex or flat flower cluster, the central flowers unfolding

D

Deciduous – Dying back; seasonal shedding of leaves or other structures; falling off.

Decumbent – Lying flat, or being prostrate, but with the tip growing upwards.

Dentate – Toothed, with outwardly projecting teeth.

Denticulate – Finely toothed. Diffuse- Loosely spreading.

Dioecious – Only one sex in a plant; with male or female flowers only.

Disk (disc) – A flattened enlargement of the receptacle of a flower or inflorescence; the head of tubular flowers, as in sunflower.

Dissected – Divided into many segments.

Drupe – A fruit with a fleshy or pulpy outer part and a bone-like inner part; a single-seeded fleshy fruit.

Drupelet – A small drupe, as one section of a blackberry.

E

Elliptic – Oval.

Entire – Without teeth, serrations, or lobes, as in leaf margins.

F

Fascicle – A cluster of leaves or other structures crowded on a short stem.

Fibrous – A mass of adventitious fine roots

Filiform – Threadlike.

Flaccid – Limp or flabby.

Follicle – A many-seeded dry fruit, derived from a single carpel, and splitting longitudinally down one side.

Fruit – The ripened ovary or ovaries with the attached parts Fuscous - Dingy-brown.

G

Glabrate – Nearly without hairs.

Glabrous – Smooth or hairless.

Glaucous – Covered with a bluish or white bloom.

Glume – small dry, membranous bract at the base of a grass spikelet

H

Hastate – Arrow-shaped with the basal lobes spreading.

Head – A dense inflorescence of sessile or nearly sessile flowers, as in Compositae

Hirsute – Having rather coarse, stiff hairs.

I

Incised – Cut into sharp lobes.

Indehiscent – Not opening at maturity.

Inflorescence – The arrangement of flowers on the flowering shoot, as a spike, panicle, head, cyme, umbel, raceme.

Involucre – Any leaflike structure protecting the reproducing structure, as in flower heads of *Compositae* and *Euphorbiaceae*.

K

Keel – Projecting, united front petals as in the flowers of *Fabaceae* (peas).

L

Lanceolate – Flattened, two or three times as long as broad, widest in the middle and tapering to a pointed apex; lance-shaped.

Leaf sheath –The lower part of a leaf, which envelopes the stem, as in grasses.

Leaflet – One of the divisions of a compound leaf.

Legume pod – A dry fruit, splitting by two longitudinal sutures with a row of seeds on the inner side of the central suture; as in family *Fabaceae* (*Leguminosae*).

Lenticular – Bean-shaped; shaped like a double convex lens.

Ligule – A membrane at the junction of the leaf sheath and leaf base of many grasses.

Linear – A long and narrow organ with the sides nearly parallel.

Lobed – Divided to about the middle or less.

M

Midrib – The central rib of a leaf or other organ; midvein.

Monoecious – Flowers unisexual, both types on the same plant.

N

Node – The part of a stem where the leaf, leaves, or secondary branches emerge.

Nutlet – A one-seeded portion of a fruit that fragments at maturity.

O

Obcordate – Inversely heart-shaped.

Oblanceolate – Inversely lanceolate.

Oblique – With part not opposite, but slightly uneven.

Oblong – Elliptical, blunt at each end, having nearly parallel sides, two to four times as long as broad.

Obovate – Inversely ovate.

Obtuse – Blunt or rounded.

Ocrea – A thin, sheathing stipule or a united pair of stipules (as in *Polygonaceae*).

Orbicular – Nearly circular in outline.

Ovate – Egg-shaped.

P

Palmate – Diverging like the fingers of a hand.

Panicle – An inflorescence, a branched raceme, with each branch bearing a raceme of flowers, usually of pyramidal form.

Pappus – A ring of fine hairs developed from the calyx, covering the fruit; acting as a parachute for wind-dispersal, as in dandelion.

Pedicel or peduncle – A short stalk.

Pedicelled – Having a short stalk, as a flower or fruit.

Peltate – More or less flattened, attached at the center on the underside.

Perennial – Growing many years or seasons.

Perfect – A flower having both stamens and carpels.

Perfoliate – Leaves clasping the stem, forming cups.

Perianth – The calyx and corolla together; a floral envelop.

Pericarp – The body of a fruit developed from the ovary wall and enclosing the seeds.

Persistent – Remaining attached after the growing season.

Petal – One of the modified leaves of the corolla; usually the colorful part of a flower.

Petiole –The unexpanded portion of a leaf; the stalk of a leaf.

Pilose – Having scattered, simple, moderately stiff hairs.

Pinnate – Leaves divided into leaflects or segments along a common axis; a compound leaf.

Pinnatifid – Pinnately cleft to the middle or beyond.

Pistillate – Female-flowered, with pistils only.

Prickle – A stiff, sharp-pointed outgrowth from the epidermis, as in Solanum.

Procumbent – Lying on the ground.

Puberulent – With very short hairs; woolly.

Pubescent – Covered with fine, soft hairs.

Punctate – With translucent dots or glands.

R

Raceme – An inflorescence, with the main axis bearing stalked flowers, these opening from the base upward.

Racemose – Like a raceme or in a raceme.

Rachis – The axis of a pinnately compound leaf ; the axis of inflorescence; the portion of a fern frond to which the pinnae are attached.

Ray – A marginal flower with a strap-shaped corolla, as in *Compositae*.

Receptacle – The end of the flower stalk, bearing the parts of the flower.

Reniform – Kidney-shaped. Reticulate - netted, as veins in leaves; with a network of fine upstanding ridges, as on the surface of spores.

Retuse – Having a bluntly rounded apex with a central notch.

Rhizome – An elongated underground stem, as in ferns.

Rootstock – An elongated underground stem, usually in higher plants.

Rosette – A cluster of leaves, usually basal, as in dandelion.

S

Sagittate – Arrowhead-shaped.

Scale – A highly modified, dry leaf, usually for protection.

Scape – A leafless or nearly leafless stem, coming from an underground part and bearing a flower or flower cluster, as in *Allium*.

Segment – A division of a compound leaf or of a perianth.

Sepal – One of the members of the calyx.

Serrate – With teeth projecting forward

Serrulate – Finely serrate

Sessile – Lacking a petiole or stalk

Sigmoid – S-shaped

Silicle – Similar to a silique, but short and broad

Silique – A dry elongated fruit divided by a partition between the two carpels.

Sinuate – With long wavy margins

Sinus – A depression or notch in a margin between two lobes.

Sorus – The brown colored fruiting structure of ferns, usually on the underside of the frond.

Spatulate – Widened at the top like a spatula.

Spike – An elongated inflorescence with sessile (stalkless) or nearly sessile flowers.

Spikelet – A small or secondary spike: the ultimate flower cluster of the inflorescence of grasses and sedges.

Spine – A short thorn-like structure.

Spinose – With spines.

Spinulose – With small, sharp spines.

Spreading – Diverging from the root and nearly prostrate.

Stamen – Male reproductive structure of a flower, consisting of the pollen bearing structure (anther) borne on a stalk or filament.

Staminate – Male- flowered, with stamens only.

Standard – The large petal that stands up at the back of the flower as in a pea flower.

Stellate – Star-shaped.

Stipule – An appendage at the base of a leaf, or other plant part.

Stolon – A basal branch rooting at the nodes.

Stramineous – Straw colored.

Striate – Marked with fine, longitudinal , parallel lines, ridges, or grooves

T

Taproot – A strong, fleshy root that grows vertically into the soil, with smaller lateral roots.

Tendril – Thread-like stem or leaf that clings to adjacent structures for support (peas)

Ternate – in 3s.

Tomentose – Densely matted with soft hairs.

Toothed – Dentate.

Trifoliate – A compound leaf with 3 leaflets (clover).

Tuber – Swollen underground stem for storing food (potato, poison hemlock), that can sprout to form new plants.

Tuberous – Forming tubers

U

Ubiquitous – Everywhere, in all types of habitat.

Umbel – Umbrella-shaped inflorescence, in which the pedicels (flower stalks) radiate from a common point like the ribs of an umbrella.

Undulate – Wavy, as the margins of leaves.

V

Veins – The vascular portions of the leaves.

Villous – Covered with short, fine hairs.

Viscid – Sticky.

W

Whorled – Three or more leaves, petals, or branches arranged in a ring at a node.

Wing – A thin, membranous extension of an organ.

Fringed sage (*Artemisia frigida*), 220

Gambels oak (*Quercus gambelii*), 273–275
Giant ragweed (*Ambrosia trifida*), 20
Goat's heat (*Tribulus terrestris*), 163
Golden banner (*Thermopsis montana;*
    T. rhombifolia), 299–300
Golden chain tree
    (*Laburnum anagyroides*), 238
Goldenrod, rayless
    (*Isocoma pluraflora; I. wrightii*), 231
Goldenweed, rayless
    (*Oonopsis engelmannii*), 311
Goose grass (*Triglochin maritima*), 14
Gray horsebrush
    (*Tetradymia canescens*), 162
Greasewood
    (*Sarcobatus vermiculatus*), 263, 268
Great blue lobelia (*Lobelia siphilitica*), 240
Greater ammi (*Ammi majus*), 151
Green false hellebore
    (*Veratrum viride*), 283–284
Gregg's catclaw accacia (*Accacia greggii*), 6
Ground cherry (*Physalis virginiana*), 118
Groundsel (*Senecio*), 157–158
Guajillo (*Accacia greggii*), 6
Gum tree (*Eucalyptus*), 2
Gumweed (*Grindelia*), 314

Hairy caltrop
    (*Kallstroemia hirsutissima*), 245
Hairy vetch (*Vicia villosa*), 169–170
Halogeton (*Halogeton glomeratus*), 263,
    266–267
Heavenly bamboo (*Nandina domestica*), 2
Hedge bindweed (*Convolvulus sepium*), 24
Hellebore (*Veratrum*), 283–284
Hemp (*Cannabis sativa*), 246
Hoary alyssum (*Berteroa incana*), 303
Honey mesquite, 94–95
Horse chestnut (*Aesculus*), 233–234
Horse nettle (*Solanum carolinense*), 115
Horsebrush (*Tetradymia*), 162
Horsetail, common (*Equisetum*), 224–225
Hound's tongue
    (*Cynoglossum officinale*), 159
Hydrangea (*Hydrangea*), 2

Indian grass (*Sorghastrum nutans*), 1, 2
Indian hemp (*Apocynum cannabinum*), 60
Indian lobelia (*Lobelia inflata*), 240
Indian paintbrush (*Castilleja*), 316

Japanese pieris (*Pieris japonica*), 131
Japanese yew (*Taxus cuspidata*), 68–69
Jequirity bean (*Abrus precatorius*), 124
Jessamine
    Carolina (*Gelsemium sempervirens*), 236
    day-blooming (*Cestrum diurnum*), 295
    –296
Jimmy weed
    (*Isocoma pluraflora; I. wrightii*), 231
Jimson weed (*Datura stramonium*), 107
Johnson grass (*Sorghum halapense*), 2, 12,
    242–243
June berry (*Amelanchier alnifolia*), 7

Kale (*Brassica oleracea*), 188
Kansas thistle (*Solanum rostratum*), 116
Kentucky coffee tree
    (*Gymnocladus dioica*), 235
Kentucky mahogany
    (*Gymnocladus dioica*), 235
Klamath weed
    (*Hypericum perforatum*), 153
Kleingrass (*Panicum coloratum*), 164
Kochia (*Kochia scoparia*), 27, 166–167

Lambert's red loco
    (*Oxytropis lambertii*), 215
Lamb's-quarter (*Chenopodium*), 22
Lantana (*Lantana camara*), 160
Larkspur (*Delphinium*), 30–36
Lead tree (*Leucaena leucocephala*), 171–172
Leafy spurge (*Euphorbia esula*), 119–120
Lechuguilla (*Agave lecheguilla*), 154
Leucaena (*Leucaena leucocephala*),171–172
Lily of the valley (*Convallaria majalis*), 61
Lima bean (*Phaseolus lunatus*), 2
Little horsebrush
    (*Tetradymia glabrata*), 162
Little mallow (*Malva parviflora*), 28
Lobelia (*Lobelia*), 240
Locoweeds (*Astragalus; Oxytropis*), 204–
    216, 278–280, 323–325
Lucky nut (*Thevetia thevetioides*), 64

Note Page numbers in *italics* indicate figures.
Page numbers followed by "*t*" indicate tables.

indolizidine alkaloids in, 104
lentiginosus, 212
   reproductive failure with, 206
miser, 216
   toxin in, 209–210
   var. *hylophilus*, 210*t*, 216
   var. *oblongifolius*, 210*t*, 216
   var. *serotinus*, 210*t*, 216
moencoppensis, 310*t*
mollissimus, 213
   congenital abnormalities with, 206
   nitro compound poisoning from, 210–211
oocalysis, 310*t*
osterhouti, 310*t*
pattersonii, 310*t*
pectinatus, 310*t*
praelongus, 210*t*, 310*t*
preussii, 310*t*
pterocarpus, 210*t*
racemosus, 310*t*
recedens, 310*t*
sabulosus, 310*t*
selenium-accumulating species of, 310*t*
selenium in, 306*t*
species containing swainsonine, 205*t*
swainsonine in, 81
tenellus, 204
tetrapterus, 210*t*
toanus, 210*t*, 310*t*
toxins in, 325*t*
   affecting milk, 323–324
whitnii, 210*t*
Atamasco. *See Zephranthes atamasco*
Ataxia, 242
Atriplex
   canescens, 315
   selenium in, 306*t*
Atropa belladona, 105
Atropine
   in Datura, 109
   for *Gymnocladus dioica* poisoning, 235
   in *Hyoscyamus niger*, 111
   isomeric form of, 110
   for larkspur poisoning, 34
   for lobelia poisoning, 240
   in nightshades, 105
   for rhododendron poisoning, 128
   in *Solanum nigrum*, 112
Atropine sulfate
   for cardiac glycoside poisoning, 59
   for yew poisoning, 69
   for *Zigadenus* poisoning, 72
Australian stringhalt syndrome, 304
Avena
   fatua
      nitrates in, 15*t*, 21
   sativa, 21, 147*t*
      nitrates in, 15*t*
Avocado. *See Persea americana*

Azalea, 129
Azaradine, 87–88

Baccharis halimifolia, 315
   symptoms of, 120*t*
Bahia oppositifolia, 2*t*
Baileya multiradiata, 96
Baneberry. *See Actaea rubra*
Baptisia, 232
Barberry family, *Podophylum peltatum*, 126
Barnyard grass. *See Echinochloa crus-galli*
Bassia hyssopifolia, 147*t*, 264*t*, 266
Batyl alcohol, 197
Bavilla. *See Halogeton glomeratus*
Be-still tree. *See Thevetia thevetioides*
Bean tree. *See Laburnum anagyroides*
Bear grass, *163*
Beard tongue. *See Penstemon*
Beargrass. *See Xerophyllum tenax*
Beetles, in photosensitizing plant control, 150
Belladonna, 105. *See also Atropa belladonna*
Bellflower family, 240
Belvedere. *See Kochia scoparia*
Berberidaceae, *Podophylum peltatum*, 126
Berteroa incana, 303
Berula pusilla, 248*t*
Beta vulgaris, 264*t*
   nitrates in, 15*t*
β-amino proprionitrile, 3
   in sweet pea, 244
β-cyanoalanine, 244
β-glucosidase, 1
Big sagebrush. *See Artemisia tridentata*
Bile acid, serum levels of, 149–150
Biliary occlusive photosensitization, 146–148
Bindweed. *See Convolvulus arvensis*
Birds, *Sesbania* poisoning in, 92
Birth defects. *See* Congenital deformities
Bishop's weed. *See Ammi majus*
Bitter nightshade. *See Solanum dulcamara*
Bitter sneezeweed. *See Helenium amarum/tenuifolium*
Bittersweet
   climbing. *See Solanum dulcamara*
   toxic, 105
Bitterweed, 100
   Hymenoxys odorata, 102
   in sneezing and vomiting, 96
Bitterwood tree. *See Quassia simarouba*

NOTES

NOTES

NOTES

NOTES

NOTES

NOTES

NOTES

NOTES

NOTES

NOTES

NOTES

NOTES

NOTES

ISBN 1-893441-11-3

We hope you enjoy this volume.

If you would like more information on Teton NewMedia products please call toll-free 877-306-9793 or visit us at www.tetonnm.com

A Guide to
Plant Poisoning
OF ANIMALS IN NORTH AMERICA
Anthony P. Knight, BVSc, MS
Richard G. Walter, MA Botany